Human Behavior Learning and Transfer

Human Behavior Learning and Transfer

Yangsheng Xu
The Chinese University of Hong Kong

Ka Keung C. Lee
The Chinese University of Hong Kong

A CRC title, part of the Taylor & Francis imprint, a member of the Taylor & Francis Group, the academic division of T&F Informa plc.

Published in 2006 by
CRC Press
Taylor & Francis Group
6000 Broken Sound Parkway NW, Suite 300
Boca Raton, FL 33487-2742

No claim to original U.S. Government works
Printed in the United States of America on acid-free paper
10 9 8 7 6 5 4 3 2 1

International Standard Book Number-10: 0-8493-7783-8 (Hardcover)
International Standard Book Number-13: 978-0-8493-7783-9 (Hardcover)
Library of Congress Card Number 2005049923

Library of Congress Cataloging-in-Publication Data

Xu, Yangsheng.
Human behavior learning and transfer / Yangsheng Xu, Ka Keung Lee.
p. cm.
ISBN 0-8493-7783-8
1. Learning, Psychology of. I. Lee, Ka Keung. II. Title.

BF318.X8 2006
153.1'54--dc22 2005049923

Taylor & Francis Group
is the Academic Division of T&F Informa plc.

Visit the Taylor & Francis Web site at
http://www.taylorandfrancis.com

and the CRC Press Web site at
http://www.crcpress.com

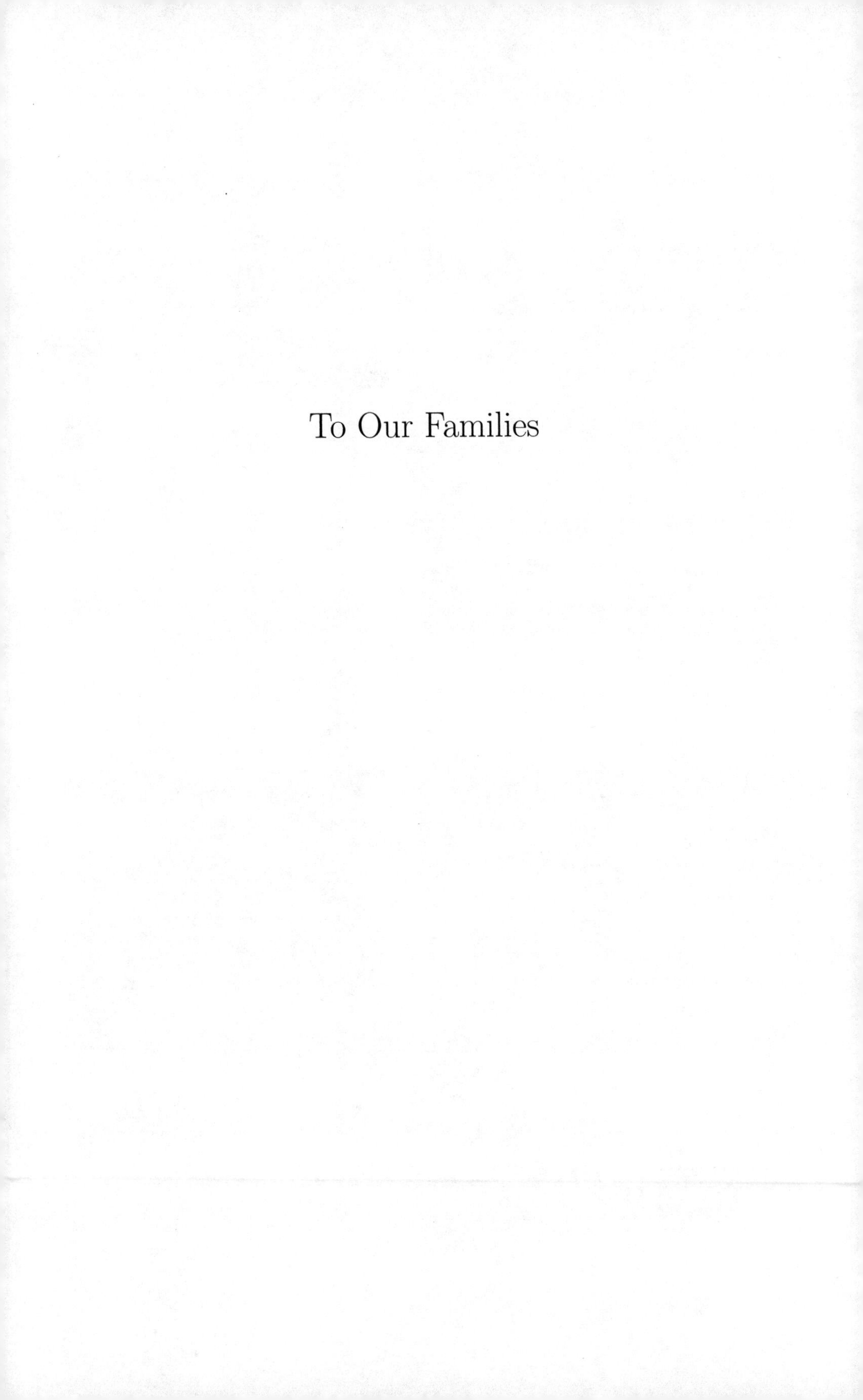

To Our Families

Preface

Over the past two decades, rapid advances in computer performance have not been matched with similar advances in the development of intelligent robots and systems. Although humans are quite adept at mastering complex and dynamic skills, we are far less impressive in formalizing our behavior into algorithmic, machine-codable strategies. Therefore, it has been difficult to duplicate the types of intelligent skills and actions we witness every day as humans, in robots and other machines. This not only limits the capabilities of individual robots, but also the extent to which humans and robots can safely interact and cooperate with one another. Nevertheless, human actions are currently our only examples of truly "intelligent" behavior. As such, there exists a profound need to abstract human skill into computational models, capable of realistic emulation of dynamic human behavior.

The main focus of this book is the modeling and transfer of human *action* and *reaction* behaviors. For human reaction skills, we apply machine learning techniques and statistical analysis towards abstracting models of human behavior. It is our contention that such models can be learnt efficiently to emulate complex human control behaviors in the feedback loop. For human action skills, the methods presented here are based on techniques for reducing the dimensionality of data sets while preserving as much useful information as possible.

The research in modeling human behaviors from human data or human demonstration has thus far not addressed a number of important issues, including modeling of continuous and discontinuous behaviors, model validation, model evaluation, model optimization, and efficient transfer of model charateristics. Chapters 2 to 8 of this book will discuss various topics involved in human reaction skill modeling. We propose and develop an efficient continuous learning framework for modeling human control strategy (HCS) that combines cascade neural networks and node-decoupled extended Kalman filtering. We then apply cascade learning towards abstracting HCS models from experimental control strategy data (Chapter 3). Next, we present a stochastic, discontinuous modeling framework for modeling discontinuous human control strategies. For validation of behavior model, we formulate a stochastic similarity measure, based on hidden Markov model analysis that is capable of comparing multi-dimensional stochastic control trajectories (Chapter 4).

For model evaluation, we propose and develop performance criteria based on inherence analysis (Chapter 5). We also propose an iterative optimization algorithm, based on simultaneous perturbed stochastic approximation

(SPSA), for improving the performance of learnt HCS models (Chapter 6).

In Chapter 7, we develop two algorithms for HCS model transferring. One of them is based on the similarity measure and SPSA. The proposed algorithms allow us to develop useful apprentice models that nevertheless incorporate some of the robust aspects of the expert HCS models. In Chapter 8, we will present a case study in which human navigational skills are transferred to a wheelchair.

In Chapters 9 to 14, issues related to human action skill modeling will be presented. In Chapter 10, we will discuss global parametric modeling of human performance data. Local modeling based on general purpose non-parametric methods is the focus of Chapter 11.

Chapter 12 presents adaptations to these non-parametric methods for the specific purpose of modeling human performance, in which we derive a s-pline smoother that can build best-fit trajectories of human performance data through phase space.

In Chapters 13 and 14, three case studies in human action modeling are presented. We will model the actions of human on three different levels, namely, facial actions, full-body actions, and walking trajectories.

We wish to acknowledge the contributors, who are the past students of the first author, on whose work our presentation is partly based. Michael Nechyba's work laid the foundation of human reaction modeling and the related work are presented in Chapters 2 to 4. Jingyan Song developed the methodology for the evaluation, optimization and trasfer of human reaction behaviors, which form the contents in Chapters 5 to 7. The work presented in Chapter 8 was mainly performed by Kyle H. N. Chow. Chapters 9 to 12, which discuss the modeling of human action skills, come primarily from the research of Christopher Lee. The first author would also like to thank his students for their support and friendship while he worked at Carnegie Mellon University and the Chinese University of Hong Kong.

Our book reviewers have contributed their advices in making critical suggestions. We are also grateful for the effort of the editors and staff at Taylor and Francis throughout the development of this book.

This book could not have happened without the support from our funding sources. The research and development work described in this book is partially supported by the grants from the Research Grants Council of the Hong Kong Special Administration Region (Projects No. CUHK4138/97E, CUHK4164/98E, CUHK 4197/00E, CUHK4228/01E, CUHK4317/02E, and CUHK4163/03E), and by the Hong Kong Innovation and Technology Fund under grant ITS/140/01.

Yangsheng Xu and Ka Keung Lee
The Chinese University of Hong Kong Spring 2005

Authors

Yangsheng Xu received his Ph.D. from University of Pennsylvania in the area of robotics in 1989. He has been with the Department of Automation and Computer-Aided Engineering at The Chinese University of Hong Kong (CUHK) since 1997, and served as department chairman from 1997 to 2004. Prof. Xu is currently a chair professor in the department and he was a faculty member at the School of Computer Science, Carnegie Mellon University (CMU) from 1989 to 1999.

Prof. Xu's research interests have been in robotics and human interface, and their applications in service, aerospace, and industry. At first he worked on designing and controling robots for space operations. He also made contributions in human control strategy modeling and applications in real-time control. His more recent efforts have been concentrated on wearable interface, intelligent surveillance, and future space systems.

He has been a principal investigator in more than 30 projects funded by both governments and industries. Based on his research work, he has fortunately published over 70 papers in journals, 130 papers in international conferences, and several book contributions and books. He has been serving or served on advisory boards or panels in various government agencies and industries in the United States, Japan, Korea, Hong Kong, and mainland China. He is a fellow of IEEE, HKIE, and IEAS.

Ka Keung Lee received his bachelor of information technology and bachelor of engineering (hons.) degrees from the Australian National University (ANU) in 1995 and 1997, respectively. From ANU, he received a full-fee international undergraduate scholarship for the years 1992-1996. He received the master of philosophy and doctor of philosophy degrees from the Department of Automation and Computer-Aided Engineering at The Chinese University of Hong Kong in 2000 and 2004, respectively.

Dr. Lee worked in the School of Computer Science of the Australian Defence Force Academy (ADFA) in 1995; was a visiting scholar at the Division of Information Technology of the Commonwealth Scientific and Research Organisation (CSIRO), Australia in 1996; and served as a software engineer in the computer harddisk industry in Hong Kong in 1997-1998. His current research interests include human action modeling, human sensation modeling, intelligent surveillance, and wearable robotics. Currently, Dr. Lee is a postdoctoral fellow in the Department of Automation and Computer-Aided Engineering at The Chinese University of Hong Kong.

Contents

1

Introduction

1.1 Motivation

Over the past two decades, rapid advances in computer performance have not been matched with similar advances in the development of intelligent robots and systems. Although humans are quite adept at mastering complex and dynamic skills, we are far less impressive in formalizing our behavior into algorithmic, machine-codable strategies. Therefore, it has been difficult to duplicate the types of intelligent skills and actions, we witness every day as humans, in robots and other machines. This not only limits the capabilities of individual robots, but also the extent to which humans and robots can safely interact and cooperate with one another. Nevertheless, human actions are currently our only examples of truly "intelligent" behavior. As such, there exists a profound need to abstract human skill into computational models, capable of realistic emulation of dynamic human behavior.

Models of human skill can transfer intelligent control behaviors to robots. This is especially critical for robots that have to operate in remote or inhospitable environments that humans cannot reach. For example, sending humans to operate in space is often expensive, dangerous, or outright impossible, while sending robots instead is comparatively cheap and involves no risk to human life. Replacing astronauts with robots is only feasible, however, if the robots are equipped with sufficient autonomous skill and intelligence; we suggest that a robot can acquire those necessary capabilities through abstracted models of human skill.

In other robotic applications, we would like robots to carry out tasks which humans have traditionally performed. For example, the Intelligent Vehicle Highway System (IVHS), currently being developed through massive initiatives in the United States, Europe, and Japan [41], [46], envisions automating much of the driving on our highways. The required automated vehicles will need significant intelligence to interact safely with variable road conditions and other traffic. Modeling human intelligence offers one way of building up the necessary skills for this type of intelligent machine.

Broadly speaking, *human skill* can be grouped into two categories: (1) *action* skills, which are open-loop, and (2) *reaction* skills, which are closed-loop, and require sensory feedback to successfully execute the skill. Tossing or kicking a ball is an example of action skill. Driving a car, on the other hand, is a classic example of reaction skill, where the human closes the feedback control loop.

Human control strategy we study in this book is a subset of this type of reaction skill. In terms of complexity, human control strategy lies between low-level feedback control and high-level reasoning, and encompasses a wide range of useful physical tasks with a reasonably well-defined numeric input/output representation.

Although a great deal of work in robotics has gone into the study of methods for modeling human *reaction* skills, much less work has so far been done on the important related problem of modeling human *action* skills. Action learning is the characterization of open-loop control signals into a system, or the characterization of the output of either an open-loop or closed-loop system. While reaction learning generates a mapping from an input space or state space to a separate action/output space, action learning characterizes the state space or action space explored during a performance. Given data collected from multiple demonstrations of task performance by a human teacher, where the state of the performer or the task-state is sampled over time during each performance, the goal is to extract from the recorded data succinct models of those aspects of the recorded performances most responsible for the successful completion of the task. Reaction learning focuses on the muscle control a dancer uses to move his or her body, for example, while action learning studies the resulting dance.

1.2 Overview

The research in modeling skill from human data or human demonstration has thus far not addressed a number of important issues. This book will discuss two main aspects of human behavior modeling: (1) Human reaction skills modeling and (2) Human action skills modeling.

For human reaction skills, this book applies machine learning techniques and statistical analysis towards abstracting models of human control strategy. It is our contention that such models can be learned efficiently to emulate complex human control behaviors in the feedback loop. For human action skills, the methods presented here are based on techniques for reducing the dimensionality of data sets while preserving as much useful information as possible.

The book is organized as follows. Chapters 2 to 8 will discuss various topics involved in human reaction skill modeling. In Chapters 9 to 14, issues related to human action skills modeling will be presented.

In Chapter 2, past work on human reaction learning will be reviewed.

In Chapter 3, we describe a graphic dynamic driving simulator which we have developed as an experimental platform for collecting, modeling, and analyzing human control data. We propose and develop an efficient continuous learning framework for modeling human control strategy that combines cascade neural networks and node-decoupled extended Kalman filtering. We then apply cascade learning towards abstracting HCS models from experimental control strategy data. We compare two training algorithms – namely, quickprop and NDEKF. We observe that the dissimilarity between the human and model-generated control trajectories is principally caused by switching discontinuities in one of the model outputs, and that, in fact, *any* continuous modeling framework would suffer equally in attempting to model that output without high-frequency noise. Therefore, we propose and develop a stochastic, discontinuous modeling framework for modeling discontinuous human control strategies. This approach models different control actions as individual statistical models, which, together with the prior probabilities of each control action, combine to generate a posterior probability for each action, given the current model inputs.

In Chapter 4, we then set out to quantify the qualitative observations of model fidelity. As a first step in validating the learned models of human control strategy, we propose and develop a stochastic similarity measure, based on Hidden Markov Model analysis that is capable of comparing multi-dimensional stochastic control trajectories. The goal of this similarity measure is to compare model-generated control trajectories with their respective human control trajectories.

In Chapter 5, we apply event analysis to arrive at some performance criteria for the HCS models. We view the driving task as a combination of events, such as, driving along a straight road, turning through a curve, and avoiding obstacles. These performance criteria, such as collision avoidance and tight turning analysis, can measure important event characteristics of the HCS model. They also measure the ability to change speeds quickly while maintaining vehicle safety and stability. All of these test the behavior of the HCS cascade neural network model outside the range of training data from which the models are learned.

In Chapter 6, we propose and develop the performance criteria based on the inherence analysis. For the real task of driving, many candidate performance criteria, such as average speed, the feeling of passenger, driving stability, driving smoothness, and fuel efficiency, exist. We term them as the criteria based on the characteristic inherence. In these inherent analysis, the feeling of passenger and the inherent frequency characteristics are more important. Both of them are based on the data analysis, and the data are collected from the whole process, not based on the event. These criteria measure the inherent characteristics of HCS model.

Performance evaluation is, however, only part of the solution for effectively applying models of human control strategy. While humans are in general very capable of demonstrating intelligent behaviors, they are far less capable of demonstrating those behaviors without occasional errors and random (noise) deviations from some nominal trajectory. Any empirical learning algorithm will necessarily incorporate those problems in the learned model, and will consequently be less optimal. Furthermore, control requirements may differ between humans and robots, where stringent power or force requirements often have to be met. A given individual's performance level, therefore, may or may not be sufficient for a particular application. Therefore, in Chapter 6, we also propose an iterative optimization algorithm, based on simultaneous perturbed stochastic approximation (SPSA), for improving the performance of learned HCS models. This algorithm leaves the learned model's structure in tact, but tunes the parameters of the HCS model in order to improve performance. It requires no analytic formulation of performance, only two experimental measurements of a user-defined performance criterion per iteration.

In Chapter 7, we develop two algorithms for HCS model transferring. One of them is based on the similarity measure and the SPSA. The proposed algorithm allows us to develop useful apprentice models that nevertheless incorporate some of the robust aspects of the expert HCS models. In this transfer learning algorithm, we propose to raise the similarity between an expert HCS model and an apprentice HCS model. Alternatively, we can think of the expert model guiding the actions of the apprentice model. The overall algorithm consists of two steps. In the first step, we let the expert model influence the eventual structure of the HCS model. Once an appropriate model structure has been chosen, we then tune the parameters of the apprentice model through simultaneously perturbed stochastic approximation, to increase the similarity between the expert HCS model and the apprentice HCS model. The other algorithm is based on model compensation, to compensate the "gap" between two HCS models by a compensation neural network model. The new compensated apprentice HCS model is a combination of original HCS model and the trained cascade neural network compensation model.

In Chapter 8, we will present a case study in which human navigational skills are transferred to a wheelchair. A novel navigation/localization learning methodology is presented to abstract and transfer the human sequential navigational skill to a robotic wheelchair by showing the platform how to respond in different local environments along a demonstrated, designated route using a lookup-table representation.

In the remaining chapters, we will discuss the methodology through which human action skills can be modeled and transferred. Chapter 9 gives an introduction to human action skill modeling and the related literature is reviewed.

In Chapter 10, we will discuss global parametric modeling of human performance data. We will present three global methods for this purpose: principal component analysis (PCA), nonlinear principal component analysis (NLPCA), and a variation on NLPCA called sequential nonlinear principal component analysis (SNLPCA).

Local modeling based on general purpose non-parametric methods is the focus of Chapter 11, which discusses the advantages that local models (i.e., as opposed to global parametric models) hold for human performance modeling. In particular, we will focus on trajectory-fitting.

Chapters 12 presents adaptations to these non-parametric methods for the specific purpose of modeling human performance, in which we derive a s-pline smoother that can build best-fit trajectories of human performance data through phase space.

In Chapters 13 and 14, three case studies in human action modeling are presented. We will model the actions of humans on three different levels, namely, facial actions, full-body actions, and walking trajectories.

Chapter 15 summarizes the conclusions and contributions of the book.

2

Introduction to Human Reaction Skill Modeling

2.1 Motivation

With increased intelligence and sophistication in robotic systems, analysis of human-robot coordination in tightly coupled human-machine systems will become increasingly relevant. In IVHS, for example, there will be ubiquitous interaction between autonomous vehicles and their human drivers/passengers. Moreover, the currently limited application domain for robots may broaden into other aspects of consumer life, where household and service robots will interact primarily with non-experts. To ensure safe coordination with humans in a shared workspace, we must incorporate appropriate models of human behavior into the world model of the robots. We can assess the quality of joint human-machine systems by including computational models of human behavior in the overall system analysis.

Realistic simulation of human behavior is required not only in human-machine systems, but also in the burgeoning field of virtual reality. As graphic displays become increasingly life-like, the dynamic behavior of the virtual world will need to match the increased visual realism. Computational models of human skill can impart the necessary sense of realism to the actions and behaviors of virtual humans in the virtual world. Consider, for example, a NASCAR video game. Rather than have preprogrammed behaviors, human driver models, abstracted from different race car drivers, could generate more diverse and human-like driving behaviors in the simulated competitors.

Finally, accurate models of human skill can contribute to improved expert training and human-computer interfacing (HCI). Consider, for example, the tasks of teleoperating robots in remote environments or learning to fly a high-performance jet. Training for both of these tasks is difficult, expensive, and time consuming for a novice [176], [215]. We can accelerate learning for the novice operator by providing on-line feedback from virtual teachers in the form of skill models, which capture the control strategies of expert operators. Through the use of human skill models, operator performance can be monitored during training or actual task execution as information is displayed through different sensor modalities and layouts.

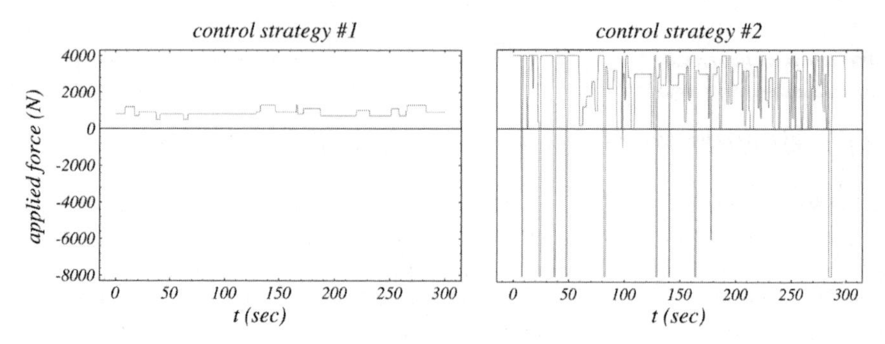

FIGURE 2.1: Control forces applied by two different individuals in a driving simulator for the same road and simulation parameters. The two control strategies are quite different.

Thus, models of human skill find application in far-ranging fields, from autonomous robot control and teleoperation to human-robot coordination and human-robot system simulation. Since scientific understanding of human intelligence is incomplete at best, however, models of human skill cannot be derived analytically. Rather, we have to model human skill through observation, or *learning*, of experimentally provided human training data. Current learning paradigms are not sufficiently rich to model the full range of human skill, from low-level muscle control to high-level reasoning and abstract thought.

Human control strategy we study in the coming chapters is a kind of reaction skill. In terms of complexity, human control strategy lies between low-level feedback control and high-level reasoning, and encompasses a wide range of useful physical tasks with a reasonably well-defined numeric input/output representation. On the one hand, a control strategy is not only defined by the "gains" or parameters in the controller, but also the structure or approach of the strategy. Consider the skill of driving a car, for example. Figure 2.1 illustrates applied force profiles over the same road for two different individuals in a driving simulator. The distinction between the two driving styles is a difference in kind rather than merely a difference in degree, similar to the structural difference between a linear feedback and a variable structure controller. Each represents a unique control strategy. On the other hand, the demonstrated skill in Figure 2.1 requires no high-level reasoning or abstract thought. Modeling such mental processes, of which humans are capable, requires an as-of-yet unavailable understanding of the human traits of self-awareness and consciousness.

2.2 Related work

The field of *intelligent control* [250] has emerged from the field of classical control theory to deal with applications too complex for classical control approaches. Broadly speaking, intelligent control combines classical control theory with *adaptation*, *learning*, or *active exploration*. Methods in intelligent control include fuzzy logic control, neural network control, reinforcement learning, and locally weighted learning for control.

Neural networks are used to map unknown nonlinear functions and have been applied in control most commonly for dynamic system identification and parameter adaptive control [15], [153], [165]. Locally weighted learning presents an alternative to neural networks, and maps unknown functions through local statistical regression of the experimental data, rather than through a functional representation [16], [17]. In reinforcement learning, an appropriate control strategy is found through dynamic programming of a discretized state space [24]. Below, we describe previous work relating to each of these methods.

2.2.1 Skill learning through exploration

Learning skill through exploration has become a popular paradigm for acquiring robotic skills. In *reinforcement learning* [24], [158], [232], [248], data is not given as direct input/output data points; rather data is specified by an input vector and an associated (scalar) reward from the environment. This reward represents the reinforcement signal, and is akin to "learning with a critic" as opposed to "learning with a teacher." The reinforcement learning algorithm is expected to explore and learn a suitable control strategy over time. References [85] and [17] give some examples of reinforcement learning control for a robot manipulator and a simulated car in a hole, respectively. Schneider [210] learns the open-loop skill of throwing through a search of the parameter space which defines all possible throwing motions. One of the advantages of learning human control strategy directly from human control data is that we avoid the need for this type of state space search to find a suitable control strategy.

Lee and Kim [139] have proposed and verified an *inductive learning* scheme, where control rules are learned from examples of perception/action data through hypothesis generation and testing. Their learning paradigm, Expert Assisted Robot Skill Acquisition (EARSA) [140], consists of two steps: (1) skill acquisition from human expert rules, and (2) skill discovery or refinement through hypothesis generation and testing.

2.2.2 Skill modeling from human data

Interest in modeling human control goes all the way back to World War II, when engineers and psychologists attempted to improve the performance of pilots, gunners, and bombardiers [94]. Early research in modeling human control is based on the *control-theory paradigm* [151], which attempts to model the human-in-the-loop as a simple feedback control system. These modeling efforts generally focussed on simple tracking tasks, where the human is most often modeled as a simple time delay in the overall human-machine system [216].

More recently, work has been done towards learning more advanced skills directly from humans. In fuzzy control schemes [131], [132], human experts are asked to specify "if-then" control rules, with fuzzy linguistic variables (e.g., "hot," "cold," etc.), which they believe guide their control actions. For example, simple human-in-the-loop control models based on fuzzy modeling have been demonstrated for automobile steering [123] and ships helmsmen [233]. Although fuzzy control systems are well suited for simple control tasks with few inputs and outputs, they do not scale well to the high-dimensional input spaces required in modeling human control strategy [15].

Robot learning from human experts has also been applied to a deburring robot. Asada and Yang [13] derive control rules directly from human input/output data, by associating input patterns with corresponding output actions for the deburring robot. In [257], Yang and Asada combine linguistic information and numeric input/output data for the overall control of the robot. Expert linguistic rules are acquired directly from a human expert to partition the control space. For each region, a corresponding linear control law is derived from the numeric demonstration data by the human expert. In [12], [144], [218], the same deburring robot is controlled through an associative neural network which maps process parameter features to action parameters from human control data. The proper tool feed rate is determined from the burr characteristics of the current process.

Lee and Chen [138] use feasible state transition graphs through self-organizing data clusters to abstract skill from human data. Skills are modeled as optimal sequences of one-step state transitions that transform the current state into the goal state. The approach is verified on demonstrated human Cartesian teleoperation skill. Yang, *et. al.* [258], [259] implement a different state-based approach to open-loop skill learning and telerobotics using Hidden Markov Models (HMMs). HMMs are trained to learn both telerobotic trajectories executed by a human operator and simple human handwriting gestures. Yamato, *et. al.* [255], [256] also train HMMs to recognize open-loop human actions.

Friedrich, *et. al.* [73] and Kaiser [113] review programming by demonstration and skill acquisition via human demonstration for elementary robot skills. Lee [133] investigates human-to-robot skill transfer through demonstration of task performance in a virtual environments. Voyles, *et. al.* [245] program a robot manipulator through human demonstration and abstraction of gesture primitives. Ude [241] learns open-loop robot trajectories from human demonstration. Skubic and Volz [223] transfer force-based assembly skills to robots from human demonstrations. Iba [105] models open-loop sensory motor skills in humans. Gingrich, *et. al.* [79] argue that learning human performance models is valuable, but offer results only for simulated, known dynamic systems.

Several approaches to skill learning in human driving have been implemented. In [65], [66], neural networks are trained to mimic human behavior for a simulated, circular racetrack. The task essentially involves avoiding other computer-generated cars; no dynamics are modeled or considered in the approach. Pomerleau [187], [188] implements real-time road-following with data collected from a human driver. A static feedforward neural network with a single hidden layer, ALVINN, learns to map coarsely digitized camera images of the road ahead to a desired steering direction, whose reliability is given through an input-reconstruction reliability estimator. The system has been demonstrated successfully at speeds up to 70 mi/h. Subsequently, a statistical algorithm called RALPH [186] has been developed for calculating the road curvature and lateral offset from the road median. Neuser, *et. al.* [172] control the steering of an autonomous vehicle through preprocessed inputs to a single-layer feedforward neural network. These preprocessed inputs include the car's yaw angle with respect to the road, the instantaneous and time-averaged road curvature, and the instantaneous and time-averaged lateral offset. Driving data is again collected from a human operator. In [155], the authors provide a control theoretic model of human driver steering control. Finally, Pentland and Liu [183] apply HMMs towards inferring a particular driver's high-level intentions, such as turning and stopping.

Other, higher level skills have also been abstracted from human performance data. Kang [115], for example, teaches a robot assembly through human demonstration. The system observes a human performing a given task, recognizes the human grasp, and maps it onto an available manipulator. In other words, a sequence of camera images, observing the human demonstration, is automatically partitioned into meaningful temporal segments. Kosuge, *et. al.* [120] also abstract high-level assembly skill from human demonstration data. The high-level sequence of motion is decomposed into discrete state transitions, based on contact states during assembly. In each state, compliant motion control implements the corresponding low-level control. Hovland, *et. al.* [99] encode human assembly skill with Hidden Markov Models. In [181], neural networks encode simple pick-and-place skill primitives from human demonstrations.

2.2.3 Neural network learning

Interest and research in neural network-based learning for control has exploded in recent years. References [7], [8], [30], [103], [153], [204] provide good overviews of neural network control over a broad range of applications. Most often, the role of the neural network in control is restricted to modeling either the nonlinear plant or some nonlinear feedback controller.

In choosing a neural network learning architecture, there are several choices to be made, including the (1) type, (2) architecture, and (3) training algorithm. Broadly speaking there are two types of neural networks: (1) feedforward and (2) recurrent networks. *Feedforward neural networks* have connections in only one direction (from inputs to outputs). As such, they are static maps, which, in order to model *dynamic systems*, require time-delayed histories of the sensor and previous control variables as input. *Recurrent networks*, on the other hand, permit connections between units in all directions, including self connections, thereby allowing the neural network to implicitly model dynamic characteristics (i.e., discover the state) of the system. Compared to static feedforward networks, the learning algorithms for recurrent networks are significantly more computationally involved, requiring relaxation of sets of differential equations [92]. Yet, Qin, *et. al.* [191] show similar error convergence in mapping simple dynamic systems with feedforward and recurrent networks, respectively. As such, we will restrict the remainder of this discussion to feedforward models only.

Research into feedforward neural networks began in earnest with the publication of the backpropagation algorithm in 1986 [202]. Since then a number of different learning architectures have been proposed to adjust the structure of the feedforward neural network (i.e., the number and arrangement of hidden units) as part of learning. These approaches can be divided into (1) destructive and (2) constructive algorithms. In destructive algorithms, oversized feedforward models are trained first, and then, after learning has been completed, "unimportant" weights, based on some relevancy criteria are pruned from the network. See, for example [29], [36], [38], [86], [162], [163], [237]. In constructive algorithms, on the other hand, neural networks are initialized in some minimal configuration and additional hidden units are added as the learning requires. Ash [14], Bartlett [23], and Hiroshe [96], for example, have all experimented with adaptive architectures where hidden units are added one at a time in a single hidden layer as the error measure fails to reduce appreciably during learning. Fahlman [59], [60] proposes a cascade learning architecture, where hidden units are added in multiple cascading layers as opposed to a single hidden layer. Cascade learning has a comparative advantage over the other adaptive learning architectures in that (1) new hidden nodes are not arranged in a single hidden layer, allowing more complex mappings, and (2) not all weights are trained simultaneously, resulting in faster convergence. With respect to constructive algorithms, destructive algorithms compare unfavorably, since initially, a lot of effort is expended training (by definition) too many weights, and the pruned networks need to be retrained multiple times, after each individual weight or unit has been pruned.

Finally we note that there is a selection of training algorithms available for feedforward neural networks. As we have already noted, the first of these was the backpropagation algorithm, which implements local gradient descent on the weights during training in order to minimize the sum-of-squared residuals. Since this training method was first proposed [202], modifications to standard backpropagation, as well as other training algorithms have been suggested. An adaptive learning rate [39], as well as an additive momentum term [185] are both somewhat effective in accelerating convergence of backpropagation in "flat" regions of the error hypersurface. Quickprop [59] incorporates local, second-order information in the weight-update algorithm to further speed up learning. Kollias and Anastassious [119] propose applying the Marquardt-Levenberg least squares optimization method, which utilizes an approximation of the Hessian matrix. Extended Kalman filtering [190], [222], where the weights in the neural network are viewed as states in a discrete-time finite dimensional system, outperform the previously mentioned algorithms in terms of learning speed and error convergence [190]. In all cases, experimental data is usually partitioned into two random sets – one for actual training, and the other for cross validation [92]; training is generally stopped once the error measure on the cross-validation set no longer decreases.

2.2.4 Locally weighted learning

Neural networks have received great attention for nonlinear learning and control applications. Another learning paradigm, *locally weighted learning*, has emerged more recently and has shown great success for a number of different control applications, ranging from devil sticking [207] to robot juggling [208]. Atkeson, *et. al.* [16] offer an excellent overview of locally weighted learning, while [17] addresses control-specific issues. *Locally weighted regression* is one instance of locally weighted learning and is similar in approach to CMAC [5] and RBF [157] neural networks in that local (linear) models are fit to nearby data. All the data is explicitly stored and organized in efficient data structures (such as k-d trees [28] or Bump trees [178], for example). When the model is queried for the output at a specified input vector, points in the database near that input vector are used to construct a local linear map.

Locally weighted regression offers several advantages over global learning paradigms (such as neural networks) [17]. First, locally weighted regression results in smooth predicted surfaces. Second, it automatically linearizes the system around a query point by providing the local linear map. Third, adding new data to the locally weighted regression model is cheap, as it merely requires inserting a new data point in the existing data base. Learning occurs in one-shot, since all the data is retained for the construction of the local linear models. Fourth, local minima are not a problem, as no gradient descent is required for learning the model. Finally, interference (e.g., the catastrophic forgetting problem) between old and new experiences does not occur.

Despite these appealing features, locally weighted regression also suffers from some shortcomings. First, computational and memory costs increase as the size of the data increases. Second, efficient data structures become less efficient as the dimensionality of the input space grows. Finally, locally

3

Learning of Human Control Strategy: Continuous and Discontinuous

In this chapter, we apply machine learning techniques and statistical analysis towards abstracting models of human control strategy. It is our contention that such models can be learned efficiently to emulate complex human control behaviors in the feedback loop.

In Section 3.1, we describe a graphic dynamic driving simulator which we have developed as an experimental platform for collecting, modeling, and analyzing human control data. We then define the class of models – static feedforward models – to which we restrict ourselves in this study. Next, in Section 3.2, we propose and develop an efficient continuous learning framework for modeling human control strategy that combines cascade neural networks and node-decoupled extended Kalman filtering. In Section 3.3, we then apply cascade learning towards abstracting HCS models from experimental control strategy data. We compare two training algorithms, namely, quickprop and NDEKF, and show that cascade/NDEKF converges orders of magnitude faster than cascade/quickprop for the human control training data. In Section 3.4, we propose and develop a stochastic, discontinuous modeling framework for modeling discontinuous human control strategies. This approach models different control actions as individual statistical models, which, together with the prior probabilities of each control action, combine to generate a posterior probability for each action, given the current model inputs. A control decision is then made stochastically, based on the posterior probabilities.

3.1 Experimental design

3.1.1 Motivation

Human control strategy, as we have defined the term, encompasses a large set of human-controlled tasks. It is neither practical nor possible to investigate all of these tasks in a finite amount of time. In the coming chapters, we therefore look towards a prototypical control application – the task of *human driving* – to collect and model control strategy data from different human subjects.

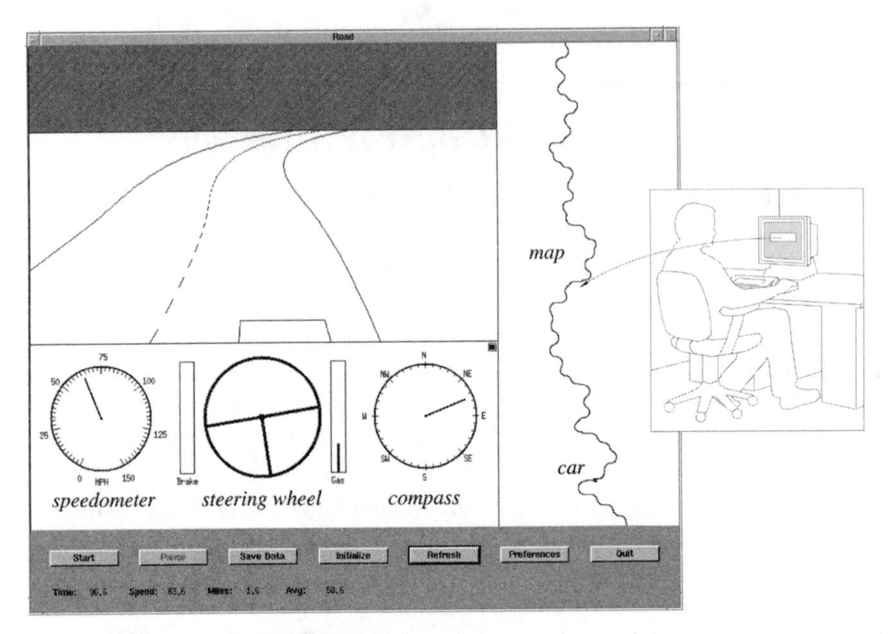

FIGURE 3.1: Our driving simulator generates a perspective view of the road for the user, who has independent control over steering, braking, and acceleration (gas).

3.1.2 Simulation environment

Dynamic driving simulator

Figure 3.1 shows the real-time, dynamic, graphic driving simulator which we have developed for collecting and analyzing human control strategy data. In the interface, the human operator has full control over the steering of the car (mouse movement), the brake (left mouse button), and the accelerator (right mouse button); the middle mouse button corresponds to slowly easing off whichever pedal is currently being "pushed." The vehicle dynamics are given in (3.1) through (3.19) below (modified from [91]):

$$\dot{\omega} = \left(l_f \phi_f \delta + l_f F_{\xi f} - l_r F_{\xi r}\right)/I \tag{3.1}$$

$$\dot{v}_\xi = \left(\phi_f \delta + F_{\xi f} + F_{\xi r}\right)/m - v_\eta \omega - (\text{sgn } v_\xi)\, c_D v_\xi^2 \tag{3.2}$$

$$\dot{v}_\eta = \left(\phi_f + \phi_r - F_{\xi f} \delta\right)/m - v_\xi \omega - (\text{sgn } v_\eta)\, c_D v_\eta^2 \tag{3.3}$$

$$\begin{bmatrix} \dot{x} \\ \dot{y} \\ \dot{\theta} \end{bmatrix} = \begin{bmatrix} \cos\theta & \sin\theta & 0 \\ -\sin\theta & \cos\theta & 0 \\ 0 & 0 & 1 \end{bmatrix} \begin{bmatrix} v_\xi \\ v_\eta \\ \omega \end{bmatrix} \tag{3.4}$$

where $\{x, y, \theta\}$ describe the Cartesian position and orientation of the car; v_ξ is the lateral velocity of the car; v_η is the longitudinal velocity of the car; and ω is the angular velocity of the car. Furthermore,

$$F_{\xi k} = \mu F_{zk} \left(\tilde{\alpha}_k - (\text{sgn}\, \delta)\, \tilde{\alpha}_k^2/3 + \tilde{\alpha}_k^3/27\right) \tag{3.5}$$

$$\sqrt{1 - \phi_k^2/\left(\mu F_{zk}\right)^2 + \phi_k^2/c_k^2}, k \in \{f, r\},$$

$$\tilde{\alpha}_k = c_k \alpha_k / \left(\mu F_{zk}\right),\ k \in \{f, r\}, \tag{3.6}$$

$$\alpha_f = \text{front tire slip angle} = \delta - \left(l_f \omega + v_\xi\right)/v_\eta, \tag{3.7}$$

$$\alpha_r = \text{rear tire slip angle} = \left(l_r \omega - v_\xi\right)/v_\eta, \tag{3.8}$$

$$F_{zk} = \left(mgl_r - \left(\phi_f + \phi_r\right) h\right) / \left(l_f + l_r\right), \tag{3.9}$$

$$F_{zr} = \left(mgl_f + \left(\phi_f + \phi_r\right) h\right) / \left(l_f + l_r\right), \tag{3.10}$$

$$\xi = \text{body-relative lateral axis},\ \eta = \text{body-relative longitudinal axis} \tag{3.11}$$

$$c_f, c_r = \text{cornering stiffness of front, rear tires} = 50{,}000\text{N/rad},\ 64{,}000\text{N/rad} \tag{3.12}$$

$$c_D = \text{lumped coefficient of drag (air resistance)} = 0.0005\text{m}^{-1} \tag{3.13}$$

$$\mu = \text{coefficient of friction} = 1, \tag{3.14}$$

$$F_{jk} = \text{frictional forces},\ j \in \{\xi, z\},\ k \in \{f, r\} \tag{3.15}$$

$$\phi_r = \text{longitudinal force on rear tires} = \begin{cases} 0 & \phi_f > 0 \\ k_b \phi_f & \phi_f < 0, k_b = 0.34 \end{cases} \tag{3.16}$$

$$m = 1500\text{kg}, I = 2500\text{kg-m}^2, l_f = 1.25\text{m}, l_r = 1.5\text{m}, h = 0.5m. \tag{3.17}$$

The controls are given by,

$$\text{-0.2rad} \leq \delta \leq 0.2\text{rad} \tag{3.18}$$

$$\text{-8000N} \leq \phi = \phi_f \leq 4000\text{N} \tag{3.19}$$

where δ is the user-controlled steering angle, and ϕ is the user-controlled longitudinal force on the front tires. Note that the separate brake and gas commands for the human are, in fact, the single ϕ variable, where the sign indicates whether the brake or the gas is active.

Because of input device constraints, the force (or acceleration) control ϕ is limited during each 1/50 second time step, based on its present value. If the gas pedal is currently being applied ($\phi > 0$), then the human operator can either increase or decrease the amount of applied force by a user-specified constant $\Delta\phi_g$ or switch to braking. Similarly, if the brake pedal is currently being applied ($\phi < 0$) the operator can either increase or decrease the applied force by a second constant $\Delta\phi_b$ or switch to applying positive force. Thus, the $\Delta\phi_g$ and $\Delta\phi_b$ constants define the responsiveness of each pedal. If we denote $\phi(k)$ as the current applied force and $\phi(k+1)$ as the applied force for the next time step, we can write in concise notation,

$$\phi(k+1) \in \{\phi(k), \min(\phi(k) + \Delta\phi_g, 4000), \tag{3.20}$$
$$\max(\phi(k) - \Delta\phi_g, 0), -\Delta\phi_b\},$$
$$\phi(k) \geq 0 \tag{3.21}$$

$$\phi(k+1) \in \{\phi(k), \max(\phi(k) - \Delta\phi_b, -8000), \tag{3.22}$$
$$\min(\phi(k) + \Delta\phi_b, 0), \Delta\phi_g\},$$
$$\phi(k) < 0 \tag{3.23}$$

Road descriptions

In the simulator, we define roads as a sequence of randomly generated segments of the form $\{l, 0\}$ (straight-line segments), and $\{r, \beta\}$ (curves), connected in a manner that ensures continuous first derivatives between segments. Here, l is the length of a given straight line segment, r is the radius of curvature of a curved segment, and β is its corresponding sweep angle, defined to be negative for left curves, and positive for right curves.

In order to make the driving task challenging, we place the following constraints on the individual segments:

$$100\text{m} \leq l \leq 200\text{m}, 100\text{m} \leq r \leq 200\text{m}, \text{and } \ 20^\circ \leq |\beta| \leq 20^\circ. \tag{3.24}$$

No segment may be followed by a segment of the same type; a curve is followed by a straight line segment with probability 0.4, and an opposite curve segment with probability 0.6. A straight line segment is followed by a left curve or right curve with equal probability. Roads are defined to be 10m wide (the car is 2m wide), and the visible horizon is set to 100m. For notational convenience, let d_ξ denote the car's lateral offset from the road median.

Figure 3.2(a), (b), and (c) shows roads #1, #2, and #3, respectively, the three 20km roads over which we collected human driving control data. Figure 3.3(a) shows road #4, the 20km road which we used as a cross-validation road for each modeling approach. Finally, Figure 3.3(b) shows road #5, the 20km road which we reserve for testing each of the modeling approaches.

3.1.3 Model class

We restrict the class of models we look at in these chapters to *static* (as opposed to *dynamic*) mappings between inputs and outputs. Because human control strategy is dynamic, we must map that dynamic system (i.e. the human control strategy) onto a static map.

In general, we can approximate any dynamic system through the difference equation [165],

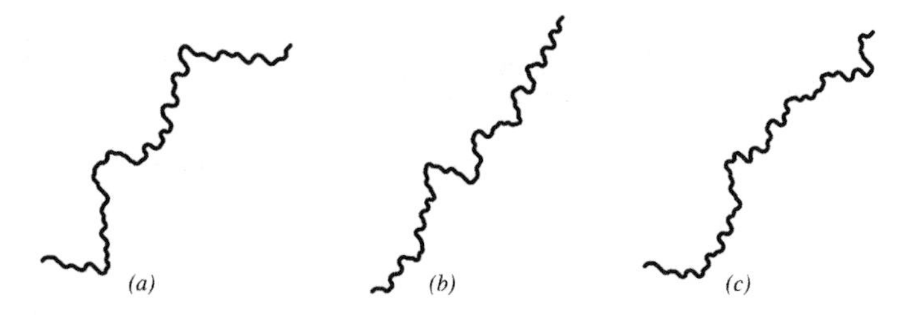

FIGURE 3.2: Data collection roads: (a) road #1, (b) road #2, (c) road #3.

FIGURE 3.3: (a) Cross-validation road #4 and (b) test road #5.

$$\begin{aligned}\bar{u}(k+1) = \; & \Gamma[\bar{u}(k), \bar{u}(k-1), \ldots, \bar{u}(k-n_u+1), \\ & \bar{x}(k), \bar{x}(k-1), \ldots, \bar{x}(k-n_x+1), \bar{z}(k)]\end{aligned} \tag{3.25}$$

where $\Gamma(\,\cdot\,)$ is some (possibly nonlinear) map, $\bar{u}(k)$ is the control vector, $\bar{x}(k)$ is the system state vector, and $\bar{z}(k)$ is a vector describing the external environment at time step k. The order of the dynamic system is given by the constants n_u and n_x, which may be infinite. Thus, a static model can abstract a dynamic system, provided that time-delayed histories of the state and command vectors are presented to the model as input, as illustrated in Figure 3.4.

For the case of the driving simulator, the HCS model will require, in general, (1) current and previous state information $\bar{x} = [v_\xi \; v_\eta \; \omega]^T$, (2) previous control information $\bar{u} = [\delta \; \phi]^T$, and a description of the road $\bar{z}$, visible from the current car position and orientation, where,

$$\bar{z} = [r_x(1) \; r_x(2) \; \cdots \; r_x(n_r) \; r_y(1) \; r_y(2) \; \cdots \; r_y(n_r)]\,,^* \tag{3.26}$$

is a $2n_r$-length vector of equivalently spaced, body-relative (x, y) coordinates (r_x, r_y) of the visible view of the road (median) ahead.

For notational convenience, we will denote the HCS model's input space for the driving simulator as,

$$\left\{v_\xi^{n_1}, v_\eta^{n_2}, \omega^{n_3}, \delta^{n_4}, \phi^{n_5}, r_x^{n_6}, r_y^{n_7}\right\}, n_i \geq 0, i \in \{1, 2, \ldots, 7\} \tag{3.27}$$

*This representation is reasonable, since computer vision algorithms such as RALPH [186] can abstract a very similar representation from real camera images.

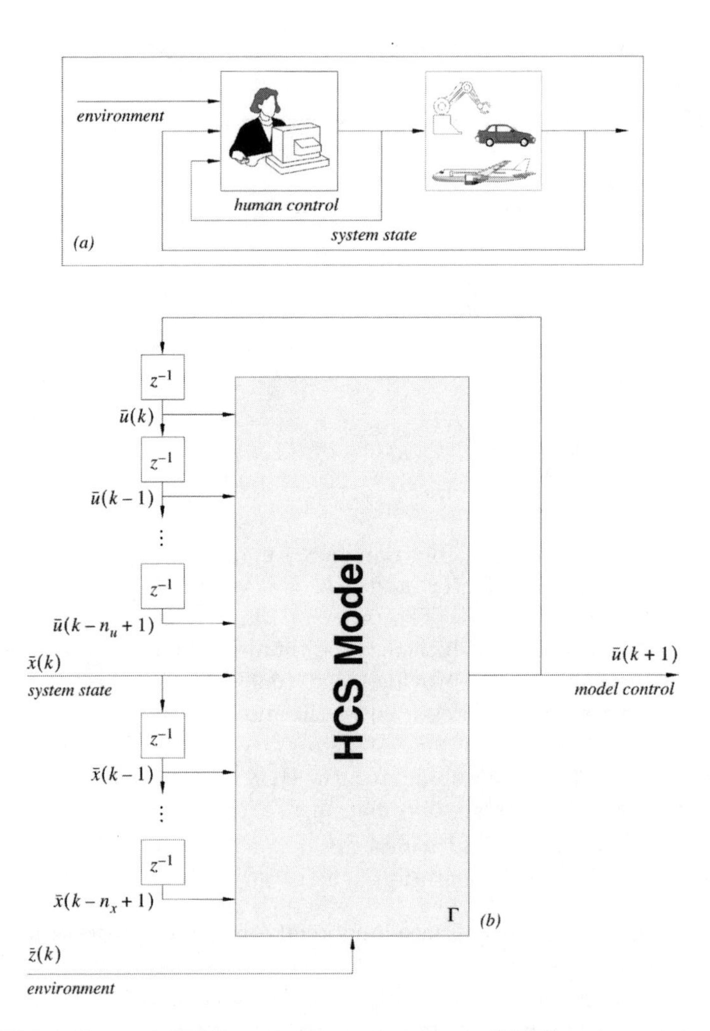

FIGURE 3.4: In modeling human control strategy (HCS), we want to replace the human controller (a) by a HCS model (b).

where at time step k,

$$\chi^{n_i} = \begin{cases} [\chi(k - n_i + 1) \ \dots \ \chi(k-1) \ \chi(k)]^T & \chi \in \{v_\xi, v_\eta, \omega, \delta, \phi\} \\ [\chi(1) \ \dots \ \chi(n_i - 1) \ \chi(n_i)]^T & \chi \in \{r_x, r_y\} \end{cases} \tag{3.28}$$

Thus, the total number of inputs n_{in} is given by,

$$n_{in} = \sum_{i=1}^{7} n_i. \tag{3.29}$$

We omit χ^{n_i} from the list in equation (3.28) if $n_i = 0$. For example, $\{\delta^3, r_x^{10}, r_y^{10}\}$ denotes a model whose input space consists of the previous three steering commands and a view of the road ahead, discretized to 10 body-relative coordinates. Unless otherwise noted, each time step k is $\tau = 1/50$ sec long. Finally, when $n_1 = n_2 = n_3$, $n_4 = n_5$, and $n_6 = n_7$, we use the shorthand notation,

$$\{\bar{x}^{n_x}, \bar{u}^{n_u}, \bar{z}^{n_r}\} = \left\{v_\xi^{n_x}, v_\eta^{n_x}, \omega^{n_x}, \delta^{n_u}, \phi^{n_u}, r_x^{n_r}, r_y^{n_r}\right\}, n_s, n_c, n_r \geq 0, \tag{3.30}$$

$$n_{in} = 3n_x + 2n_u + 2n_r \tag{3.31}$$

to denote the input space in equation (3.27).

3.2 Cascade neural networks with Kalman filtering

3.2.1 Cascade neural networks

In learning human control strategies, we wish to approximate the functional mapping between sensory inputs and control action outputs which guide an individual's actions. Function approximation, in general, consists of two parts: (1) the selection of an appropriate functional form, and (2) the adjustment of free parameters in the functional model to optimize some criterion. For most neural networks used today, the learning process consists of (2) only, since a specific functional form is selected prior to learning; that is, the network architecture is usually fixed before learning begins.

We believe, however, that both (1) and (2) above have a place in the learning process. Thus, for modeling human control strategy, we look towards the flexible cascade learning architecture [61], which adjusts the structure of the neural network as part of learning. The cascade learning architecture combines the following two notions: (1) a cascade architecture, in which hidden units are automatically added one at a time to an initially minimal network, and (2) the learning algorithm which creates and installs new hidden units as the learning requires in order to reduce the RMS error (e_{RMS}) between the network's outputs and the training data.

As originally formulated in [61], *cascade neural network* training proceeds in several steps. Initially, there are no hidden units in the network, only direct input-output connections. These weights are trained first, thereby capturing any linear relationship between the inputs and outputs. With no further significant decrease in the RMS error between the network outputs and the training data (e_{RMS}), a first hidden unit is added to the network from a pool of *candidate* units. Using the quickprop algorithm [59] – an improved version of the standard backprop algorithm – these candidate units are trained independently and in parallel with different random initial weights.

Again, after no more appreciable error reduction occurs, the best candidate unit is selected and installed in the network. Once installed, the hidden-unit input weights are frozen, while the weights to the output units are retrained. By freezing the input weights for all previous hidden units, each training cycle is equivalent to training a three-layer feedforward neural network with a single hidden unit. This allows for much faster convergence of the weights during training than in a standard multi-layer feedforward network where many hidden-unit weights are trained simultaneously. The process is repeated until the algorithm succeeds in reducing e_{RMS} sufficiently for the training set or the number of hidden units reaches a specified maximum number. Figure 3.5 below illustrates, for example, how a two-input, single-output network grows as two hidden units are added. Note that a new hidden unit receives as input connections from the input units as well as all previous hidden units (hence the name "cascade"). A cascade network with n_{in} input units (including the bias unit), n_h hidden units, and n_o output units, has n_w connections where,

$$n_w = n_{in}n_o + n_h(n_{in} + n_o) + (n_h - 1)n_h/2 \tag{3.32}$$

Recent theorems by Cybenko [50] and Funahashi [74], which hold that standard layered neural networks are universal function approximators also hold for the cascade network topology, since any multi-layer feedforward neural network with k hidden units arranged in m layers, fully connected between consecutive layers, is a special case of a cascade network with k hidden units and some weight connections equal to zero.

Thus, the cascade architecture relaxes *a priori* assumptions about the functional form of the model to be learned by dynamically adjusting the network size. We can relax these assumptions further by allowing new hidden units to have variable activation functions [168], [169]. In fact, Cybenko [50] shows that sigmoidal functions are not the only possible activation functions which allow for universal function approximation. There are other nonlinear functions, such as *sine* and *cosine* for example, which are complete in the space of n-dimensional continuous functions. In the pool of candidate units, we can assign a different nonlinear activation function to each unit, rather than just the standard sigmoidal function. During candidate training, the algorithm will select for installment whichever candidate unit reduces e_{RMS} for the

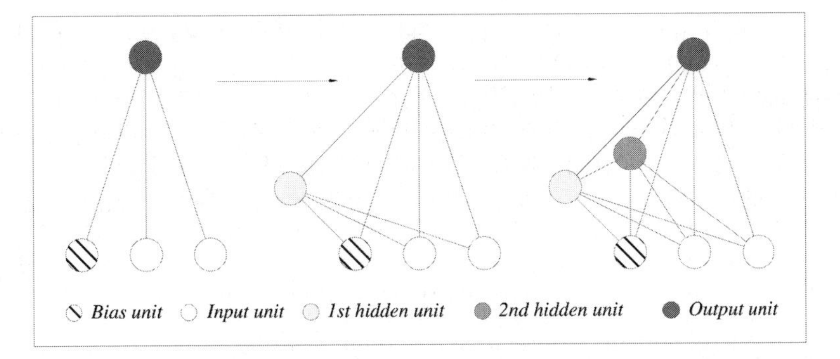

FIGURE 3.5: The cascade learning architecture adds hidden units one at a time to an initially minimal network. All connections in the diagram are feedforward.

training data the most. Hence, the unit with the most appropriate activation function at that point during training is selected. Typical alternatives to the sigmoidal activation function are the Gaussian function, Bessel functions, and sinusoidal functions of various frequency [168].

3.2.2 Node-decoupled extended Kalman filtering

While quickprop is an improvement over the standard backpropagation algorithm for adjusting the weights in the cascade network, it can still require many iterations until satisfactory convergence is reached [59], [222]. Thus, we modify standard cascade learning by replacing the quickprop algorithm with node-decoupled extended Kalman filtering (*NDEKF*), which has been shown to have better convergence properties and faster training times than gradient-descent techniques for fixed-architecture multi-layer feedforward networks [190]. As we demonstrate later, the combination of cascade learning and NDEKF alleviates critical problems that each exhibits by itself, and better exploits the main strengths of both algorithms.

Learning architecture

In general extended Kalman filtering (*GEKF*) [222], an $n_w \times n_w$ conditional error covariance matrix P, which stores the interdependence of each pair of n_w weights in a given neural network, is explicitly generated. NDEKF reduces this computational and storage complexity by – as the name suggests – decoupling weights by node, so that we consider only the interdependence of weights feeding into the same unit (or node). This, of course, is a natural formulation for cascade learning, since we only train the input-side weights of one hidden unit and the output units at any one time; we can partition the weights by unit into $n_o + 1$ groups – one group for the current hidden unit, n_o

groups for the output units. In fact, by iteratively training one hidden unit at a time and then freezing that unit's weights, we minimize the potentially detrimental effect of the node-decoupling.

Denote ω_k^i as the input-side weight vector of length n_w^i at iteration k, for unit $i \in \{0, 1, \ldots, n_o\}$, where $i = 0$ corresponds to the current hidden unit being trained, and $i \in \{1, ..., n_o\}$ corresponds to the ith output unit, and,

$$n_w^i = \begin{cases} n_{in} + n_h - 1 & i = 0 \\ n_{in} + n_h & i \in \{1, \ldots, n_o\} \end{cases} \tag{3.33}$$

The NDEKF weight-update recursion is then given by,

$$\omega_{k+1}^i = \omega_k^i + \{(\psi_k^i)^T (A_k \xi_k)\} \phi_k^i \tag{3.34}$$

where ξ_k is the n_o-dimensional error vector for the current training pattern, ψ_k^i is the n_o-dimensional vector of partial derivatives of the network's output unit signals with respect to the ith unit's net input, and,

$$\phi_k^i = P_k^i \zeta_k^i \tag{3.35}$$

$$A_k = \left[I + \sum_{i=0}^{n_o} \{(\zeta_k^i)^T \phi_k^i\} \left[\psi_k^i (\psi_k^i)^T\right] \right]^{-1} \tag{3.36}$$

$$P_{k+1}^i = P_k^i - \{(\psi_k^i)^T (A_k \psi_k^i)\} \left[\phi_k^i (\phi_k^i)^T\right] + \eta_Q I \tag{3.37}$$

$$P_0^i = (1/\eta_P) I \tag{3.38}$$

where ζ_k^i is the n_w^i-dimensional input vector for the ith unit, and P_k^i is the $n_w^i \times n_w^i$ approximate conditional error covariance matrix for the ith unit. We include the parameter η_Q in (3.37) to alleviate singularity problems for error covariance matrices [190]. In (3.34) through (3.37), {}'s, ()'s, and []'s evaluate to scalars, vectors, and matrices, respectively.

The ψ_k^i vector is easy to compute within the cascade framework. Let O_i be the value of the ith output node, Γ_O be its corresponding activation function, net_{Oi} be its net activation, Γ_H be the activation function for the current hidden unit being trained, and net_H be its net activation. Then,

$$\frac{\partial O_i}{\partial net_{Oj}} = 0, \forall i \neq j \tag{3.39}$$

$$\frac{\partial O_i}{\partial net_{Oi}} = \Gamma_O'(net_{Oi}), i \in \{1, \ldots, n_o\} \tag{3.40}$$

$$\frac{\partial O_i}{\partial net_H} = w_{Hi} \cdot \Gamma_O'(net_{Oi}) \cdot \Gamma_H'(net_H) \tag{3.41}$$

where w_{Hi} is the weight connecting the current hidden unit to the ith output unit.

Throughout the remainder of the paper, we will use the shorthand notation explained in Table 3.1 for different neural network training methodologies.

Table 3.1: Notation

Symbol	*Methodology*	*Training algorithm*
Fq	*Fixed architecture*[a]	*quickprop*
Cq	*Cascade learning*[b]	*quickprop*
Fk	*Fixed architecture*	*NDEKF*
Ck	*Cascade learning*	*NDEKF*

a. All weights are trained simultaneously.
b. Hidden units are added and trained one at a time.

3.3 HCS models: continuous control

3.3.1 Cascade with quickprop learning

Here, we present modeling results for the cascade/quickprop (*Cq*) learning architecture. In subsequent sections, we will compare these results to cascade/NDEKF learning (*Ck*).

Experimental data

Appendix A describes driving control data from six different individuals – (1) Larry, (2) Curly, (3) Moe, (4) Groucho, (5) Harpo, and (6) Zeppo across three different roads, roads #1, #2, and #3 in Figures 3.2(a), (b), and (c), respectively. For notational convenience, let $X^{(i,j)}$, $i \in \{1,2,3,4,5,6\}$, $j \in \{1,2,3\}$, denote the run from person (i) on road $\#j$, sampled at 50Hz.

Model inputs and outputs

We defer to the next section a detailed discussion of the input space choices made for the HCS models described in this section. This is necessary in part because our input selection procedure is dependent on the cascade/NDEKF learning results. For now, we simply note what the choices are for each model. In Table 3.2, n_x, n_u, and n_r, as defined in equation (3.30), completely characterize the input space for the results presented below. As we shall see later in the next section, model performance remains similar over a wide range of input spaces, especially once a sufficient number of inputs are given. Here, "sufficient number" means that there are enough time-delayed values of each state and control variable such that the model is able to build necessary derivative dependencies between the inputs and outputs. The outputs of

Table 3.2: Input space for each Cq model

Run[a]		*Input space*		n_h[b]	*Figure*
		$n_x = n_u$	n_r		
Larry's second	$X^{(1,2)}$	3	10	12	Figure 4-1
Moe's first	$X^{(3,1)}$	3	10	14	Figure 4-2
Groucho's first	$X^{(4,1)}$	6	10	19	Figure 4-3
Harpo's second	$X^{(5,2)}$	5	10	9	Figure 4-4

a. See Appendix A for a detailed description of each human control data set.
b. Number of hidden units in final Cq model.

each model are, of course, the next steering angle and acceleration command $\{\delta(k+1), \phi(k+1)\}$.

Cq training

For each model $\Gamma^{(i,j)}$, we process the training data as follows. First, we excise from the complete run $X^{(i,j)}$ those segments where the human operator temporarily runs off the road ($d_\xi > 5$m). Let $[t, t+t_l]$ denote an interval of time, in seconds, that a human operator veers off the road. Then, we cut the data corresponding to time interval $[t-1, t+t_l]$ from the training data. In other words, we not only remove the actual off-road data from the training set, but also the second of data leading up to the off-road interval. This ensures that the HCS model does not learn control behaviors that are potentially destabilizing.

Next, we normalize each input dimension of the HCS model, such that no input in the training data falls outside the interval $[-1, 1]$. Finally we randomize the input-output training vectors and select half for training, while reserving the other half for testing. All the runs $X^{(i,j)}$ are approximately equal to or longer than 10 minutes in length. Thus, at 50Hz, typical training and testing data sets will consist of approximately 15,000 data points each.

Training proceeds until the RMS error in the test data set no longer decreases. We use eight candidate units, and allow up to 500 epochs for candidate as well as for output training. Table 3.2 reports the final number of hidden units n_h for the models presented below.

HCS models

Figures 3.6, 3.7, 3.8, and 3.9 illustrate some representative Cq learning results for four different runs: $X^{(1,2)}$ (Larry's second run), $X^{(3,1)}$ (Moe's first run), $X^{(4,1)}$ (Groucho's first run), and $X^{(5,2)}$ (Harpo's second run). Part (a) of each Figure plots the original human control data, while part (b) of each Figure plots the corresponding model control over the test road (#5 in Figure 3.3(b)).

Before comparing these results to Ck, we make the following two obser-

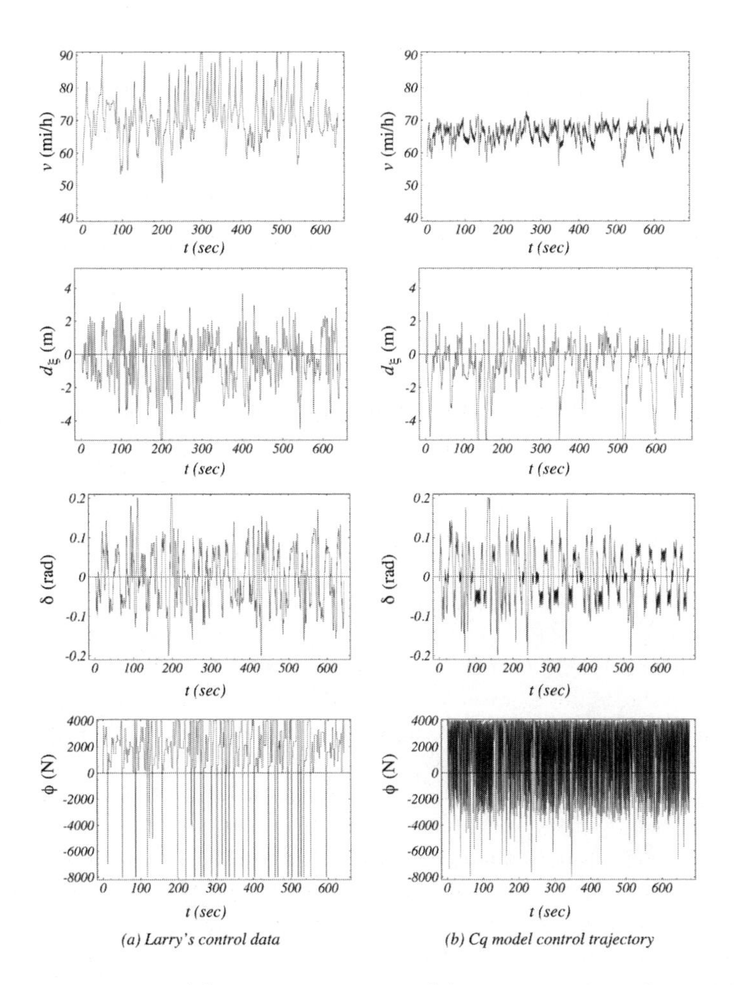

FIGURE 3.6: Larry's (a) training data and (b) corresponding Cq model data.

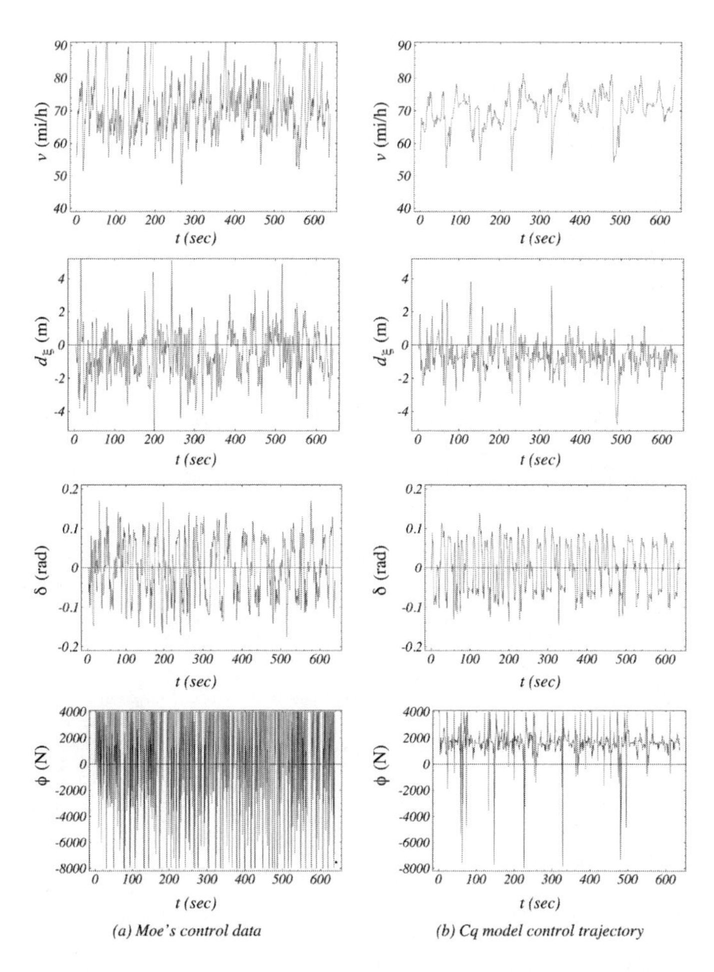

FIGURE 3.7: Moe's (a) training data and (b) corresponding Cq model data.

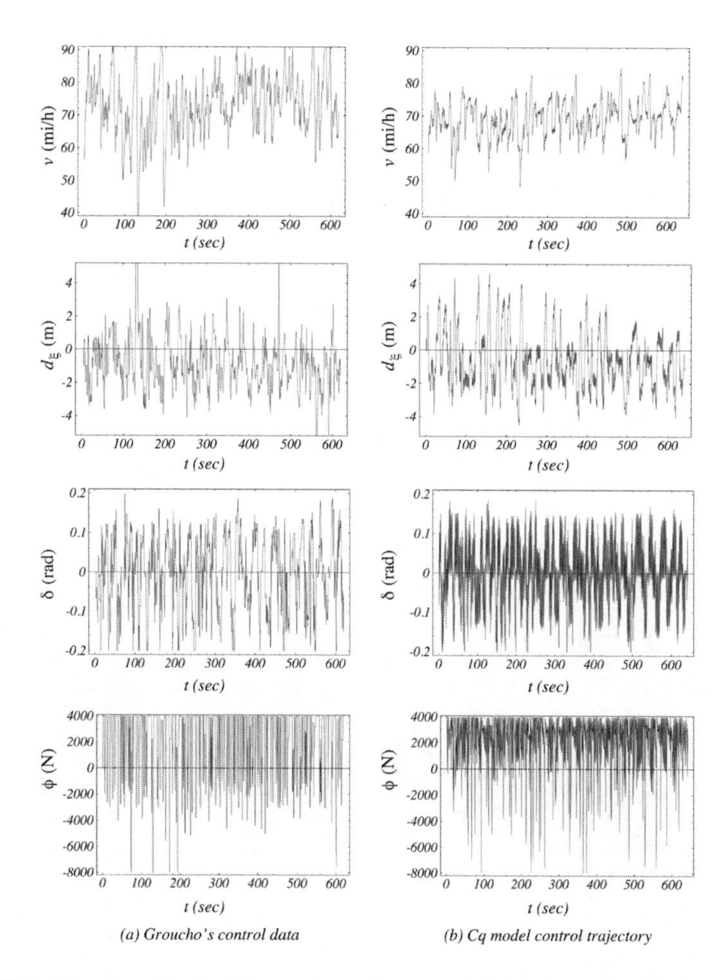

FIGURE 3.8: Groucho's (a) training data and (b) corresponding Cq model data.

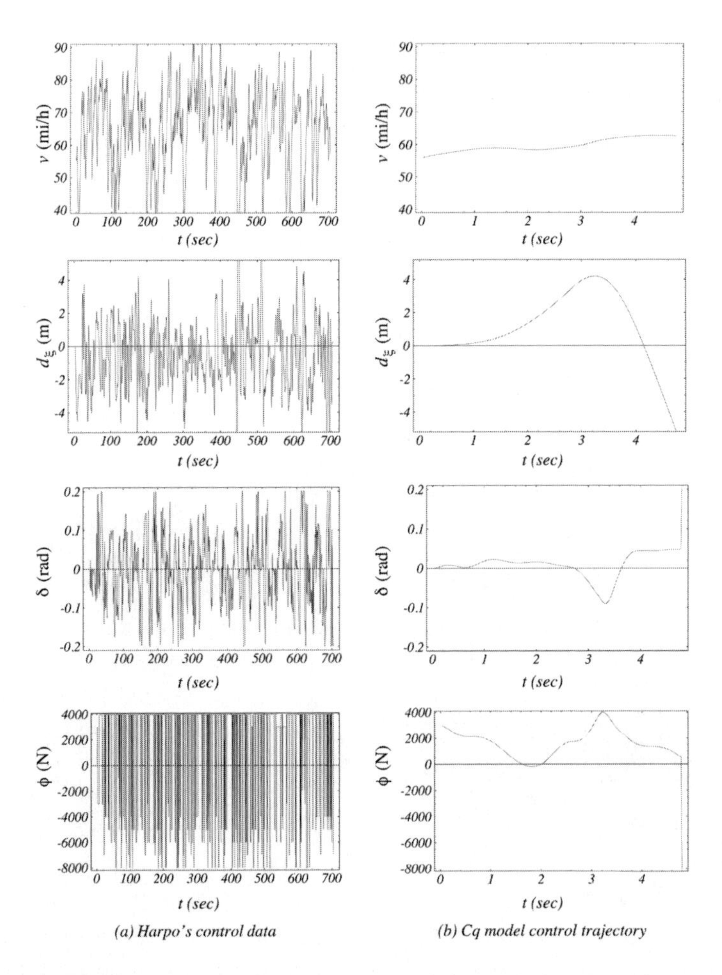

FIGURE 3.9: Harpo's (a) training data and (b) corresponding Cq model data (unstable).

vations. First, the discontinuous switching in the acceleration control ϕ induces substantial high-frequency noise in the Cq model control. This noise is especially evident in Larry's and Groucho's models (Figures 3.6 and 3.8, respectively). Second, Harpo's Cq model does not converge to a stable control strategy, as it loses complete track of the road after less than 5 seconds.

3.3.2 Cascade with NDEKF learning

Model inputs and outputs

For the neural networks trained in this section, we follow a simple experimental procedure for selecting the input space of each HCS model. Let model $\Gamma_k^{(i,j)}$ correspond to a HCS model trained with Ck on training data from run $X^{(i,j)}$ (i.e., person (i) on road $\#j$) and with input space,

$$\{\bar{x}^k, \bar{u}^k, \bar{z}^{10}\}, k \in \{1, 2, \ldots, 20\}, \tag{3.42}$$

as defined in equation (3.30). Also let,

$$\max(d_{\xi,k}^{(i,j)}), k \in \{1, 2, \ldots, 20\}, \tag{3.43}$$

denote the maximum lateral offset for model $\Gamma_k^{(i,j)}$ over the 20km validation road (#4) shown in Figure 3.3(a). Then, we select model $\Gamma_l^{(i,j)}$ for testing over the 20km testing road (#5) shown in Figure 3.3(b)[†]such that,

$$\max(d_{\xi,l}^{(i,j)}) < \max(d_{\xi,k}^{(i,j)}), \forall k \neq l. \tag{3.44}$$

In other words, we choose the model with the largest stability margin over the validation road. Figure 3.10 plots $\max(d_{\xi,k}^{(i,j)})$ as a function of k for the runs listed in Table 3.2. We observe that for $k \geq 3$, model performance, as measured by the maximum lateral offset, does not change significantly. Thus, when the model is presented with enough time-delayed values of each state and control variable, the model is able to build what appear to be necessary derivative dependencies between model inputs and outputs $\{\delta(k+1), \phi(k+1)\}$.

Ck training

For each model $\Gamma_k^{(i,j)}$, we process the training data as described in the previous section on Cq training. Training for a particular Ck neural network proceeds until the RMS error in the test data set no longer decreases. We use one candidate unit, and allow up to five epochs for candidate as well as for output training. After some experimentation, we settled on the following training parameter choices when training on human control data:

$$\eta_Q = 0.0, \eta_P = 0.000001 \tag{3.45}$$

[†]We note that roads #4 and #5 (Figure 3.3) are different from the data collection roads #1, #2, and #3 (Figure 3.2), but share similar statistical attributes.

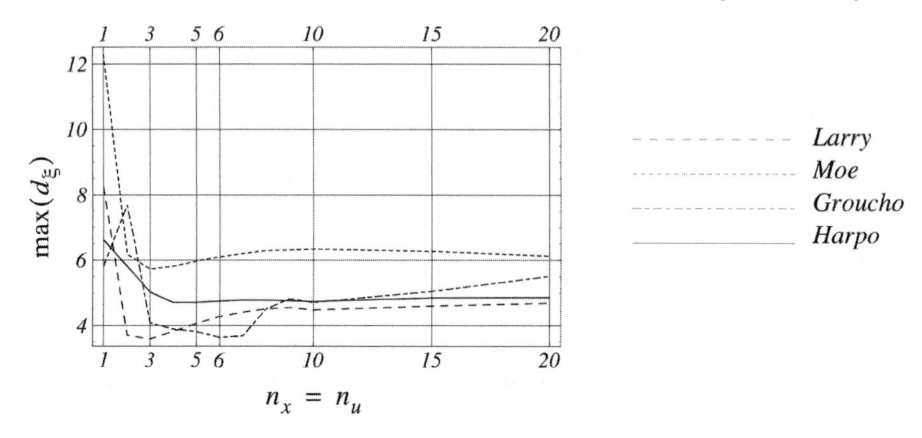

FIGURE 3.10: Maximum lateral offset d_ξ over validation road #4 as we vary the size of the input space.

HCS models

Figures 3.11, 3.12, 3.13, and 3.14 illustrate representative *Ck* learning results for the same four runs for which we report *Cq* results – namely, $X^{(1,2)}$ (Larry's second run), $X^{(3,1)}$ (Moe's first run), $X^{(4,1)}$ (Groucho's first run), and $X^{(5,2)}$ (Harpo's second run). Once again part (a) of each Figure plots the original human control data, while part (b) of each Figure plots the corresponding model control over the test road (#5).

We note that the model control trajectories in Figures 3.11 through 3.14 are for *linear Ck* models; that is, models with no hidden units. Despite the discontinuous acceleration command ϕ, these linear models are able to abstract convergent control strategies – even for Harpo's data – that keep the simulated car on the test road at approximately the same average speed and lateral distance from the road median as the corresponding human controllers. At the same time, we should point out that the *linear* Cq networks do *not* form stable controllers.

Because the *Ck* models are linear, they do not exhibit the type of high-frequency noise that we observed in the nonlinear *Cq* models. Only when we allow the *Ck* models to add nonlinear hidden units, will high-frequency noise manifest itself in the *Ck* models. Figure 3.15, for example, illustrates what happens to Larry's *Ck* model control when one hidden unit (sigmoidal) is added to the linear model. Thus, in general, the linear *Ck* models do not benefit from the addition of nonlinear hidden units. In the next section, we discuss in much greater detail the implications of this result on the stability of the HCS models, the convergence properties of the *Cq* algorithm, the models' fidelity to the source training data, and the capacity of continuous function approximators to model switching control behaviors.

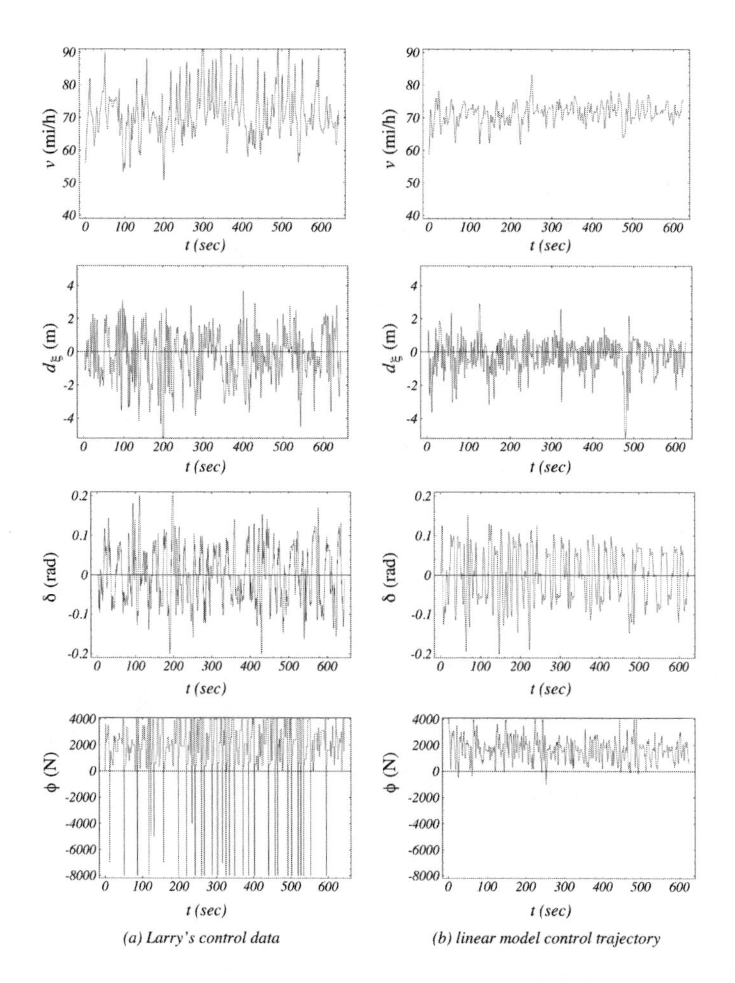

FIGURE 3.11: Larry's (a) training data and (b) corresponding linear model data.

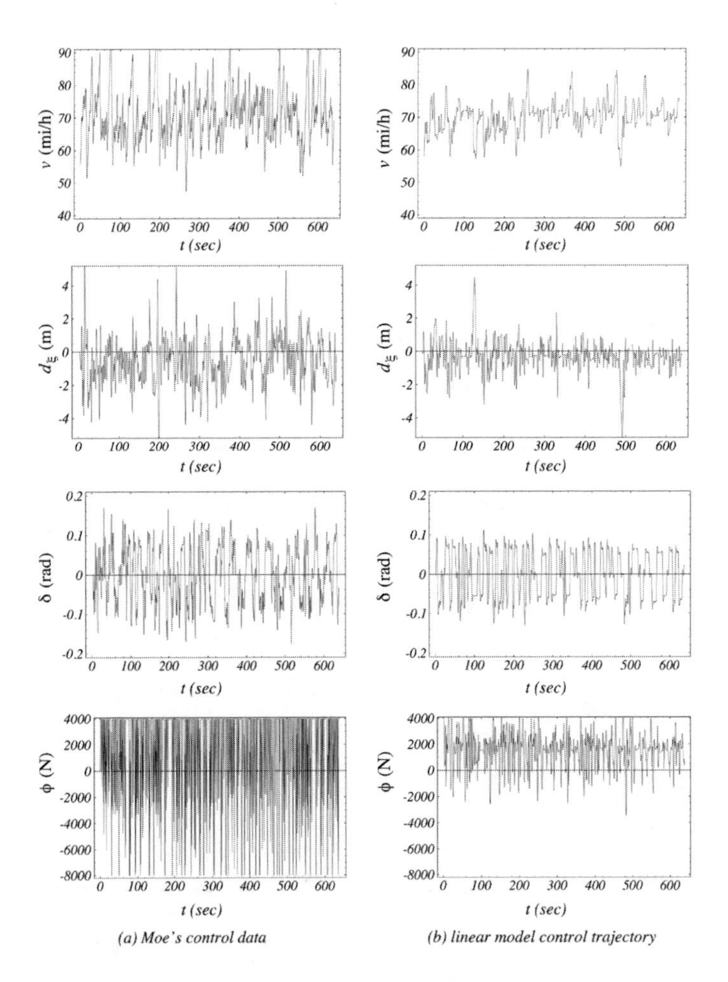

FIGURE 3.12: Moe's (a) training data and (b) corresponding linear model data.

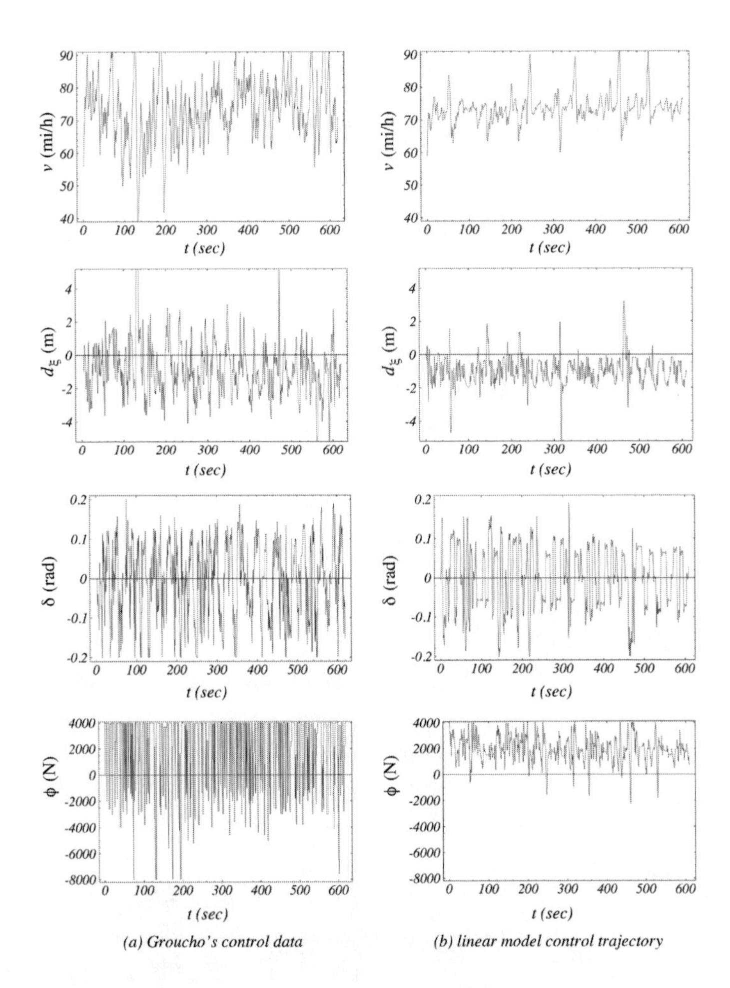

FIGURE 3.13: Groucho's (a) training data and (b) corresponding linear model data.

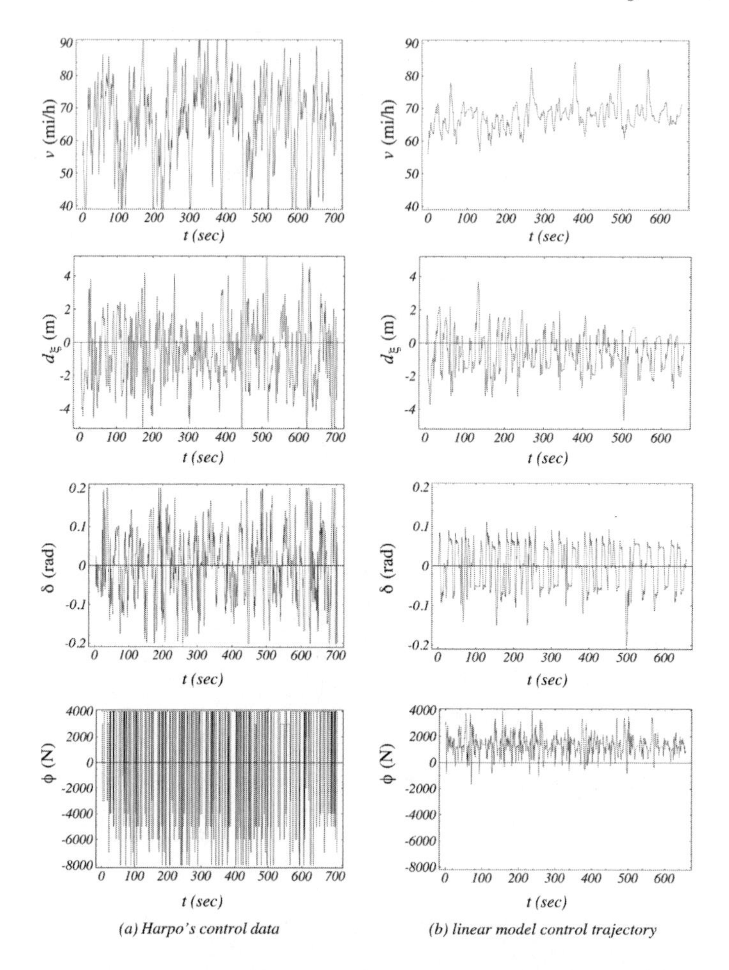

FIGURE 3.14: Harpo's (a) training data and (b) corresponding linear model data.

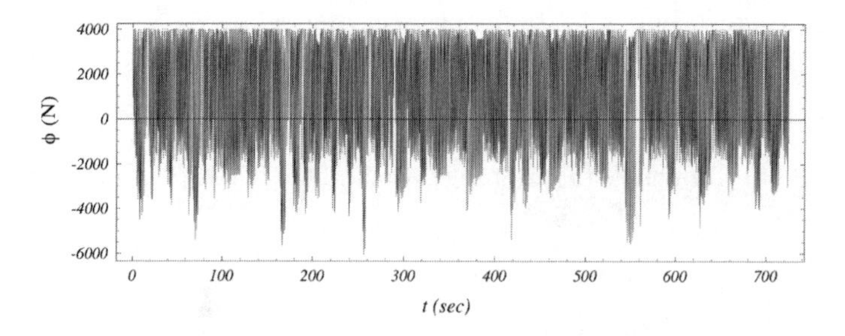

FIGURE 3.15: Larry's *Ck* model control with one hidden unit added. Note that the additional hidden unit causes significant (not really beneficial) high-frequency noise.

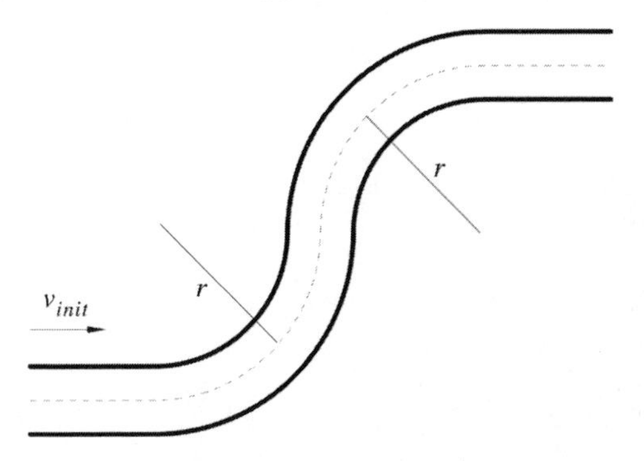

FIGURE 3.16: This s-curve test road is used to generate stability profiles for the human control strategy models, as v_{init} and r are varied.

3.3.3 Analysis

Model stability

In this section, we experimentally assess the stability of our *Ck* and *Cq* HCS models. While we have already seen one example of instability (Harpo's *Cq* model), we would like to determine for what range of initial conditions and road curvatures each model is stable (i.e., stays on the road). To do this for a given HCS model, we record the maximum lateral offset for that model ($\max(d_\xi)$) as it attempts to negotiate the s-curve shown in Figure 3.16 for different radii r and initial velocities v_{init}. Figures 3.17, 3.18, 3.19, and 3.20 plot these stability profiles for Larry, Moe, Groucho, and Harpo, respectively, where $90\text{m} \leq r \leq 250\text{m}$ and $20\text{mi/h} \leq v_{init} \leq 100\text{mi/h}$. ‡

From these Figures, we observe that, in general, the *Ck* models behave in a stable manner for a greater range of initial and environmental conditions than do the *Cq* models. Moreover, the control behaviors of the *Ck* models vary more smoothly with changes in r and v_{init}. Thus, the *Cq* models, with their many additional hidden units, do not appear to learn anything beneficial with the increased nonlinearity of the larger models.

Learning convergence

Now, we examine the difference in learning convergence between the *Cq* and *Ck* learning architectures. As we have noted previously, the linear *Ck* networks converge in less than one epoch to approximately the same RMS error as the *Cq* networks after thousands of epochs and multiple hidden units. Consider Figures 3.21 and 3.22, for example. In Figure 3.21, we show how the RMS error over Groucho's entire training and test data sets decreases in the first

‡From Harpo's stability profile, it becomes apparent why his *Cq* model fails on road #5. Road #5 begins with an s-curve whose radii are 117m and 123m, respectively.

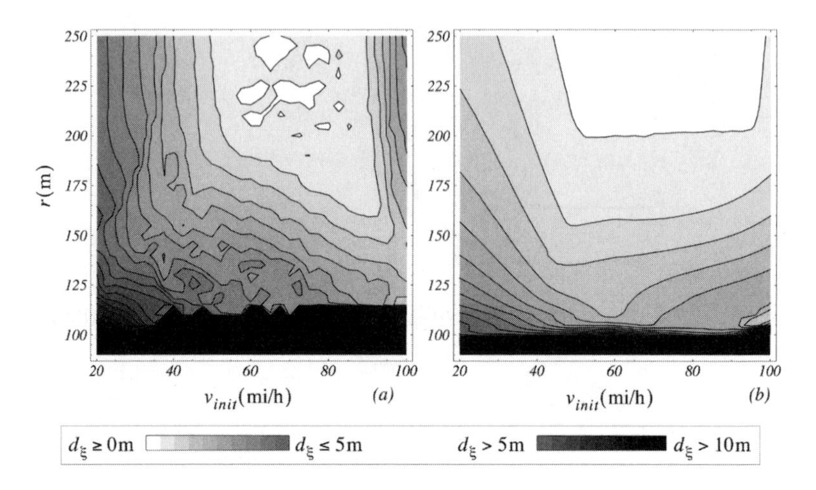

FIGURE 3.17: Larry's stability profiles through the s-curve for (a) the *Cq* model and (b) the *Ck* model (lighter colors are better).

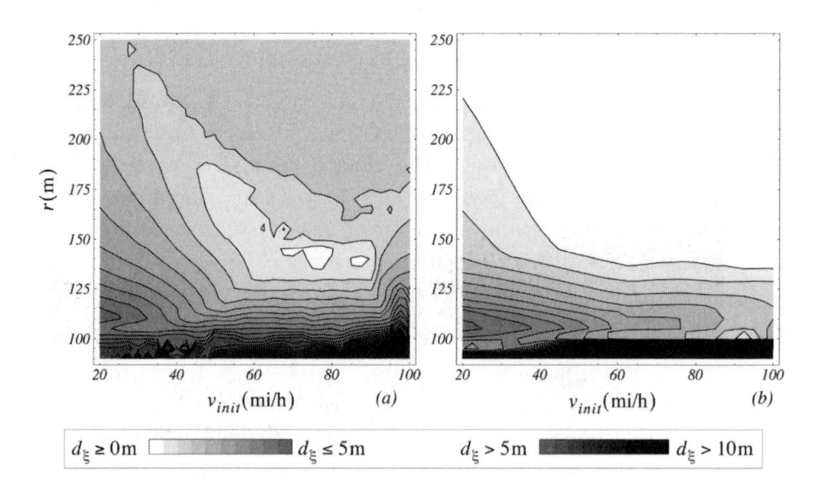

FIGURE 3.18: Moe's stability profiles through the s-curve for (a) the *Cq* model and (b) the *Ck* model (lighter colors are better).

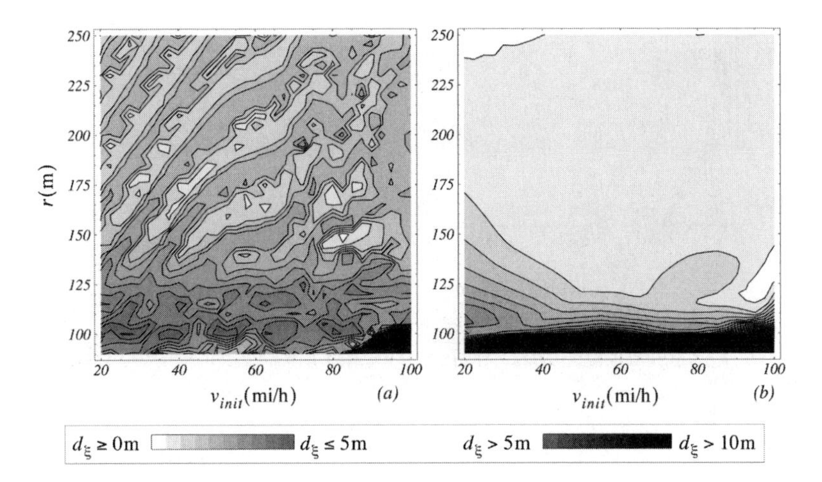

FIGURE 3.19: Groucho's stability profiles through the s-curve for (a) the *Cq* model and (b) the *Ck* model (lighter colors are better).

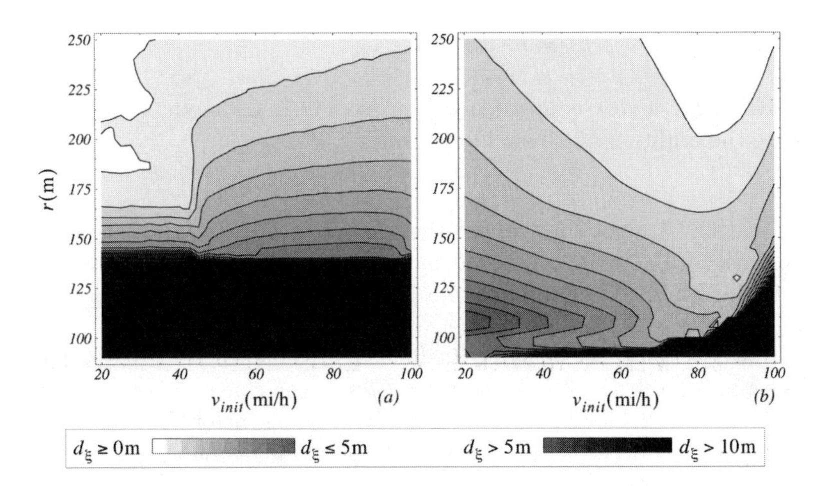

FIGURE 3.20: Harpo's stability profiles through the s-curve for (a) the *Cq* model and (b) the *Ck* model (lighter colors are better).

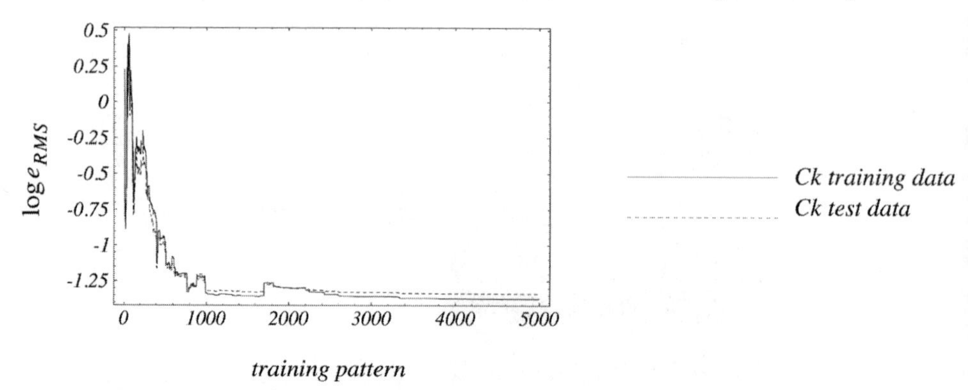

FIGURE 3.21: Error convergence per *training pattern* for Groucho's linear *Ck* model.

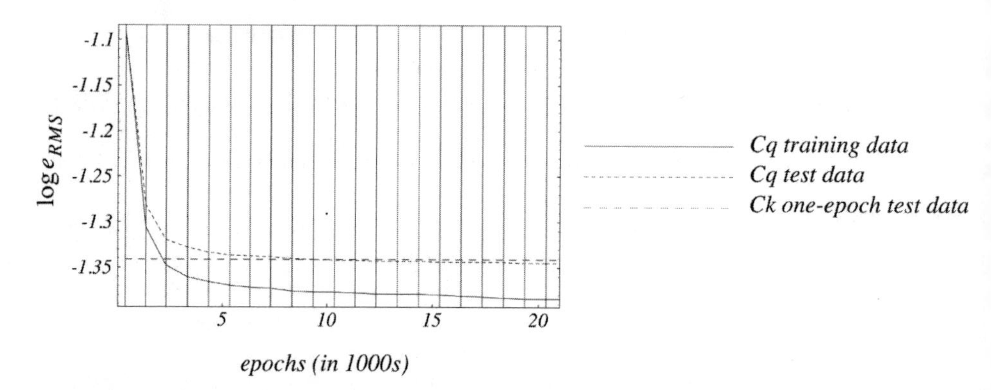

FIGURE 3.22: Error convergence for Groucho's *Cq* model. Vertical lines indicate the addition of a new hidden unit.

epoch of the *Ck* algorithm. Even though the entire training set consists of approximately 15,000 input-output patterns, the *Ck* algorithm converges very close to the final RMS error after only 1/3 of the training data set is presented *once*. By contrast, Figure 3.22 illustrates the convergence of the *Cq* algorithm for Groucho's data. Note that its linear convergence as measured by the RMS error is substantially worse than *Ck*'s linear convergence, and that *Cq* requires about 12 hidden units and 11,000 epochs before converging to an equivalent test RMS error. This is true despite repeated attempts to optimize the learning parameters for the *Cq* algorithm.

It is apparent that the *Cq* algorithm encounters significant convergence problems when presented with the correlated, time-delayed inputs of our HCS models. This becomes even more apparent if we increase n_x and n_u beyond the values in Table 3.2. As an example, consider Harpo's model. Figure 3.23 plots the stability profile for the *Ck* model, where $n_x = n_u = 20$. Although

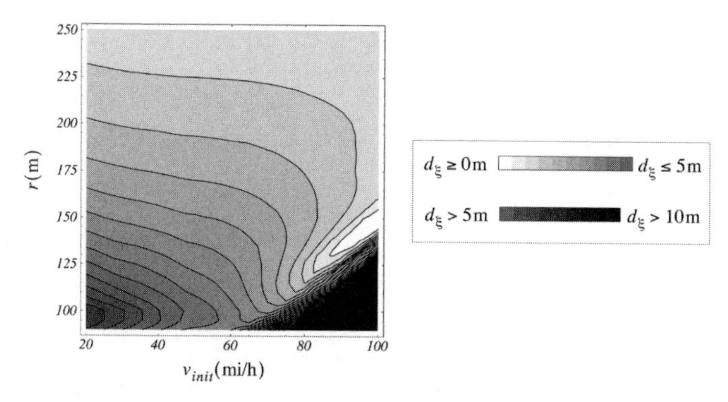

FIGURE 3.23: Harpo's stability profiles through the s-curve for the *Ck* model with $n_x = n_u = 20$. The *Cq* model is unstable for all r, v_{init}.

the dimensionality of the input space is increased from 45 inputs ($n_x = n_u = 5$, $n_r = 10$) to 120 inputs, the *Ck* algorithm preserves the stable control strategy of the original model. The *Cq* algorithm, on the other hand, does not converge to a stable control strategy for any values $r \subset [90\text{m}, 250\text{m}]$, $v_{init} \subset [20\text{mi/h}, 100\text{mi/h}]$.

Thus, although the reduction of e_{RMS} as hidden units are added to the *Cq* models suggests that substantial *nonlinear* modeling is occurring, this is not the case, since virtually *all* the reduction in e_{RMS} can be captured by a *linear* *Ck* model. In fact, any training algorithm that does not explicitly factor in the interdependence of weights in the neural network model is doomed to a similar fate, due to the correlated nature of the time-delayed state and control inputs, as well as the correlation between visible road coordinates.

Discussion

While thus far we have argued that the *Ck* algorithm shows better convergence to stabler HCS models, we have not yet addressed how closely each of the learned models approximates its corresponding training data. Examining Figures 3.6 through 3.9 and Figures 3.11 through 3.14, we make the *qualitative* observation that none of the models' control strategies closely mirror those of the corresponding human data. Neither the *Cq* nor the *Ck* learning algorithm appears to be able to model the driving control strategies with a high degree of fidelity to the source training data.

The principal source of this inability appears to be the discontinuous switching of the acceleration control ϕ. To better appreciate why this is the case, we would like to visualize how different input vectors in the training data map to different acceleration outputs $\phi(k+1)$. As an example, consider Groucho's control strategy data and let $n_x = n_u = 6$, $n_r = 10$, as before. For these input space parameters, the input training vectors $\zeta(k)$ are of length 50. Since it is impossible to visualize a 50-dimensional input space, we decompose each of

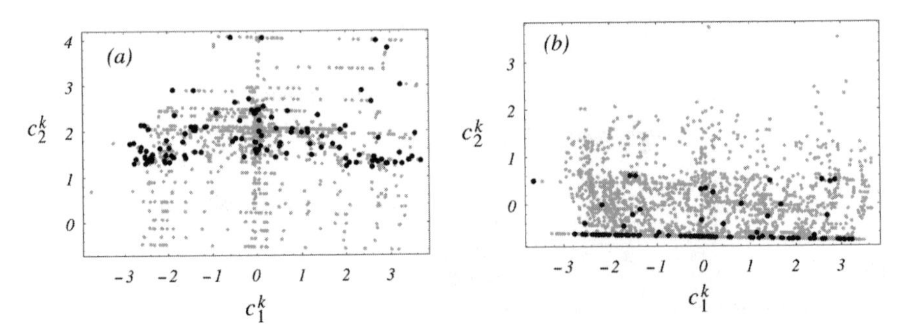

FIGURE 3.24: Switching actions (black) significantly overlap other actions (grey) when the current applied force is (a) negative (brake), and (b) positive (gas).

the input vectors $\zeta(k)$ in the training set into the principal components (PCs) [189] over Groucho's entire data set, such that,

$$\zeta(k) = c_1^k\gamma_1 + c_2^k\gamma_2 + \cdots + c_{50}^k\gamma_{50}, \tag{3.46}$$

where γ_i is the principal component corresponding to the ith largest eigenvalue σ_i. Now, for Groucho's control data we have that,

$$|\sigma_2/\sigma_1| = 0.44, |\sigma_i/\sigma_1| \leq 0.05, i \in \{3, 4, \ldots, 50\}, \tag{3.47}$$

so that we coarsely approximate the input vectors $\zeta(k)$ as,

$$\zeta(k) \approx c_1^k\gamma_1 + c_2^k\gamma_2. \tag{3.48}$$

By plotting the PC coefficients (c_1^k, c_2^k) in 2D space, we can now visualize the approximate relative location of the input vectors $\zeta(k)$. Figure 3.24(a) and (b) show the results for $\phi(k) < 0$ (brake), and $\phi(k) \geq 0$ (gas), respectively. In each plot, we distinguish points by whether or not $\phi(k+1)$ indicates a discontinuity (i.e., a switch between braking and accelerating) such that,

$$\phi(k) < 0 \text{ and } \phi(k+1) > 0 \;\; \text{[Figure 3.24(a)]} \tag{3.49}$$

$$\phi(k) > 0 \text{ and } \phi(k+1) < 0 \;\; \text{[Figure 3.24(b)]} \tag{3.50}$$

Those points that involve a switch are plotted in black, while a representative sample (20%) of the remaining points are plotted in grey.

We immediately observe from Figure 3.24 that – at least in the low-dimensional projection of the input vectors – the few training vectors that involve a switch overlap the many other vectors that do not. In other words, very similar inputs $\zeta(k)$ can lead to radically different outputs $\phi(k+1)$. Consequently, Groucho's acceleration control strategy may not be easily expressible in a functional form, let alone a smooth functional form. This poses an impossible learning challenge not just for cascade neural networks, but *any* continuous function approximator. In theory, no continuous function approximator will be capable of modeling the switching of the acceleration control ϕ (Figure 3.25).

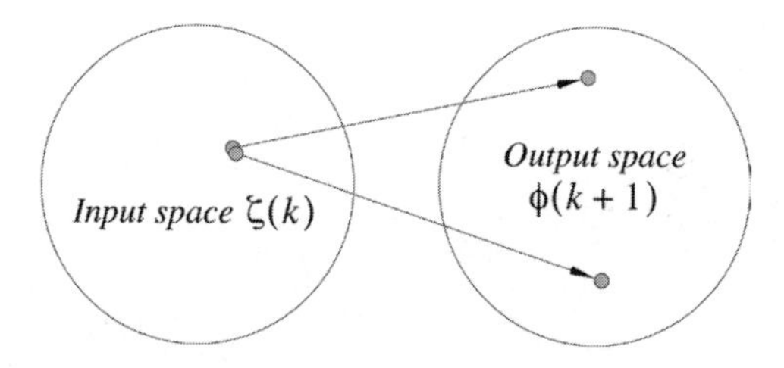

FIGURE 3.25: Switching causes very similar inputs to be mapped to radically different outputs.

3.4 HCS models: discontinuous control

To adequately model control behaviors that involve discontinuities in the input-output mapping, in this section we propose a stochastic, discontinuous learning algorithm. The proposed algorithm models possible control actions as individual statistical models. During run-time execution of the algorithm, a control action is then selected stochastically, as a function of both prior probabilities and posterior evaluation probabilities. We show that the resulting controller overcomes the shortcomings of continuous modeling approaches in modeling discontinuous control strategies, and that the resulting model s-trategies appear to exhibit a higher degree of fidelity to the human training data.

3.4.1 Hybrid continuous/discontinuous control

Figure 3.26 provides an overview of the proposed modeling approach. As before, we use a *Ck* model for the steering control δ. Now, however, we model the acceleration control ϕ in a discontinuous, statistical framework. The following sections describe this framework in much greater detail.

General statistical framework

First, let us derive a general statistical framework for faithfully modeling discontinuous control strategies. For now, we make the following assumptions. First, assume a control task where at each time step k, there is a choice of one of N different control actions A_i, $i \in \{1, \dots, N\}$. Second, assume that we have sets of input vector training examples $\bar{\zeta}_i^j$, $j \in \{1, 2, \dots, n\}$, where each $\bar{\zeta}_i^j$ leads to control action A_i at the next time step. Finally, assume that we can train statistical models λ_i, which maximize,

$$\prod_{j=1}^{n_i} P(\lambda_i | \bar{\zeta}_i^j), i \in \{1, \dots, N\}. \tag{3.51}$$

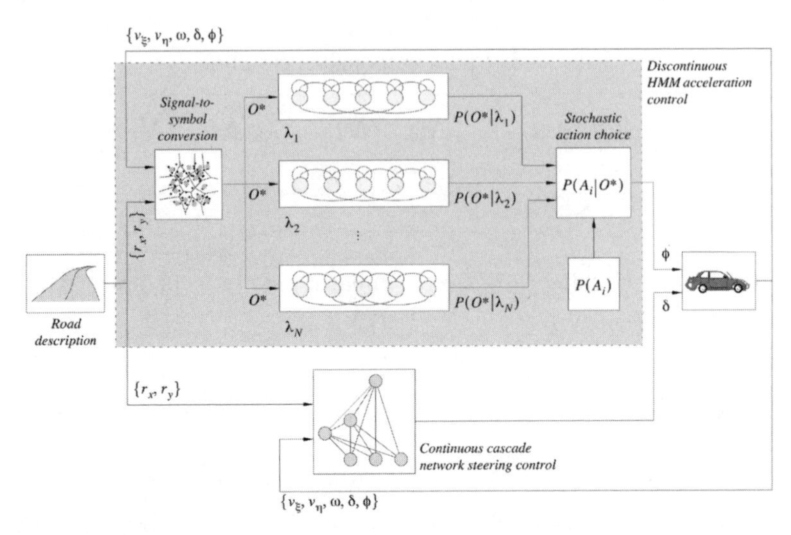

FIGURE 3.26: Overall hybrid discontinuous/continuous controller. Steering is controlled as before by a cascade model, while the discontinuous acceleration command is controlled by the HMM-based, stochastic framework (shaded box).

Given an unknown input vector $\bar{\zeta}^*$, we would like to choose an appropriate, corresponding control action A^*. Since model λ_i corresponds to action A_i, we define,

$$P(\bar{\zeta}^*|A_i) \equiv P(\bar{\zeta}^*|\lambda_i). \tag{3.52}$$

By Bayes Rule,

$$P(A_i|\bar{\zeta}^*) = \frac{P(\bar{\zeta}^*|A_i)P(A_i)}{P(\bar{\zeta}^*)}, \tag{3.53}$$

where,

$$P(\bar{\zeta}^*) \equiv \sum_{i=1}^{N} P(\bar{\zeta}^*|A_i)P(A_i), \tag{3.54}$$

serves as a normalization factor, $P(A_i)$ represents the *prior* probability of selecting action A_i and $P(A_i|\bar{\zeta}^*)$ represents the *posterior* probability of selecting action A_i given in the input vector $\bar{\zeta}^*$.

We now propose the following stochastic control strategy for A^*. Let,

$$A^* = A_i \text{ with probability } P(A_i|\bar{\zeta}^*), \tag{3.55}$$

so that, at each time step k, the control action A^* is generated stochastically as a function of the current model inputs ($\bar{\zeta}^*$) and the prior likelihood of each action.

Table 3.3: Pedal responsiveness choices

Individual	$\Delta\phi_g$ *(N)*	$\Delta\phi_b$ *(N)*
Larry	100	1000
Moe	100	300
Groucho	100	200
Harpo	1000	1000

Action definitions

As we point out in equations (3.20) and (3.22), the acceleration command ϕ is limited at each time step k to the following actions. When $\phi(k) \leq 0$ (the gas is currently active),

$$A_1\colon\ \phi(k+1) = \phi(k), \tag{3.56}$$

$$A_2\colon\ \phi(k+1) = \min(\phi(k) + \Delta\phi_g, 4000), \tag{3.57}$$

$$A_3\colon\ \phi(k+1) = \max(\phi(k) - \Delta\phi_g, 0), \tag{3.58}$$

$$A_4\colon\ \phi(k+1) = -\Delta\phi_b, \tag{3.59}$$

and when $\phi(k) < 0$ (the brake is currently active),

$$A_5\colon\ \phi(k+1) = \phi(k), \tag{3.60}$$

$$A_6\colon\ \phi(k+1) = \max(\phi(k) - \Delta\phi_b, -8000), \tag{3.61}$$

$$A_7\colon\ \phi(k+1) = \min(\phi(k) + \Delta\phi_b, 0), \tag{3.62}$$

$$A_8\colon\ \phi(k+1) = \Delta\phi_g. \tag{3.63}$$

Actions A_1 and A_5 correspond to no action for the next time step; actions A_2 and A_6 correspond to pressing harder on the currently active pedal; actions A_3 and A_7 correspond to easing off the currently active pedal; and actions A_4 and A_8 correspond to switching between the gas and brake pedals. The constants $\Delta\phi_g$ and $\Delta\phi_b$ are set by each human operator to the pedal responsiveness level he or she desires. Table 3.3 lists those choices for our four test individuals.

Statistical model choice

In part because of our familiarity with Hidden Markov Models, and because of the capacity of HMMs to model arbitrary statistical distributions, we choose discrete-output HMMs to be the trainable statistical models λ_i. Consequently, we must convert the multi-dimensional, real-valued model input space to discrete symbols.

For a particular data set X let,

$$\bar{\zeta}(k) = \left[v_{\xi}^{n_x} \; v_{\eta}^{n_x} \; \omega^{n_x} \; \delta^{n_u} \; \phi^{n_u} \; r_x^{n_r} \; r_y^{n_r} \right]^T, \tag{3.64}$$

denote the normalized model input vector at time step k corresponding to control action $\phi(k+1)$, where $\{v_{\xi}^{n_x}, v_{\eta}^{n_x}, \omega^{n_x}, \delta^{n_u}, \phi^{n_u}, r_x^{n_r}, r_y^{n_r}\}$ are defined in equation (3.28). Also, let V be a matrix, whose rows are the $\bar{\zeta}(k)$ vectors. Using the LBG VQ algorithm [142], we generate a codebook Q_L of size L that minimizes the quantization distortion $D(V, Q_L)$ defined in equation (4.67). To complete the signal-to-symbol conversion, the discrete symbols $o(k)$ corresponding to the minimum distortion of $\bar{\zeta}(k)$ are given by,

$$o(k) = T_{VQ}^{v}\left(\bar{\zeta}(k), Q_L\right), \tag{3.65}$$

where $T_{VQ}^{v}(\,\cdot\,)$ is defined in equation (4.69). Finally, let us define the observation sequence $O(k)$ of length n_O to be,

$$O(k) = \{o(k - n_O + 1), o(k - n_O + 2), \ldots, o(k)\}. \tag{3.66}$$

Now, suppose that we want to provide the Hidden Markov Models λ_i with m time-delayed values of the state and control variables as input. There are at least two ways to achieve this. We can either (1) let $n_x = n_u = m$, $n_O = 1$, or (2) let $n_x = n_u = 1$, $n_O = m$. In the first case, we vector quantize the entire input vector into a single observable, and base our action choice on that single observable. This necessarily forces the HMMs λ_i to single-state models, such that each model is completely described by its corresponding output probability vector B_i. Alternatively, we can vector quantize shorter input vectors but provide a longer sequence of observables $n_O > 1$ for HMM training and evaluation.

While in theory both choices start from identical input spaces, the single-observable, single-state case works better in practice. There are two primary reasons for this. Because the amount of data we have available for training comes from finite-length data sets, and is therefore necessarily limited in length, we must be careful that we do not overfit the models λ_i. Assuming fully forward-connected, left-to-right models λ_i, increasing the number of states from n_s to $(n_s + 1)$ increases the number of free (trainable) parameters by $n_s + L$, where L is the number of observables. Thus, having too many states in the HMMs substantially increases the chance of overfitting, since there may be too many degrees of freedom in the model. Conversely, by minimizing the number of states, the likelihood of overfitting is minimized.

A second reason that the single-observable, single-state case performs better relates to the vector quantization process. To understand how, consider that each input vector $\bar{\zeta}(k)$ minimally includes $2n_r$ road inputs $\{r_x^{n_r}, r_y^{n_r}\}$. If we let $n_r = 10$, then for $n_x = n_u = 1$, 80% of the input dimensions will be road-related, while only 20% will be state-related. Thus, the vector quantization will most heavily minimize the distortion of the road inputs, while in comparison neglecting the potentially crucial state and previous command inputs. With larger values of n_x and n_u, the vector quantization process relies more equally on the state, previous control, and road inputs, and therefore forms more pertinent feature (prototype) vectors for control.

Thus, for a VQ codebook Q_L, input vector $\bar{\zeta}^*$, $n_x = n_u = m$, $n_r = 10$, and $n_O = 1$,

$$O^* = l = T_{VQ}^v\left(\bar{\zeta}^*, Q_L\right), \text{ and} \tag{3.67}$$

$$P(A_i|\bar{\zeta}^*) = P(A_i|O^*) \propto P(O^*|\lambda_i)P(A_i) = b_i(l)P(A_i), \tag{3.68}$$

where $b_i(l)$ denotes the lth element in the λ_i model's output probability vector B_i. If we interpret the codebook vectors $\bar{q}_l$ as states S_l, then the discontinuous controller can be viewed as a learned stochastic policy that maps states S_l to actions A_i, where,

$$P(A_i|S_l) = b_i(l)P(A_i)/\sum_{i=1}^{N} b_i(l)P(A_i). \tag{3.69}$$

Prior probabilities

In order to calculate $P(A_i|S_l)$, we need to assign values to the prior probabilities $P(A_i)$. One approach is to estimate $P(A_i)$ by the frequency of occurrence of action A_i in a particular control data set X. For $\phi(k) \geq 0$,

$$P(A_i) = \begin{cases} n_i/\sum_{k=1}^{4} n_k & i \in \{1,2,3,4\} \\ 0 & i \in \{5,6,7,8\} \end{cases} \tag{3.70}$$

where n_i denotes the number of times action A_i was executed in the data set X; similarly, for $\phi(k) < 0$,

$$P(A_i) = \begin{cases} 0 & i \in \{1,2,3,4\} \\ n_i/\sum_{k=5}^{8} n_k & i \in \{5,6,7,8\} \end{cases} \tag{3.71}$$

Task-based modifications

While the assignment of priors in equations (3.70) and (3.71) are the best estimates for $P(A_i)$ from the data X, they are sometimes problematic when dealing with marginally stable training data. Consider, for example, Figure 3.27, where we plot a small part of Groucho's first run. We observe that Groucho's trajectory takes him close to the edge of the road; what keeps him from driving off the road is the switch from the gas to the brake at time t_s.

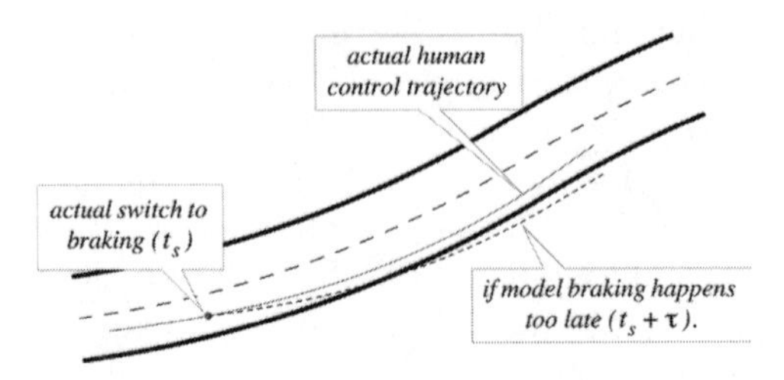

FIGURE 3.27: Instability can result if the stochastic controller switches to braking too late.

Now, because the action selection criterion in equation (3.55) is stochastic, it is possible that the stochastic controller will only brake at time $t_s + \tau$, even if it correctly models that time t_s is the most likely time for a control switch. Braking at time $t_s + \tau$, however, may be too late for the car to maintain contact with the road.

Consequently, we would like to improve the stability margins of the stochastic control model. The stability of the system (i.e., the simulated car) is directly related to the kinetic energy T being pumped into the system,

$$T \propto \sum_k \phi(k) \tag{3.72}$$

so that the expected value of T, $E[T]$, is proportional to,

$$E[T] \propto \sum_k E[\phi(k)]. \tag{3.73}$$

Hence, for increased stability margins, we want to adjust the stochastic model to generate $\phi'(k)$, where,

$$E[\phi'(k)] < E[\phi(k)] \tag{3.74}$$

We can realize condition (3.74) by slight increases in the priors for those actions that decrease $E[\phi(k)]$ – namely, A_3 or A_4. To stay within probabilistic constraints, we offset these changes by slight decreases in the priors A_2 or A_1, respectively, so that the modified priors are given as either,

$$P'(A_3) = P(A_3) + \varepsilon_s \text{ and } P'(A_2) = P(A_2) - \varepsilon_s, \text{ or} \tag{3.75}$$

$$P'(A_4) = P(A_4) + \varepsilon_s \text{ and } P'(A_1) = P(A_1) - \varepsilon_s, \tag{3.76}$$

where ε_s determines the degree to which we decrease $E[\phi(k)]$. As we shall observe later, for some human control data $P(A_3) = 0$. In that case we choose modification (3.76), so as not to introduce a control action that was never observed in the human control strategy. When $P(A_3)$ is substantial, then we choose modification (3.75).

While instability is a common failure mode of the unmodified stochastic controller, another very rare failure mode leads to exactly the opposite: the brake is engaged too long by the stochastic controller and the simulated car comes to a screeching halt. This problem is very similar to the instability problem, in that a switch – in this case from braking to accelerating – occurs too late. Once the car is stopped, the distortion for the vector-quantized input vector $\bar{\zeta}(k)$ is large for all VQ codebook vectors $\bar{q}_l$. It will be smallest, however, for those codebook vectors where the previous acceleration commands ϕ are less than zero. Hence, once the car is stopped, the brake remains engaged for a long time. Although the stochastic selection criterion in (3.55) ensures that eventually the simulated car will once again switch from braking, we would like to prevent the car from stopping altogether. Unlike the instability problem, this is significantly easier to monitor, since the velocity v directly predicts when a stopping event is about to occur. Consequently, we modify the statistical controller so that,

$$P(A_8|v < v_{min}) = 1, P(A_i|v < v_{min}) = 0, i \in \{1, 2, \ldots, 7\}, \qquad (3.77)$$

where v_{min} is chosen to reflect the range of velocities in the human control data. Over repeated trials, condition (3.77) is invoked on average approximately one time per 20km test run.

3.4.2 Experimental results

Model training

In order to make a fair comparison of the HCS models in this section with those of the previous section, we select the same input space parameters n_x, n_u, and n_r as those listed in Table 3.2. Furthermore, we let $n_O = n_s = 1$, so that the HMMs λ_i are single-state models. As we have already argued, we get significantly better performance from the more constrained models than we do if we let $n_O > 1$ and $n_s > 1$. We vector quantize the training data for each run to $L = 512$ levels. Also, for each run we choose the stabilization parameter ε_s to ensure stability over the validation road #4 (Figure 3.3(a)), and then test the resulting modified controller over test road #5 (Figure 3.3(b)). Table 3.4 summarizes the stabilization parameter and minimum velocity choices for each model. Finally, for the steering control δ, we select the same linear *Ck* as in the previous section.

HCS models

Figures 3.28, 3.29, 3.30, and 3.31 illustrate representative hybrid controller results for the same four runs for which we report cascade network results – namely, $X^{(1,2)}$ (Larry's second run), $X^{(3,1)}$ (Moe's first run), $X^{(4,1)}$ (Groucho's first run), and $X^{(5,2)}$ (Harpo's second run). Once again part (a) of each Figure plots the original human control data, while part (b) of each Figure plots the corresponding model control over the test road (#5).

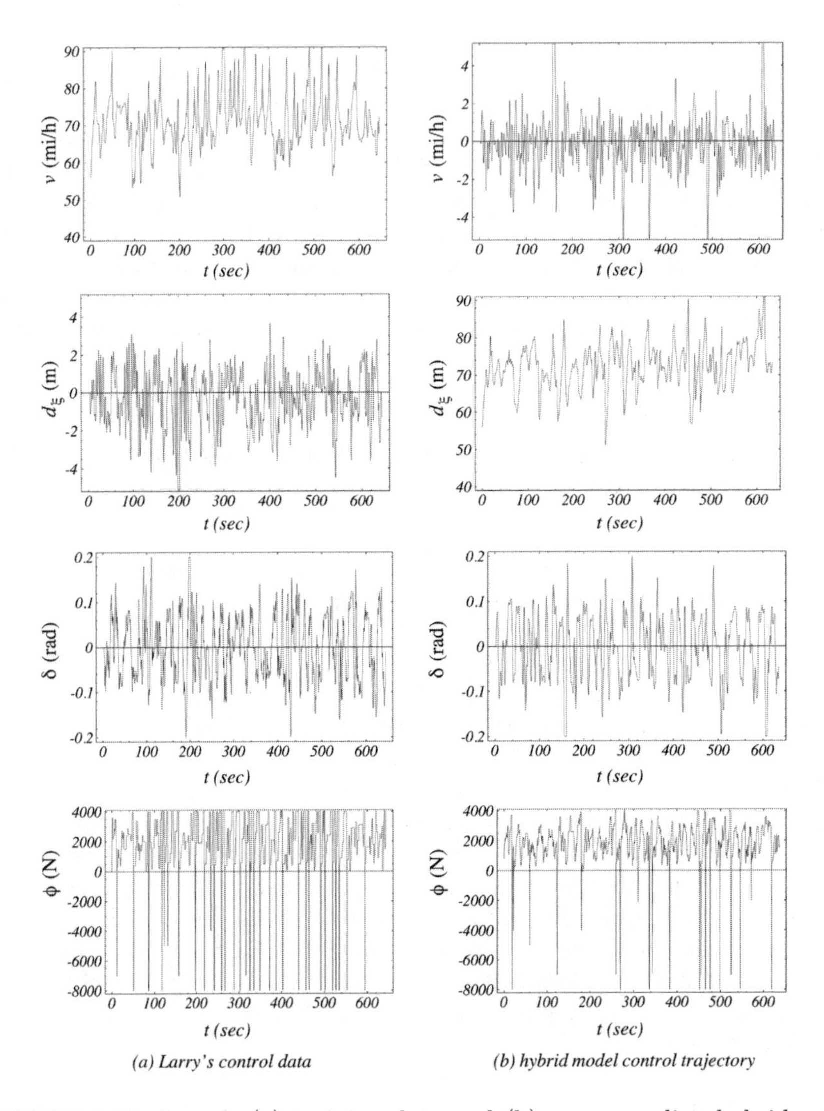

FIGURE 3.28: Larry's (a) training data and (b) corresponding hybrid controller data.

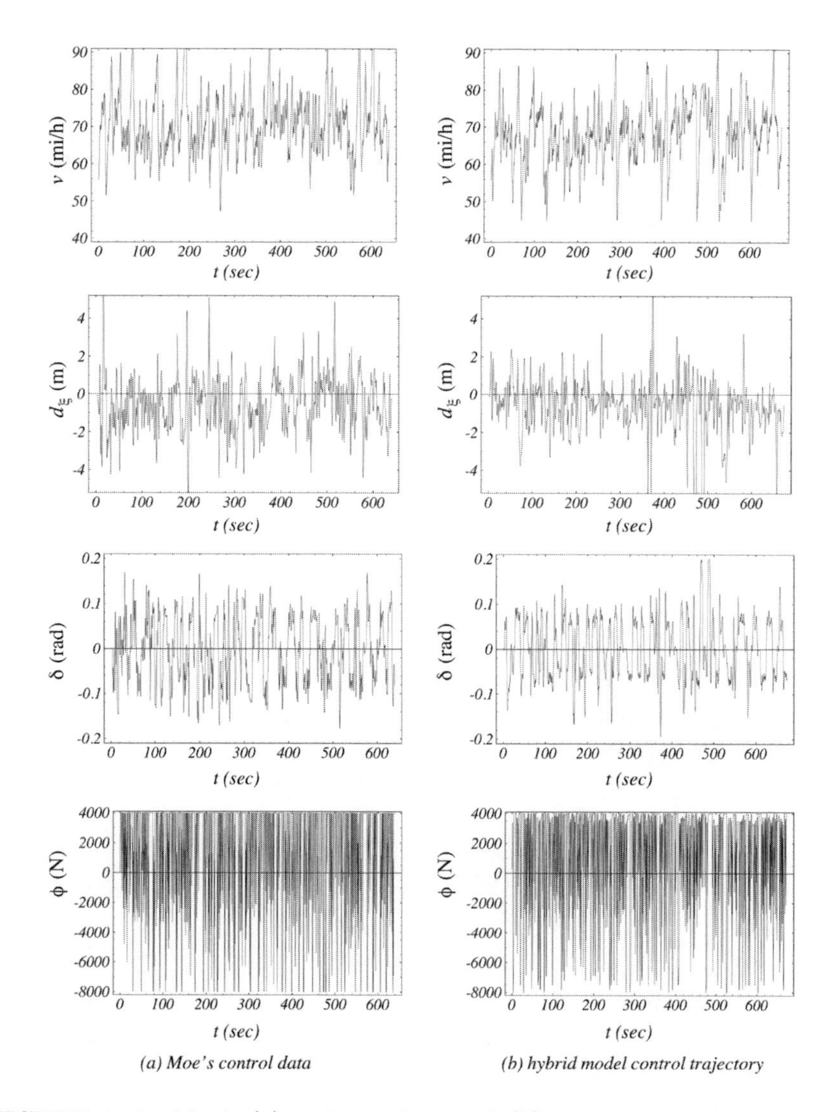

FIGURE 3.29: Moe's (a) training data and (b) corresponding hybrid controller data.

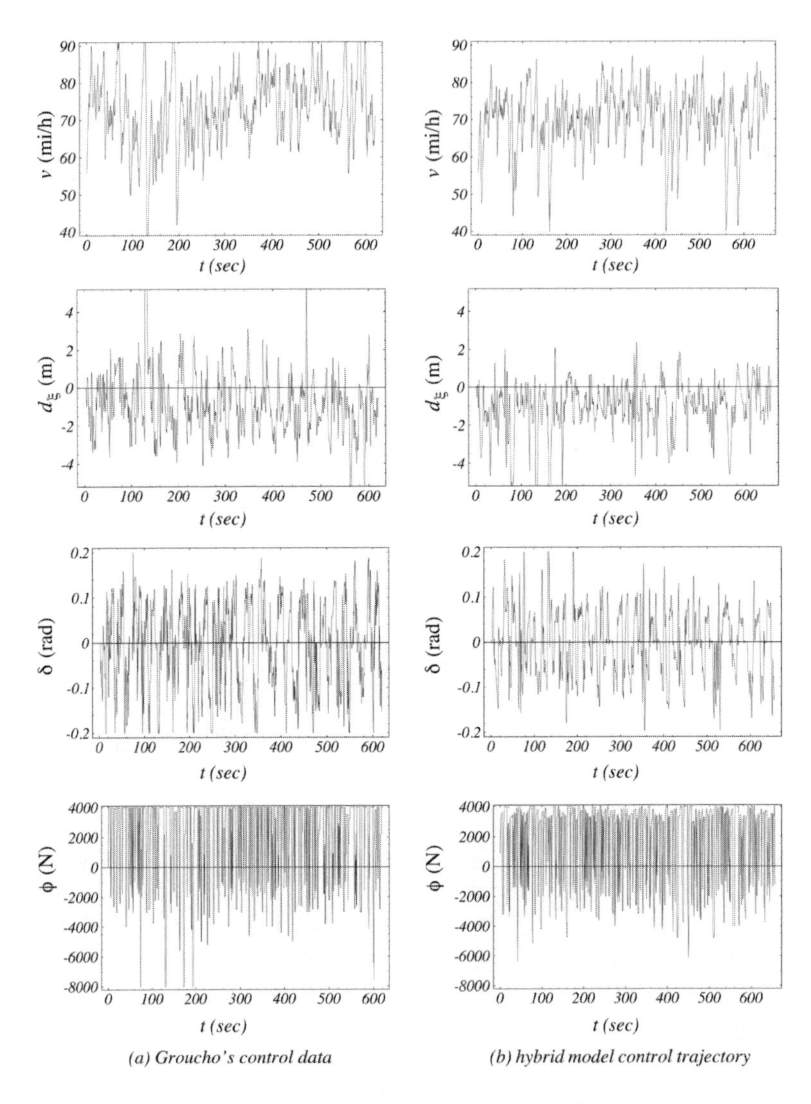

FIGURE 3.30: Groucho's (a) training data and (b) corresponding hybrid controller data.

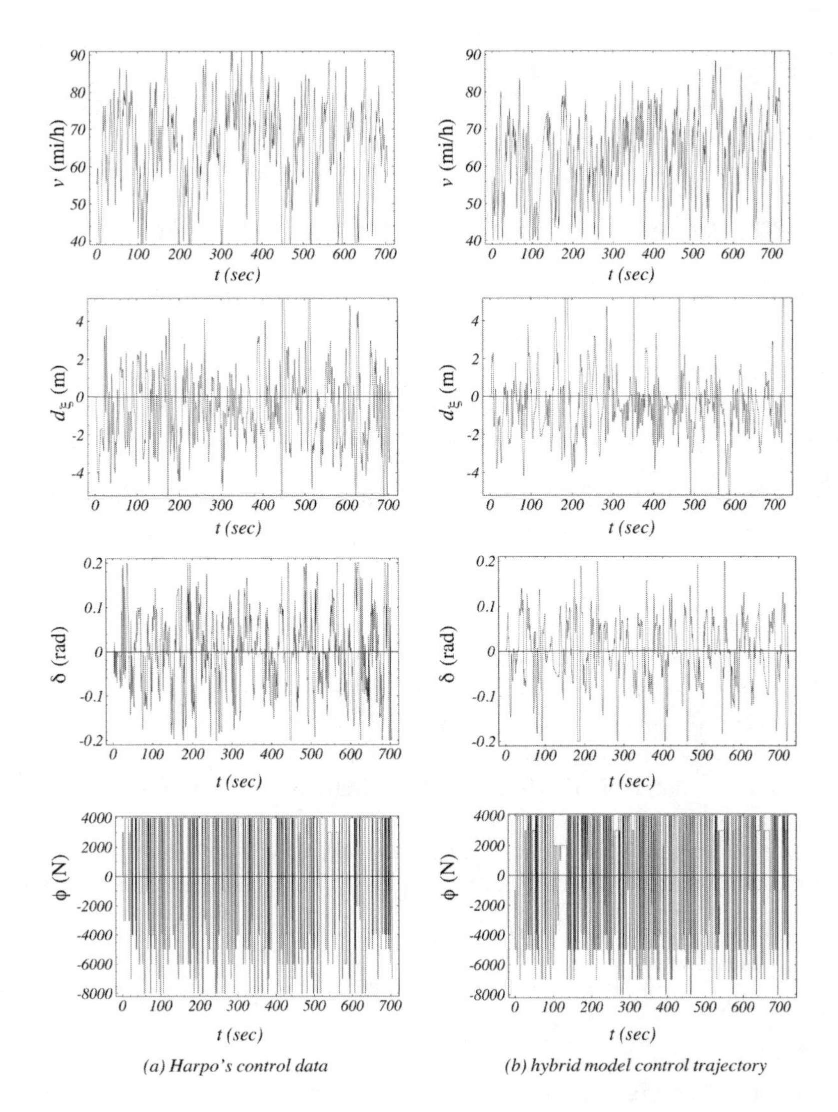

FIGURE 3.31: Harpo's (a) training data and (b) corresponding hybrid controller data.

Table 3.4: Hybrid controller design choices

Individual	v_{min} *(mi/h)*	ε_s	$P(A_i)$ *modified*
Larry	50	0.010	$\{P(A_2), P(A_3)\}$
Moe	45	0.005	$\{P(A_1), P(A_4)\}$
Groucho	40	0.005	$\{P(A_1), P(A_4)\}$
Harpo	40	0.005	$\{P(A_1), P(A_4)\}$

Comparing these modeling results to the *Cq* and *Ck* results in Figures 3.6 through 3.9 and Figures 3.10 through 3.14, respectively, we ask ourselves, which controllers, the continuous cascade network controllers or the discontinuous stochastic controllers, perform better? The answer to that question depends on what precisely is meant by "better."

If we evaluate the two modeling strategies based on absolute performance criteria, such as range of stability, the cascade network controllers probably perform better. Whereas the linear model controllers rarely, if ever, run off the road, the hybrid controllers temporarily run off the test road more often (for $w = 10$m). Simply put, the linear controllers appear more stable than their hybrid counterparts.

If, on the other hand, we evaluate the two modeling approaches on how closely they approximate the corresponding operator's control strategy, then the verdict likely changes. As we have already noted, the *Ck* models' control trajectories do not look anything like their human counterparts' trajectories, due to the continuous models' inability to faithfully model the discontinuous acceleration command ϕ. Qualitatively, the hybrid continuous/discontinuous controllers appear to approximate more closely their respective training data.

3.4.3 Analysis

Sample curve control

Here, we examine the control behavior of the hybrid models in somewhat greater detail – through a road sequence composed of (1) a 75m straight-line segment, and (2) a subsequent 150m-radius, 120° curve. Since, this particular road sequence appears in road #1 (Figure 3.2(c)), we can directly compare the actual human control strategy with the corresponding hybrid-model control.

Figures 3.32, 3.33, 3.34, and 3.35 plot the driving control for Larry, Moe, Groucho, and Harpo and their respective hybrid models. In each figure, the vertical lines indicate the start of the turn for the human (dashed) and the corresponding hybrid-model control data (solid).

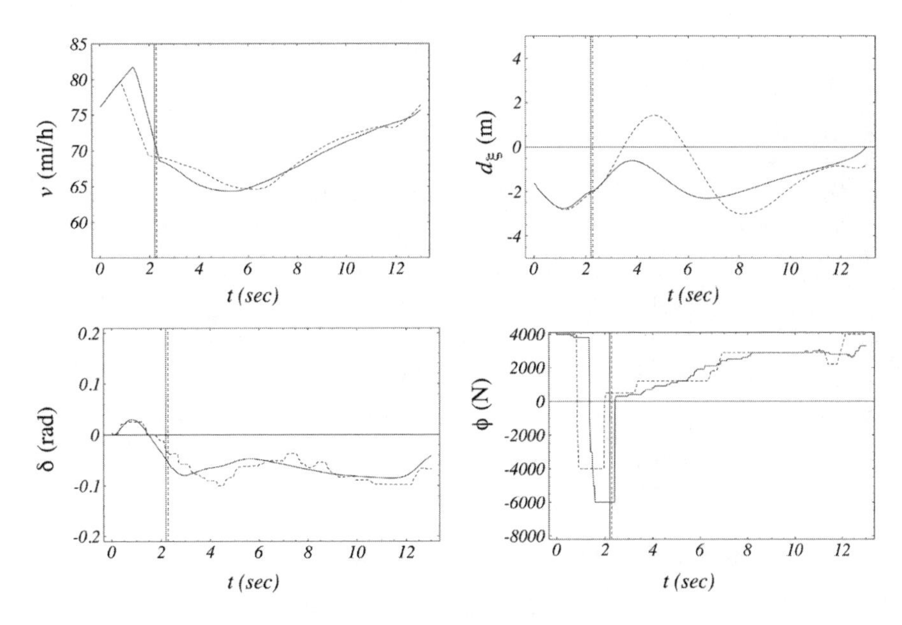

FIGURE 3.32: Larry's (dashed) and his hybrid model's (solid) control through a given turn.

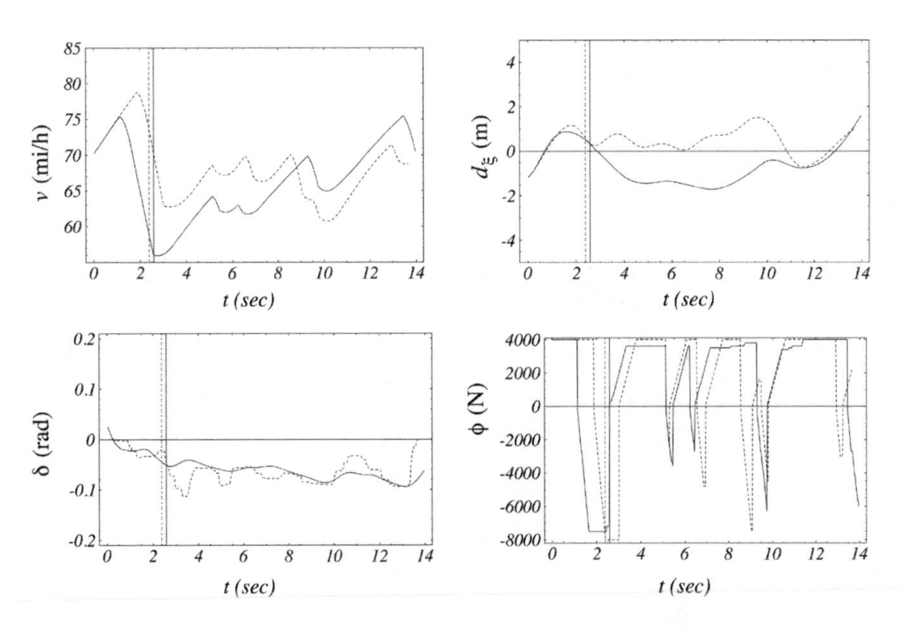

FIGURE 3.33: Moe's (dashed) and his hybrid model's (solid) control through a given turn.

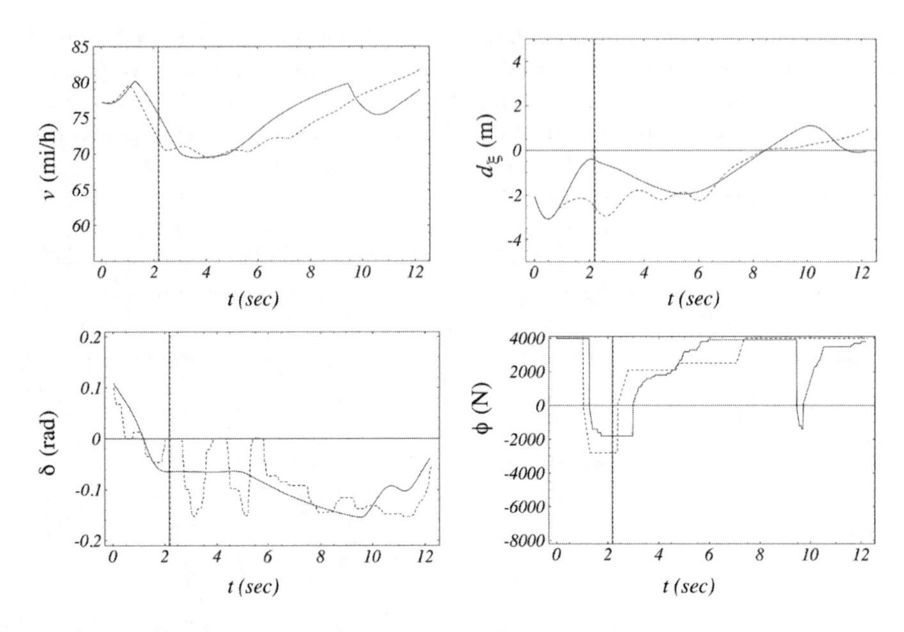

FIGURE 3.34: Groucho's (dashed) and hybrid model's (solid) control through a given turn.

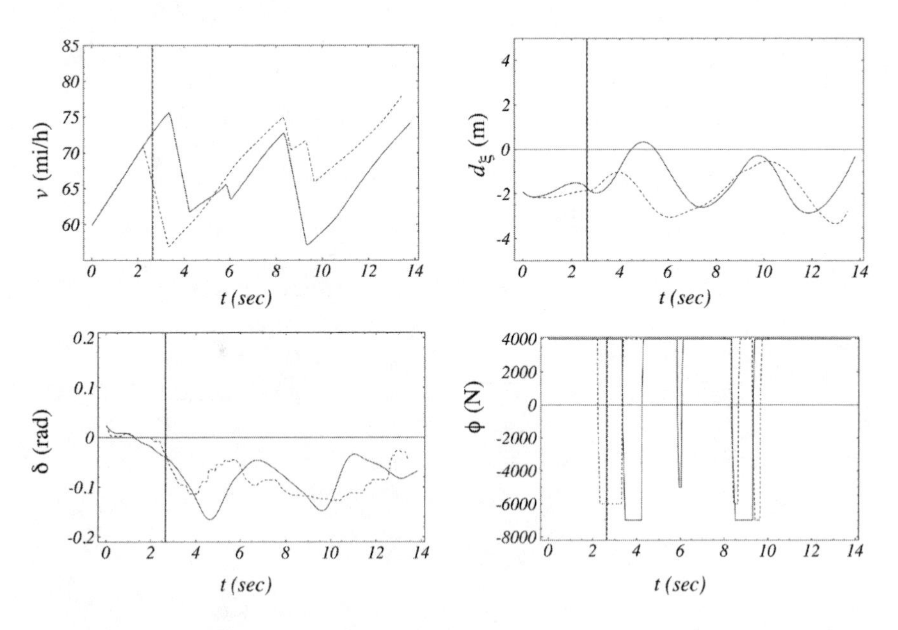

FIGURE 3.35: Harpo's (dashed) and his hybrid model's (solid) control through a given turn.

From these figures we make a few general observations. First, each hybrid model completes the curve in almost exactly the same time as its respective human. The largest difference in completion times – 2.5% (0.34 sec) – occurs for Harpo; in two of the other cases (Larry and Groucho), the time difference between the actual and hybrid controls is less than 0.5%.

Second, since the models' steering is handled through a linear, continuous mapping, the steering δ profiles for the models vary more smoothly than their human counterparts. Consequently, the lateral offset from the road median (d_ξ) also differs between the model and human control trajectories.[§]

Finally, while the applied force ϕ profiles between the humans and corresponding hybrid models are not identical, they are, in fact, similar. For Larry, the hybrid model's initial brake heading into the curve occurs approximately 1/2 second after Larry's initial brake. Thus, the model is slightly faster when first braking; to compensate for the higher speed, the model brakes slightly harder, and thereafter tracks Larry's applied force profile closely.

Groucho's case is similar to Larry's. The initial brake for the model occurs approximately 1/4 second after Groucho's initial brake. In this case, however, the model compensates not by braking harder, but by braking longer. Thereafter, the main difference between Groucho's control and the model is a quick brake maneuver while in the turn to compensate for the model's somewhat larger acceleration in the turn. Harpo's model also initially brakes after Harpo (by about 1 second), and consequently is also forced to brake more while in the turn to compensate for the higher speed going into the turn.

Moe is perhaps the most interesting of the four cases. This time, the model initially brakes approximately 1/2 second before Moe himself does, albeit with somewhat less force than Moe. Thereafter, the hybrid model closely emulates Moe's strategy of rapid switching between the brake and the accelerator while in the turn.

In summary, we observe that each hybrid controller, while not replicating the human's control strategy exactly, does a good job of emulating its respective human's turn maneuver. The following section examines the underlying reason for this success.

Probability profile

The most important reason behind the success of the hybrid controller is that it is able to successfully model the switching behavior between the gas and brake pedals as a probabilistic event, since the *precise time* that a switch occurs is not that important (as we observed in the previous section). What

[§]Although lateral offset from the road median is an important criterion for distinguishing between drivers in *real* driving, we shall see later that in our type of *simulated* driving, drivers pay little attention to d_ξ as long as they maintain contact with the road. Since drivers are inconsistent in their lateral lane position from one curve to the next, we cannot expect that a model will very closely track lateral lane position in any specific instance.

is more important is that the switch take place in some *time interval* around the time that the human operator would have executed the switch. Consider, for example, Figure 3.36, which plots the posterior probabilities $P(A_i|O)$ for a small segment of Groucho's hybrid model control. We see that switches between the gas and brake pedals (actions A_4 and A_8), while never very likely for any individual time step, are modeled as intervals where,

$$P(A_4|O) = p > 0, \text{ or } P(A_8|O) = p > 0. \tag{3.78}$$

The probability that a switch will occur after m time steps given the constant probability p is given by,

$$1 - (1 - p)^m \tag{3.79}$$

Figure 3.37 plots this probability as a function of time (at 50 Hz) for $p = 0.1$ and $p = 0.05$. Thus, we see that even for small values of p, the likelihood of a switch rises quickly as a function of time.

Because we train *separate* models λ_i for each action A_i, the hybrid modeling approach does not encounter the same one-to-many mapping problem, illustrated in Figures 3.24 and 3.25, that the continuous cascade networks encounter. The relatively few occurrences of switching in each control data set are sufficient training data, since the switching models λ_4 and λ_8 see *only* that data during training. Including the priors $P(A_i)$ in the action selection criterion (3.55) then ensures that the model is not overly biased towards switching.

Modeling extension

Suppose the acceleration control ϕ were not constrained by equations (3.20) and (3.22), and thus were not as readily expressible through discrete actions. For example, suppose that the separate gas and brake commands could change by an arbitrary amount for each time step, not just by $\Delta\phi_g$ and $\Delta\phi_b$. How would this change the proposed control framework?

Figure 3.38 suggests one possible solution. Initially, we train two separate continuous controllers, the first corresponding to $\phi(k) \geq 0$, and the second corresponding to $\phi(k) < 0$. Since these controllers would not be required to model switches between braking and accelerating, the control outputs will vary continuously and smoothly with model inputs; hence a continuous function approximator should be well suited for these two modeling tasks.

Then, we train four statistical models $\tilde{\lambda}_i$, corresponding to actions $\tilde{A}_i$, $i \in \{1, 2, 3, 4\}$, where actions $\tilde{A}_1$ and $\tilde{A}_2$ correspond to *no switch* at the next time step for $\phi(k) \geq 0$ and $\phi(k) < 0$, respectively, and actions $\tilde{A}_3$ and $\tilde{A}_4$ correspond to *a switch* at the next time step for $\phi(k) \geq 0$ and $\phi(k) < 0$, respectively. This discontinuous action model would then regulate which of the continuous models is active at each time step k. Although the discontinuous controller's function in this scheme is reduced, it does preserve the critical role of the discontinuous controller in properly modeling the switching behavior,

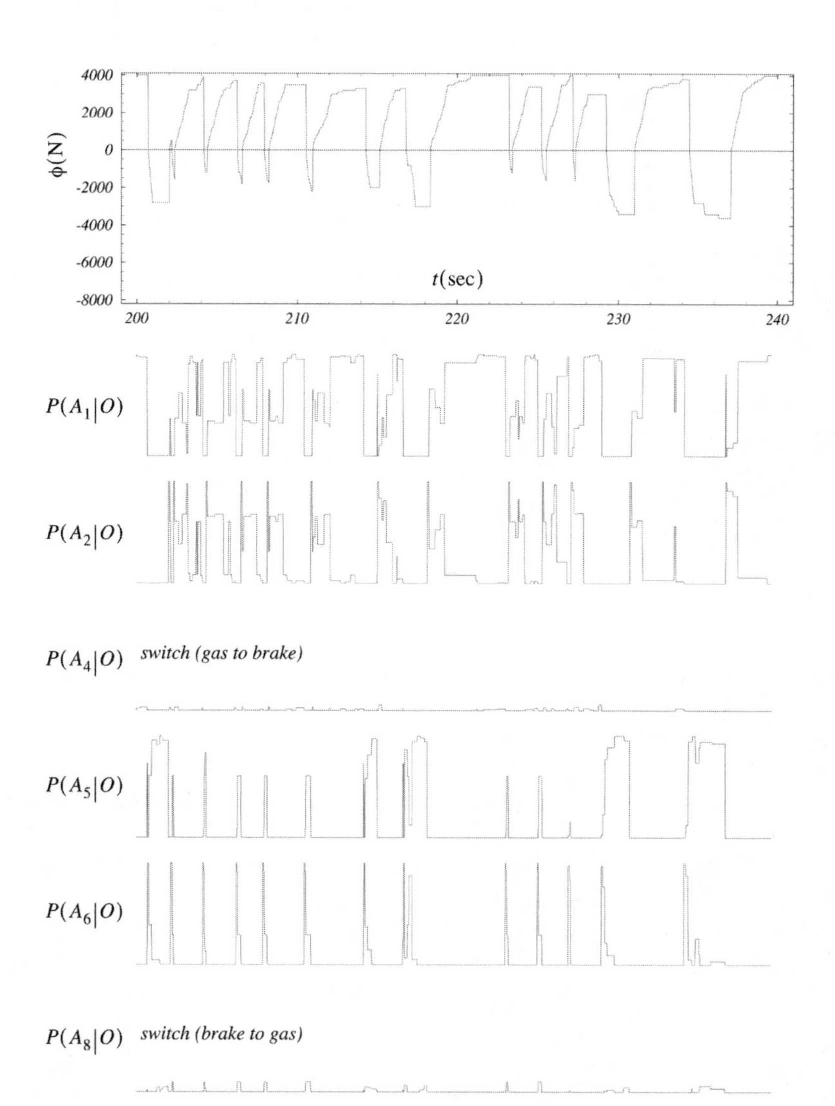

FIGURE 3.36: Posterior probabilities for Groucho's model control ($P(A_3) = P(A_7) = 0$).

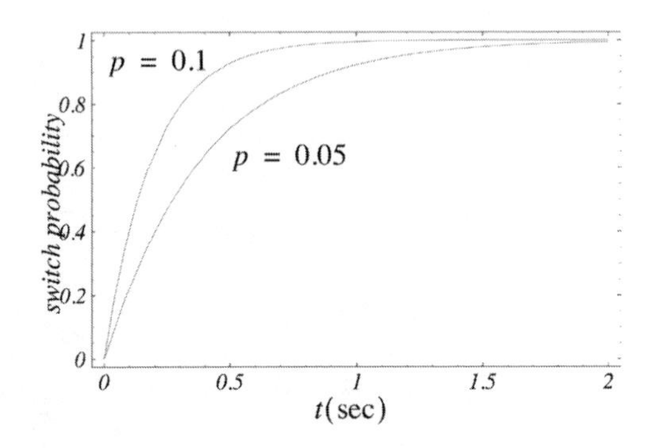

FIGURE 3.37: Probability of a switch after t seconds (at 50 Hz) when the probability of a switch at each time step is p.

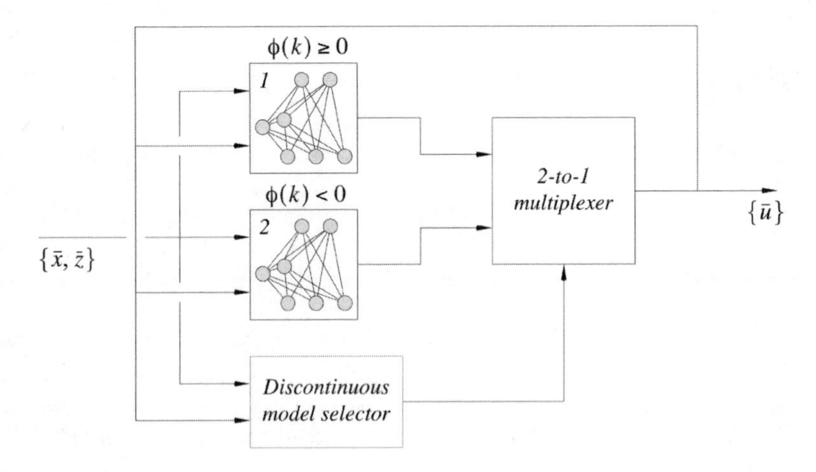

FIGURE 3.38: Alternative architecture for discontinuous strategies.

without the introduction of high-frequency noise. In fact, Figure 3.38 offers a modeling architecture which is applicable whenever discrete events or actions disrupt the continuous mapping from inputs to outputs.

Of course, the proposed statistical framework does have some limitations in comparison to functional modeling approaches. Because we vector quantize the input space, the stable region of operation for the hybrid controller is strictly limited to the input space spanned by the VQ codes. In fact, we observe from the modeling results that the continuous *Ck* models are more stable than the hybrid discontinuous/continuous models. A second limitation of the approach is the inclusion of the prior probabilities $P(A_i)$ in the stochastic selection criterion (3.55). By including the priors, we are assuming environmental conditions similar to the training environment. Radically different environmental conditions during testing presumably change the values of the priors, and therefore make the action selection criterion less appropriate.

In this chapter, we have developed a new neural network learning methodology for real-valued function approximation and dynamic system identification. In summary, we note that while we limit the volume of results presented to representative examples, the conclusions we draw with regard to the *Cq* and *Ck* algorithms have been confirmed for other human control data sets and over countless and repeated learning trials. Thus, this chapter has shown that (1) the *Ck* algorithm converges faster and more reliably than *Cq* in modeling human control strategies; (2) the *Ck* models exhibit stability over a greater range of initial and environmental conditions; (3) as long as sufficient data is provided as input, the precise input representation affects performance only marginally; and (4) the continuous *Ck* and *Cq* algorithms abstract control strategies that are qualitatively dissimilar to the original human control strategies. In the next chapter, we derive an alternative, discontinuous modeling framework which attempts to overcome this limitation.

We have developed a discontinuous modeling framework for abstracting discontinuous human control strategies, and have compared the proposed approach to competing continuous function approximators. Which control approach is preferred ultimately depends on the specific application for the HCS model. If the model is being developed towards the eventual control of a real robot or vehicle, then the continuous modeling approach might be preferred as a good starting point. Continuous models extrapolate control strategies to a greater range of inputs, show greater inherent stability, and lend themselves more readily to theoretical performance analysis. If, on the other hand, the model is being developed in order to simulate different human behaviors in a virtual reality simulation or game, then the discontinuous control approach might be preferred, since fidelity to the human training data and random variations in behavior would be the desired qualities of the HCS model. Thus, depending on the application, we believe a need exists for both types of modeling approaches.

In the next chapter, we develop a stochastic similarity measure as the first step in a post-training model-validation.

4

Validation of Human Control Strategy Models

In the previous chapter, we investigated different machine learning techniques for abstracting models of human control strategy. Each of these methods learns, to varying degrees, stable HCS models from the experimental data. As we observed, however, the different modeling techniques – *Cq*, *Ck*, and discontinuous learning – generate control trajectories that are qualitatively quite different from one another, despite training from identical human control data. This is true not only because the modeling capacity of each approach differs, but also because modeling errors can feed back on themselves to generate state and command trajectories that are uncharacteristic of the source process. Therefore, for feedback control tasks, such as human driving, we suggest that post-training model validation is not only desirable, but essential to establish the degree to which the human and model-generated trajectories are similar.

In this chapter, we first demonstrate the need for model validation with some illustrative examples. * We then propose a stochastic similarity measure – based on Hidden Markov Model analysis – for comparing stochastic, dynamic, multi-dimensional trajectories. This similarity measure can then be applied towards validating a learned model's fidelity to its training data by comparing the model's dynamic trajectories in the feedback loop to the human's dynamic trajectories.

4.1 Need for model validation

The main strength of modeling by learning is that no explicit physical model is required; this also represents its biggest weakness, however. On the one hand, we are not restricted by the limitations of current scientific knowledge, and are able to model human control strategies for which we have not yet developed adequate biological or psychological understanding. On the other hand, the lack of scientific justification detracts from the confidence that we can show in these learned models. This is especially true when the unmodeled process

Table 4.1: Sample input-output training data

Input			*Output*
$u(k-1)$	$u(k)$	$x(k)$	$u(k+1)$
-0.1	0.1	0.4	0.349
0.1	0.1	0.5	0.599
-0.3	0.2	0.3	0.123
0.3	0.2	0.4	0.673
0.2	0.0	0.5	0.650
0.0	0.2	0.3	0.348

is (1) dynamic and (2) stochastic in nature, as is the case for human control strategy. For a dynamic process, model errors can feed back on themselves to produce trajectories which are not characteristic of the source process or are even potentially unstable. For a stochastic process, a static error criterion (such as RMS error), based on the difference between the training data and predicted model outputs, is inadequate to gauge the fidelity of a learned model to the source process. Yet, for the static modeling techniques studied in this book, some static error measure usually serves as the test of convergence. While this measure is very useful during training, it offers no guarantees, theoretical or otherwise, about the dynamic behavior of the learned model in the feedback control loop.

To illustrate this problem, we consider two examples. In the first example, suppose that we wish to learn a dynamic process represented by the simple difference equation,

$$u(k+1) = 0.75u(k) + 0.24u(k-1) + x(k) \tag{4.1}$$

where $u(k)$, $x(k)$ represent the output and input of the system, respectively, at time step k. For the input/output training data in Table 4.1, at least three different linear models yield the same RMS error (6.16×10^{-3}) over the training set:

$$\#1:\ u(k+1) = 0.76u(k) + 0.25u(k-1) + x(k) \tag{4.2}$$

$$\#2:\ u(k+1) = 0.76u(k) + 0.23u(k-1) + x(k) \tag{4.3}$$

$$\#3:\ u(k+1) = 0.74u(k) + 0.23u(k-1) + x(k) \tag{4.4}$$

The dynamic trajectories for these models, however, differ markedly. As an example, consider the time-dependent input,

$$x(k) = 0.1\sin(k\pi/100) \tag{4.5}$$

Table 4.2: Dominant poles for each difference equation

system	*model #1*	*model #2*	*model #3*
$(0.992, 0)$	$(1.008, 0)$	$(0.992, 0)$	$(0.976, 0)$

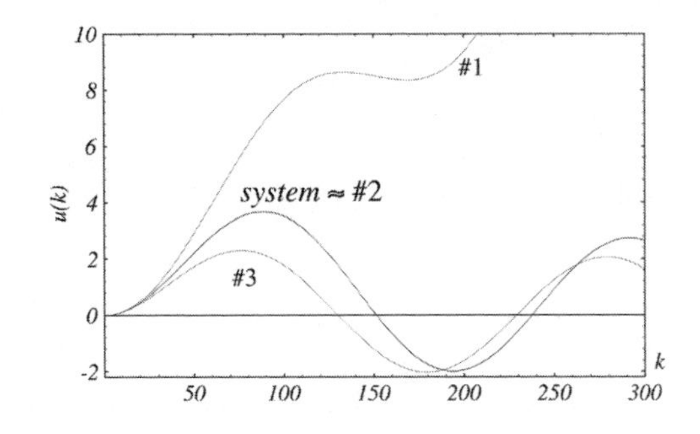

FIGURE 4.1: The three models result in dramatically different (even unstable) trajectories.

and initial conditions $u(-1) = u(0) = 0$. Figure 4.1 plots the system as well as the model trajectories for $0 \leq k \leq 300$. We see that model #1 diverges to an unstable trajectory; model #3 remains stable, but approximates the system with significantly poorer accuracy; and model #2 matches the system's response very closely. These responses are predicted by the dominant pole for each difference equation (Table 4.2). Except for the unstable model (#1), each model's dominant pole lies inside the unit circle, thus ensuring stability.

It is apparent from this example that a biased estimator of a marginally stable system may well result in an unstable model, despite RMS errors which appear to be equivalent to those of better models. Not only biased models are a problem, however. Static models of the type shown in Figure 3.4 can achieve deceptively low RMS errors by confusing *causation* with *correlation* between the model's input and output spaces. In our second example, we illustrate this problem with some human driving data collected from Groucho.

In the driving simulator (Figure 3.1), we ask Groucho to drive over road #1 shown in Figure 3.2(a). We simplify the problem by fixing the acceleration command at $\phi = 300$N, keeping the velocity around 40mph, and requiring only control of the steering angle δ. Now, we train two linear models Γ_1 and Γ_2 with input representations $\{r_x^{10}\}$ and $\{\delta^3\}$, respectively. Note that model Γ_2 receives no road information as input, and is therefore guaranteed to be unstable.

For Γ_1, the RMS error over the data set converges to 9.70×10^{-3}, while for Γ_2, the RMS error converges to 0.71×10^{-3}, an order of magnitude smaller. All that the Γ_2 model has "learned," however, is a *correlation* between the

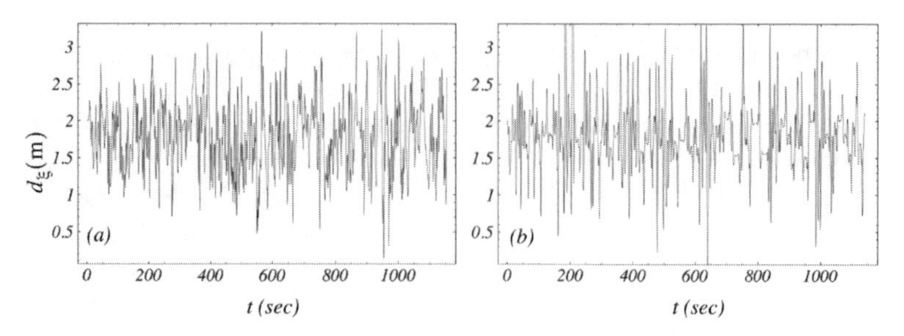

FIGURE 4.2: Lateral offset (e.g., distance from road median) over time, for (a) Groucho's control strategy, and (b) the first model's control trajectory.

previous values of δ and the next value of δ. In fact, the full model,

$$\delta(k+1) = 0.9997\delta(k) + 0.6647\delta(k-1) - 0.6655\delta(k-2) \tag{4.6}$$

simplifies to (approximately),

$$\delta(k+1) = \delta(k). \tag{4.7}$$

Despite the larger RMS error over the training data, model Γ_1, on the other hand, converges to a stable control strategy, as is shown in Figure 4.2. It learned a *causal* relationship between the curvature of the road ahead and the steering command $\delta(k+1)$.

The two examples above are the extremes. Not all models will be either a good or an unstable approximation of the human control data. In general, similarity between model-generated trajectories and the human control data will vary continuously for different models, from completely dissimilar to n-early identical. Furthermore, for stochastic systems (such as humans), one cannot expect equivalent trajectories for the system and the learned model, given equivalent initial conditions. Therefore, we require a stochastic similarity measure, with sufficient representational power and flexibility to compare multi-dimensional, stochastic trajectories.

4.2 Stochastic similarity measure

Similarity measures or metrics have been given considerable attention in computer vision [22], [31], [249], image database retrieval [109], and 2D or 3D shape analysis [126], [219]. These methods, however, generally rely on the special properties of images, and are therefore not appropriate for analyzing sequential trajectories.

Many parametric methods have been developed to analyze and predict time-series data. One of the more well known, autoregressive-moving average (ARMA) modeling [35], predicts the current signal based on a *linear* combination of previous time histories and Gaussian noise assumptions. Since we have already observed in the previous chapter that a linear model is insufficient to qualitatively replicate switching, nonlinear control strategies, ARMA models may form a poor foundation upon which to develop a similarity measure. Other work has focused on classifying temporal patterns using Bayesian statistics [58], wavelet and spectral analysis [230], neural networks (both feedforward and recurrent) [92], [227], and Hidden Markov Models (see discussion below). Much of this work, however, analyzes only short-time trajectories or patterns, and, in many cases, generates only a binary classification, rather than a continuously valued similarity measure. Prior work has not addressed the problem of comparing long, multi-dimensional, stochastic trajectories, especially of human control data. Thus, we propose to evaluate *stochastic similarity* between two dynamic, multi-dimensional trajectories using *Hidden Markov Model (HMM)* analysis.

4.2.1 Hidden Markov models

Rich in mathematical structure, HMMs are trainable statistical models, with two appealing features: (1) no *a priori* assumptions are made about the statistical distribution of the data to be analyzed, and (2) a degree of sequential structure can be encoded by the Hidden Markov Models. As such, they have been applied for a variety of stochastic signal processing. In speech recognition, where HMMs have found their widest application, human auditory signals are analyzed as speech patterns [101], [192]. Transient sonar signals are classified with HMMs for ocean surveillance in [125]. Radons, *et. al.* [194] analyze 30-electrode neuronal spike activity in a monkey's visual cortex with HMMs. Hannaford and Lee [87] classify task structure in teleoperation based on HMMs. In [255], [256], HMMs are used to characterize sequential images of human actions. Finally, Yang and Xu apply Hidden Markov Models to open-loop action skill learning [258] and human gesture recognition [259].

A Hidden Markov Model consists of a set of n states, interconnected through probabilistic transitions; each of these states has some output probability distribution associated with it. Although algorithms exist for training HMMs with both discrete and continuous output probability distributions, and although most applications of HMMs deal with real-valued signals, discrete HMMs are preferred to continuous or semi-continuous HMMs in practice, due to their relative computational simplicity (orders of magnitude more efficient) and lesser sensitivity to initial random parameter settings [193]. In the following sections, we will describe how we use discrete HMMs for analysis of real-valued signals by converting the data to discrete symbols through pre-processing and vector quantization. It then follows with methods for minimizing the detrimental effects of discretization.

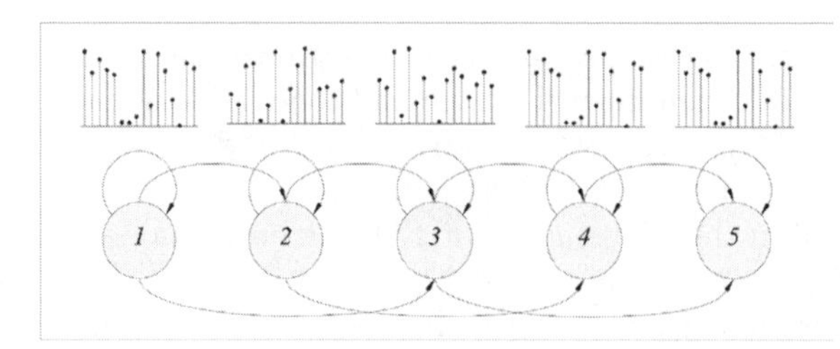

FIGURE 4.3: A 5-state Hidden Markov Model, with 16 observable symbols in each state.

A discrete HMM is completely defined by the following triplet [192],

$$\lambda = \{A, B, \pi\} \tag{4.8}$$

where A is the probabilistic $n_s \times n_s$ state transition matrix, B is the $L \times n_s$ output probability matrix with L discrete output symbols $l \in \{1, 2, \ldots, L\}$, and π is the n-length initial state probability distribution vector for the HMM. Figure 4.3, for example, represents a 5-state HMM, where each state emits one of 16 discrete symbols.

We define the notion of *equivalent* HMMs for two HMMs λ_1 and λ_2 such that,

$$\lambda_1 \sim \lambda_2, \text{ iff. } P(O|\lambda_1) = P(O|\lambda_2), \forall O. \tag{4.9}$$

Note that λ_1 and λ_2 need not be identical to be equivalent. The following two HMMs are, for example, equivalent, but not identical:

$$\begin{aligned} \lambda_1 &= \left\{ [1], [0.5\ 0.5]^T, [1] \right\} \\ \lambda_2 &= \left\{ \begin{bmatrix} 0.5 & 0.5 \\ 0.5 & 0.5 \end{bmatrix}, \begin{bmatrix} 1 & 0 \\ 0 & 1 \end{bmatrix}, [0.5\ 0.5]^T \right\} \end{aligned} \tag{4.10}$$

Finally, we note that for an observation sequence O of discrete symbols and an HMM λ, we can locally maximize $P(\lambda|O)$ using the well-known Baum-Welch algorithm [192], [26]. We can also evaluate $P(O|\lambda)$ using the computationally efficient forward-backward algorithm.

4.2.2 Similarity measure

Here, we derive a stochastic similarity measure, based on discrete-output H-MMs. Assume that we wish to compare observation sequences from two s-tochastic processes Γ_1 and Γ_2. Let $\bar{O}_i = \{O_i^{(k)}\}$, $k \in \{1, 2, \ldots, n\}$, $i \in \{1, 2\}$, denote the set of n_i observation sequences of discrete symbols generated by process Γ_i. Each observation sequence is of length $T_i^{(k)}$, so that the total

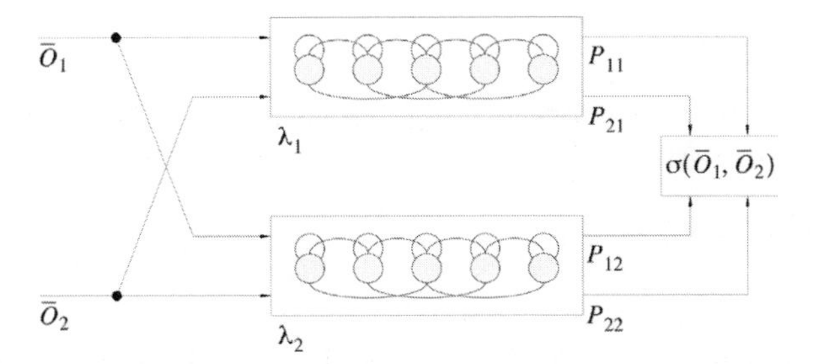

FIGURE 4.4: Four normalized probability values make up the similarity measure.

number of symbols in set $\bar{O}_i$ is given by,

$$\bar{T}_i = \sum_{k=1}^{n_i} T_i^{(k)}, i \in \{1, 2\}. \tag{4.11}$$

Also let $\lambda_j = \{A_j, B_j, \pi_j\}$, $i \in \{1, 2\}$, denote a discrete HMM locally optimized with the Baum-Welch algorithm to maximize,

$$P(\lambda_j|\bar{O}_j) = \prod_{k=1}^{n_j} P(\lambda_j|O_j^{(k)}), j \in \{1, 2\}, \tag{4.12}$$

and let,

$$P(\bar{O}_i|\lambda_j) = \prod_{k=1}^{n_i} P(O_i^{(k)}|\lambda_j), i, j \in \{1, 2\}, \tag{4.13}$$

$$P_{ij} = P(\bar{O}_i|\lambda_j)^{1/\bar{T}_i} i, j \in \{1, 2\}, \tag{4.14}$$

denote the probability of the observation sequences $\bar{O}_i$ given the model λ_j, normalized with respect to the sequence lengths $\bar{T}_i$.†

Using the definition in (4.14), Figure 4.4 illustrates our overall approach to evaluating similarity between two observation sequences. Each observation sequence is first used to train a corresponding HMM; this allows us to evaluate P_{11} and P_{22}. We then cross-evaluate each observation sequence on the other HMM (i.e., $P(\bar{O}_1|\lambda_2)$, $P(\bar{O}_2|\lambda_1)$) to arrive at P_{12} and P_{21}. Given these four normalized probability values, we define the following similarity measure between $\bar{O}_1$ and $\bar{O}_2$:

$$\sigma(\bar{O}_1, \bar{O}_2) = \sqrt{\frac{P_{21}P_{12}}{P_{11}P_{22}}} \tag{4.15}$$

†In practice, we calculate P_{ij} as $10^{\log P(\bar{O}_i|\lambda_j)/\bar{T}_i}$ to avoid problems of numerical underflow for long observation sequences.

4.2.3 Properties

In order for the similarity measure to obey certain important properties, we restrict the HMMs λ_1 and λ_2 to have the same number of states such that,

$$n_{s,1} = n_{s,2} \tag{4.16}$$

where $n_{s,j}$, $j \in \{1,2\}$ denotes the number of states in model λ_j.

Now, let us assume that the P_{ii} are global (rather than just a local) maxima.[‡] We define model λ_i to be a *global maximum* if and only if,

$$P(O_i|\lambda_i) \geq P(O_i|\lambda), \forall\lambda, n_s = n_{s,i}, \tag{4.17}$$

where n_s is the number of states in model λ. With this assumption, we have that,

$$1/L \leq P_{ii}, \text{ and} \tag{4.18}$$

$$0 \leq P_{ij} \leq P_{ii}.^{\S} \tag{4.19}$$

The lower bound for P_{ii} in (4.18) is realized for single-state discrete HMMs, and a uniform distribution of symbols in $\bar{O}_i$. From (4.15) to (4.19), we derive the following properties for $\sigma(\bar{O}_1, \bar{O}_2)$:

$$\text{Property \#1: } \sigma(\bar{O}_1, \bar{O}_2) = \sigma(\bar{O}_2, \bar{O}_1) \text{ (by definition)} \tag{4.20}$$

$$\text{Property \#2: } 0 \leq \sigma(\bar{O}_1, \bar{O}_2) \leq 1 \tag{4.21}$$

$$\text{Property \#3: } \sigma(\bar{O}_1, \bar{O}_2) = 1 \text{ if (a) } \lambda_1 \sim \lambda_2 \text{ or (b) } \bar{O}_1 = \bar{O}_2 \tag{4.22}$$

Below, we illustrate the behavior of the similarity measure for some simple HMMs. First, for the class of single-state, discrete HMMs given by,

$$\lambda_j = \{A_j, B_j, \pi_j\} = \left\{[1], \left[b_{j1}\ b_{j2}\ \ldots\ b_{jL}\right]^T, [1]\right\} \tag{4.23}$$

[‡]Theoretically, the Baum-Welch algorithm guarantees that P_{ii} is a local maximum only. In practice, this is not a significant concern, however, since the Baum-Welch algorithm converges to near-optimal solutions for discrete-output HMMs, when the algorithm is initialized with random model parameters [192],[193]. We have verified this near-optimal convergence property experimentally in two ways. First, for a given set of observation sequences $\bar{O}$, we trained n different HMMs $\{\lambda_1, \lambda_2, \ldots, \lambda_n\}$ from different initial random parameter settings. We then observed that the probabilities $P(\bar{O}|\lambda_i)$, $i \in \{1, 2, \ldots, n\}$, were approximately equivalent. Second, for a given model λ, we *generated* a set of observation sequences $\hat{O}$. We then trained a second Hidden Markov Model $\hat{\lambda}$ (with initial random model parameters) on $\hat{O}$. Finally, we observed that $P(\hat{O}|\hat{\lambda}) \approx 1$ provided that $\hat{O}$ is sufficiently large. Both procedures suggest that the Baum-Welch algorithm does indeed converge to optimal or near-optimal solutions in practice.

[§]Note that without condition (4.16), equation (4.19) does not necessarily hold.

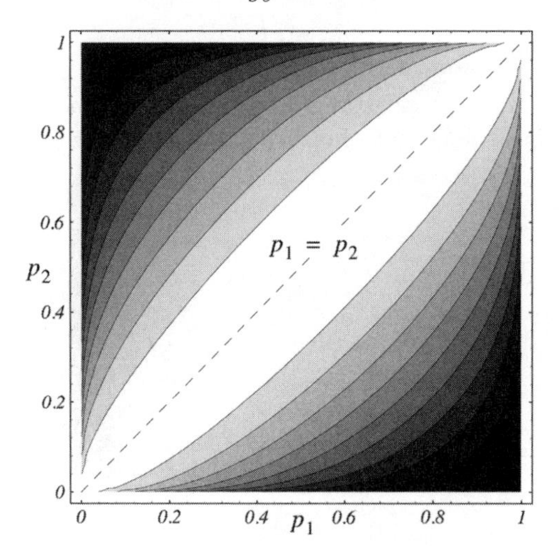

FIGURE 4.5: Similarity measure for two binomial distributions. Lighter colors indicate higher similarity.

the similarity measure reduces to,¶

$$\sigma(\bar{O}_1, \bar{O}_2) = \prod_{k=1}^{L} \left(\frac{b_{1k}}{b_{2k}} \right)^{\frac{(b_{2k} - b_{1k})}{2}} \tag{4.24}$$

which reaches a maximum when $b_{1k} = b_{2k}$, or simply, $B_1 = B_2$, and that maximum is equal to one. Figure 4.5 shows a contour plot for $B_1 = [p_1 \;\; 1 - p_1]^T$, $B_2 = [p_2 \;\; 1 - p_2]^T$, and $0 < p_1, p_2 < 1$.

Second, we give an example of how the proposed similarity measure changes, not as a function of different symbol distributions, but rather as a function of varying HMM structure. Consider the following Hidden Markov Model,

$$\lambda(\alpha) = \left\{ \begin{bmatrix} \frac{1+\alpha}{2} & \frac{1-\alpha}{2} \\ \frac{1-\alpha}{2} & \frac{1+\alpha}{2} \end{bmatrix}, \begin{bmatrix} 1 & 0 \\ 0 & 1 \end{bmatrix}, [0.5 \; 0.5]^T \right\}, 0 \leq \alpha < 1 \tag{4.25}$$

¶Note that:

$$P_{ij} = \left(\prod_{k=1}^{L} (b_{jk})^{(\bar{T}_i b_{ik})} \right)^{1/\bar{T}_i} = \prod_{k=1}^{L} (b_{jk})^{b_{ik}}$$

$$\sigma(\bar{O}_1, \bar{O}_2) = \sqrt{\frac{P_{21}P_{12}}{P_{11}P_{22}}} = \left[\frac{\left(\prod_{k=1}^{L} (b_{1k})^{b_{2k}}\right)\left(\prod_{k=1}^{L} (b_{2k})^{b_{1k}}\right)}{\left(\prod_{k=1}^{L} (b_{1k})^{b_{1k}}\right)\left(\prod_{k=1}^{L} (b_{2k})^{b_{2k}}\right)} \right]^{\frac{1}{2}}$$

$$= \prod_{k=1}^{L} \left(\frac{b_{1k}}{b_{2k}} \right)^{\frac{b_{2k}}{2}} \prod_{k=1}^{L} \left(\frac{b_{2k}}{b_{1k}} \right)^{\frac{b_{1k}}{2}} = \prod_{k=1}^{L} \left(\frac{b_{1k}}{b_{2k}} \right)^{\frac{(b_{2k} - b_{1k})}{2}}$$

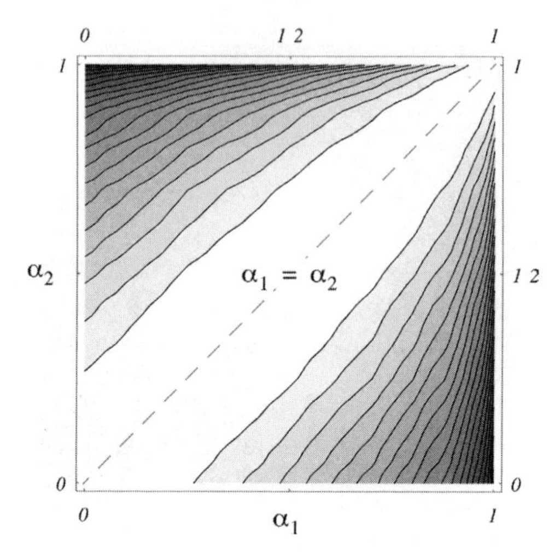

FIGURE 4.6: The similarity measure changes predictably as a function of HMM structure.

and corresponding observation sequences, $O(\alpha)$, stochastically generated from model $\lambda(\alpha)$. For all $\alpha \in [0, 1]$, $O(\alpha)$ will have an equivalent aggregate distribution of symbols 0 and 1 – namely 1/2 and 1/2. As α increases, however, $O(\alpha)$ will become increasingly structured. For example,

$$\lambda(0) = \left\{ [1], [0.5\ 0.5]^T, [1] \right\}, \text{ (equivalent to unbiased coin toss)} \tag{4.26}$$

$$\lim_{\alpha \to 1} O(\alpha) = \{\ldots, 1, 1, 1, 0, 0, 0, \ldots, 0, 0, 0, 1, 1, 1, \ldots\} \tag{4.27}$$

Figure 4.6 graphs $\sigma[O(\alpha_1), O(\alpha_2)]$ as a contour plot for $0 \leq \alpha_1, \alpha_2 < 1$, where each observation sequence $O(\alpha)$ of length $T = 10,000$ is generated stochastically from the corresponding HMM $\lambda(\alpha)$.[‖] Greatest similarity is indicated for $\alpha_1 = \alpha_2$, while greatest dissimilarity occurs for $(\alpha_1 \to 1, \alpha_2 = 0)$, and $(\alpha_1 = 0, \alpha_2 \to 1)$.

4.2.4 Distance measure

In some cases, it may be more convenient to represent the similarity between two sets of observation sequences through a *distance measure* $d(\bar{O}_1, \bar{O}_2)$, rather than a similarity measure. Given the similarity measure $\sigma(\bar{O}_1, \bar{O}_2)$, such a measure is easily derived. Let,

$$d(\bar{O}_1, \bar{O}_2) = -\log \sigma(\bar{O}_1, \bar{O}_2) = \frac{1}{2}[\log(P_{11}P_{22}) - \log(P_{21}P_{12})] \tag{4.28}$$

[‖]This procedure only approximates our similarity measure definition, since $\lambda(\alpha)$ is only optimal for $O(\alpha)$ as $T \to \infty$.

such that,

$$d(\bar{O}_1, \bar{O}_2) = d(\bar{O}_2, \bar{O}_1), \quad (4.29)$$

$$d(\bar{O}_1, \bar{O}_2) \geq 0, \quad (4.30)$$

$$d(\bar{O}_1, \bar{O}_2) = 0 \text{ if (a) } \lambda_1 \sim \lambda_2 \text{ or (b) } \bar{O}_1 = \bar{O}_2. \quad (4.31)$$

The distance measure $d(\bar{O}_1, \bar{O}_2)$ *between two sets of observation sequences* defined in (4.28) is closely related to the dual notion of distance *between two Hidden Markov Models*, as proposed in [112].

Let $\hat{O}_i$ denote a set of random observation sequences of total length $\hat{T}_i$ *generated by* the HMM $\hat{\lambda}_i$, and let,

$$\hat{P}_{ij} = P(\hat{O}_i | \hat{\lambda}_j)^{1/\hat{T}_i} \quad (4.32)$$

Then, [112] defines the following distance measure between two Hidden Markov Models, $\hat{\lambda}_1$ and $\hat{\lambda}_2$:

$$d(\hat{\lambda}_1, \hat{\lambda}_2) = \frac{1}{2}[\log(\hat{P}_{11}\hat{P}_{22}) - \log(\hat{P}_{21}\hat{P}_{12})]. \quad (4.33)$$

Unlike the observation sequences $\bar{O}_i$, the sequences $\hat{O}_i$ are not unique, since they are stochastically generated from $\hat{\lambda}_i$. Hence, $d(\hat{\lambda}_1, \hat{\lambda}_2)$ is uniquely determined only in the limit as $\hat{T}_i \to \infty$. Likewise for $d(\bar{O}_1, \bar{O}_2)$, the HMMs λ_1 and λ_2 are not unique, since P_{11} and P_{22} are in general guaranteed to be only local, not global maxima. Hence, $d(\bar{O}_1, \bar{O}_2)$ is uniquely determined only when P_{11} and P_{22} represent global maxima.

While in general, $d(\hat{\lambda}_1, \hat{\lambda}_2)$ and $d(\bar{O}_1, \bar{O}_2)$ are not equivalent, the discussion above suggests sufficient conditions for which the two notions – distance between HMMs and distance between observation sequences – do converge to equivalence. Specifically, $d(\bar{O}_1, \bar{O}_2) = d(\hat{\lambda}_1, \hat{\lambda}_2)$ if,

$$(1) \ \lambda_1 \sim \hat{\lambda}_1, \lambda_2 \sim \hat{\lambda}_2, \quad (4.34)$$

$$(2) \ P_{11}, P_{22} \text{ are global maxima, and} \quad (4.35)$$

$$(3) \ \hat{T}_i \to \infty. \quad (4.36)$$

4.2.5 Data preprocessing

Assume that we wish to analyze the similarity of N control data sets, each of which is either a human control data set or a model-generated data set. Denote these data sets as $X^n = [\bar{x}_1^n \ \bar{x}_2^n \ \ldots \bar{x}_D^n]$, $n \in \{1, 2, \ldots, N\}$, where,

$$\bar{x}_d^n = \left[x_{1d}^n \ x_{2d}^n \ \ldots \ x_{t_n d}^n\right]^T, d \in \{1, 2, \ldots, D\}, \quad (4.37)$$

denotes the dth t_n-length column vector for data set X^n. Since we use discrete-output HMMs in our similarity measure, we need to convert these multi-dimensional, real-valued data sets to sequences of discrete symbols O_n. We follow two steps in this conversion: (1) data preprocessing and (2) vector quantization, as illustrated in Figure 4.7. The primary purpose of the data preprocessing (described below) is to extract meaningful feature vectors for the vector quantizer. For our case, the preprocessing proceeds in three steps: (1) normalization, (2) spectral conversion, and (3) power spectral density (PSD) estimation.

In the normalization step, we want to scale the columns in each data set, so that each dimension takes on the same range of values, namely $[-1, 1]$. Note that the scale factor for a given dimension has to be the same across data sets X^n. Let,

$$U = N^m(X, \bar{s}) = [(\bar{x}_1/s_1)\ (\bar{x}_2/s_2)\ \ldots\ (\bar{x}_D/s_D)] \tag{4.38}$$

define a matrix-to-matrix *normalization* transform for a $t \times D$ matrix X and a D-length scale vector,

$$\bar{s} = [s_1\ s_2\ \ldots\ s_D]^T, s_d > 0, d \in \{1, 2, \ldots, D\}. \tag{4.39}$$

To perform the normalization on our data sets X^n, we choose the $\bar{s}$ vector,

$$s_d = \max_{\forall n,t} |x_{td}^n|, d \in \{1, 2, \ldots, D\}, \tag{4.40}$$

such that the normalized data sets U^n,

$$U^n = N^m(X^n, \bar{s}) = [\bar{u}_1^n\ \bar{u}_2^n\ \ldots\ \bar{u}_D^n]^T, n \in \{1, 2, \ldots, N\}, \tag{4.41}$$

$$\bar{u}_d^n = \left[u_{1d}^n\ u_{2d}^n\ \ldots\ u_{t_n,d}^n\right], d \in \{1, 2, \ldots, D\}, \tag{4.42}$$

satisfy,

$$|u_{td}^n| \le 1, \forall n, d, t. \tag{4.43}$$

After normalization, we perform spectral conversion on the columns of the normalized data sets U^n. For each column, we segment the data into possibly overlapping window frames, and apply either the Discrete Fourier Transform (DFT) or the Discrete Walsh Transform (DWT) to each frame.**

The Discrete Fourier Transform $T_F^v(\,\cdot\,)$ maps a k-length real vector $\bar{y} = [y_1\ y_2\ \ldots\ y_k]^T$ to a k-length complex vector $\bar{z}$ and is defined as,

$$\bar{z} = T_F^v(\bar{y}) = [F_0(\bar{y})\ F_1(\bar{y})\ \ldots\ F_{k-1}(\bar{y})]^T, \text{ where} \tag{4.44}$$

$$F_p(\bar{y}) = \sum_{q=0}^{k-1} y_{q+1} e^{2\pi i p q/k}, p \in \{0, 1, \ldots, k-1\}. \tag{4.45}$$

**In practice, we calculate the DFT and DWT through the fast Fourier transform (FFT) and the fast Walsh transform (FWT), the $\mathcal{O}(k \log k)$ algorithmic counterparts of the DFT and DWT, respectively. This restricts k to be of the form 2^m, $m \in \{1, 2, \ldots\}$.

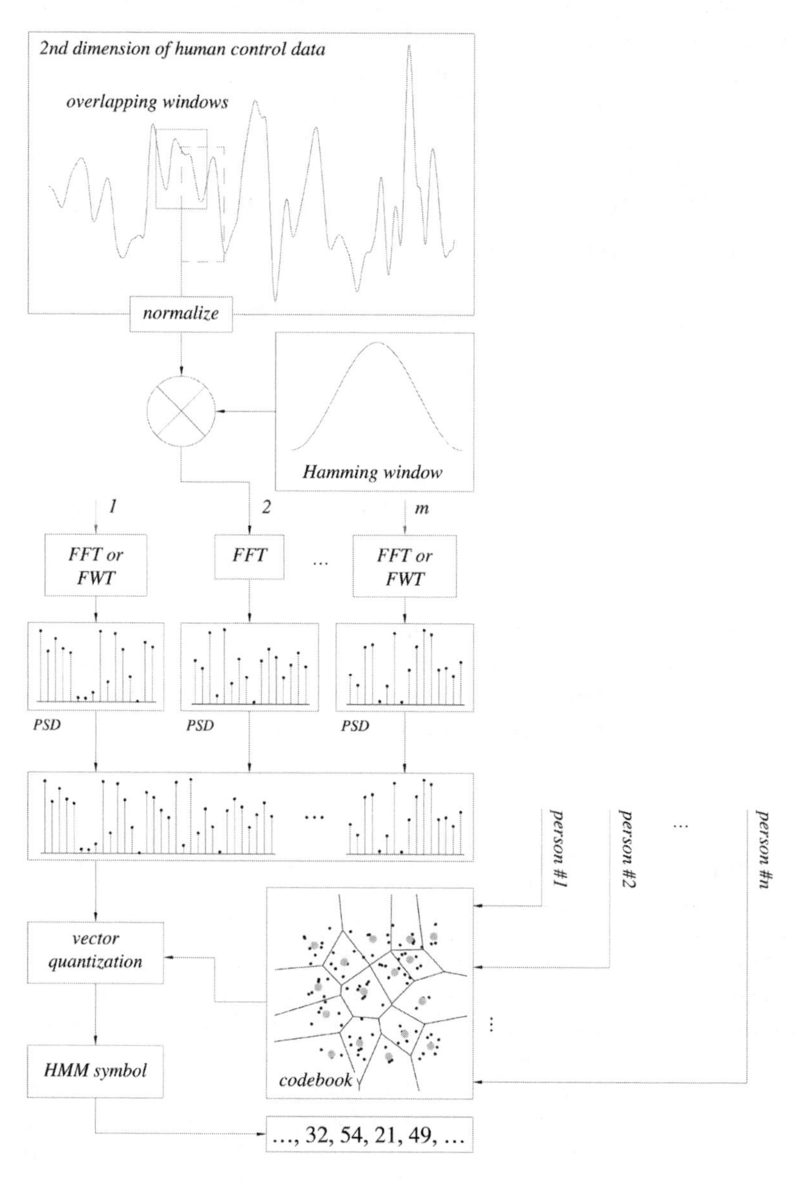

FIGURE 4.7: Conversion of multi-dimensional human control data to a sequence of discrete symbols.

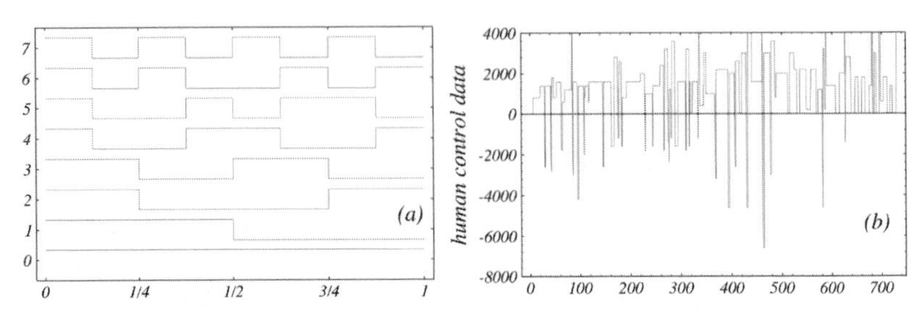

FIGURE 4.8: (a) The first eight Walsh-ordered Walsh functions, and (b) some sample human control data.

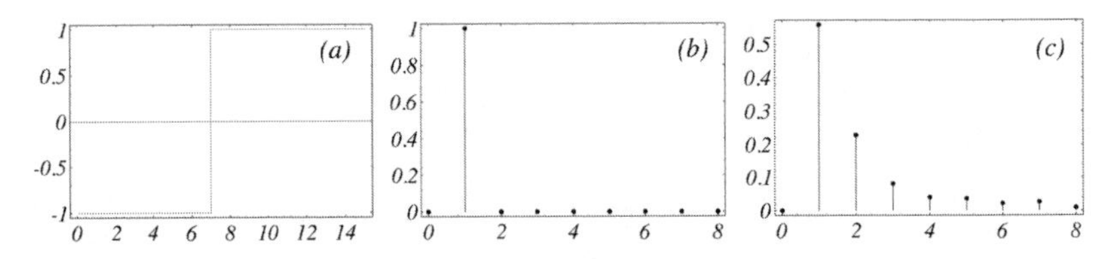

FIGURE 4.9: (a) Sample square wave and its corresponding (b) Walsh and (c) Fourier PSDs.

Prior to applying the Fourier transform, we filter each frame through a Hamming window in order to minimize spectral leakage caused by the data windowing [189]. The Hamming transform $T_H^v(\,\cdot\,)$ maps a k-length real vector $\bar{y} = [y_1\ y_2\ \ldots\ y_k]^T$ to a k-length real vector $\bar{h}$ and is defined as,

$$\bar{h} = T_H^v(\bar{y}) = [H_1 y_1\ H_2 y_2\ \ldots\ H_k y_k]^T\text{, where} \tag{4.46}$$

$$H_p = 0.54 - 0.46\cos\left[\frac{2\pi(p-1)}{k-1}\right], p \in \{1, 2, \ldots, k\} \text{ (see Figure 4.7)} \tag{4.47}$$

For notational convenience let $T_{HF}^v(\bar{y}) = T_F^v[T_H^v(\bar{y})]$.

Instead of sinusoidal basis functions, the Discrete Walsh Transform decomposes a signal based on the orthonormal Walsh functions [195]. The first eight Walsh-ordered Walsh functions are shown in Figure 4.8(a). In Figure 4.8(b), we show an example of human control data which can be characterized better through the Walsh transform, rather than the Fourier transform, due to its discontinuous profile. Consider, for example, the power spectral densities (PSDs) for the square wave in Figure 4.9(a). The Walsh PSD in Figure 4.9(b) is a more concise feature vector than the corresponding Fourier PSD in Figure 4.9(c).

The Discrete Walsh Transform (DWT) $T_W^v(\,\cdot\,)$ maps a k-length real vector $\bar{y} = [y_1\ y_2\ \ldots\ y_k]^T$ to a k-length real vector $\bar{w}$ and is defined as,

$$\bar{w} = T_W^v(\bar{y}) = [W_0(\bar{y})\ W_1(\bar{y})\ \ldots\ W_{k-1}(\bar{y})]^T\text{, where} \tag{4.48}$$

$$W_p(\bar{y}) = \sum_{q=0}^{k-1} y_{q+1}\omega(q,p), p \in \{0, 1, \ldots, k-1\}, \tag{4.49}$$

$$\omega(q,p) = (-1)^{\left\{b(q,k-1)b(p,0)+\sum_{i=1}^{k-1}[b(q,k-i)+b(q,k-i-1)]b(p,i)\right\}}, \text{ and} \tag{4.50}$$

$$b(p,i) \text{ is the } i\text{th bit in the binary representation of } p. \tag{4.51}$$

Now, let us define the power spectral density (PSD) estimates for the Hamming-Fourier ($T^v_{HF}(\,\cdot\,)$) and Walsh transforms ($T^v_W(\,\cdot\,)$). For $\bar{f} = T^v_{HF}(\bar{y}) = [f_0\ f_1\ \ldots\ f_{k-1}]^T$, the Fourier PSD is given by,

$$P^v_{HF}(\bar{f}) = \left[P_{HF0}(\bar{f})\ P_{HF1}(\bar{f})\ \ldots\ P_{HF(k/2)}(\bar{f})\right], \text{ where,} \tag{4.52}$$

$$P_{HF0}(\bar{f}) = \frac{1}{H_{ss}}|f_0|^2, \tag{4.53}$$

$$P_{HFp}(\bar{f}) = \frac{1}{H_{ss}}\left(|f_p|^2 + |f_{k-p}|^2\right), p \in \{1, 2, \ldots, k/2-1\}, \tag{4.54}$$

$$P_{HF(k/2)}(\bar{f}) = \frac{1}{H_{ss}}|f_{(k/2)}|^2, \text{ and} \tag{4.55}$$

$$H_{ss} = k\sum_{q=1}^{k} H_k^2. \tag{4.56}$$

For $\bar{w} = T^v_W(\bar{y}) = [w_0\ w_1\ \ldots\ w_{k-1}]^T$, the Walsh PSD is given by,

$$P^v_W(\bar{w}) = \left[P_{W0}(\bar{w})\ P_{W1}(\bar{w})\ \ldots\ P_{W(k/2)}(\bar{w})\right], \text{ where,} \tag{4.57}$$

$$P_{W0}(\bar{w}) = |w_0|^2, \tag{4.58}$$

$$P_{Wp}(\bar{w}) = |w_{2p-1}|^2 + |w_{2p}|^2, p \in \{1, 2, \ldots, k/2-1\}, \tag{4.59}$$

$$P_{W(k/2)}(\bar{w}) = |w_{(k/2)}|^2. \tag{4.60}$$

Finally, for notational convenience, define two unity transforms,

$$P^v_G(\bar{y}) = T^v_G(\bar{y}) = \bar{y}, \tag{4.61}$$

and let,

$$\bar{y}_{[\tau,k]} = [y_\tau\ y_{\tau+1}\ \ldots\ y_{\tau+k-1}]^T \tag{4.62}$$

be the k-length segment, beginning at element τ, for the t-length vector $\bar{y}$. Using equation (4.62), let us define the vector-to-matrix transform $T^{vm}_{(\varphi,[k_1,k_2])}(\,\cdot\,)$,

$$T^{vm}_{(\varphi,[k_1,k_2])}(\bar{y}) = \begin{bmatrix} P^v_\varphi\left\{T^v_\varphi\left(\bar{y}_{[1,k_1]}\right)\right\} \\ P^v_\varphi\left\{T^v_\varphi\left(\bar{y}_{[k_2+1,k_1]}\right)\right\} \\ P^v_\varphi\left\{T^v_\varphi\left(\bar{y}_{[2k_2+1,k_1]}\right)\right\} \\ \vdots \end{bmatrix}, \varphi \in \{F, HF, W, G\}. \tag{4.63}$$

Furthermore, given a matrix $U = [\bar{u}_1\ \bar{u}_2\ \ldots\ \bar{u}_D]$, let us define the matrix-to-matrix transform $T^{mm}_{(\bar{\varphi},[k_1,k_2])}(\,\cdot\,)$,

$$T^{mm}_{(\bar{\varphi},[k_1,k_2])}(U) = \left[T^{vm}_{(\varphi_1,[k_1,k_2])}(\bar{u}_1)\ T^{vm}_{(\varphi_2,[k_1,k_2])}(\bar{u}_2)\ \ldots\ T^{vm}_{(\varphi_D,[k_1,k_2])}(\bar{u}_D)\right],$$
$$\bar{\varphi} = [\varphi_1\ \varphi_2\ \ldots\ \varphi_D], \varphi_d \in \{F, HF, W, G\}, d \in \{1, 2, \ldots, D\}. \tag{4.64}$$

The spectral conversion and PSD estimation can now be concisely expressed as,

$$V^n = T^{mm}_{(\bar{\varphi},[\kappa_1,\kappa_2])}(U^n), \tag{4.65}$$

where the input matrix U^n has dimensions $t_n \times D$, and the output matrix V^n has dimensions $T_n \times K$,

$$T_n = \text{floor}\left(\frac{t_n - \kappa_1}{\kappa_2} + 1\right), K = \sum_{\varphi_d \in \{F,HF,W\}} (\kappa_1/2 + 1) + \sum_{\varphi_d = G} \kappa_1. \tag{4.66}$$

The integer constants $\kappa_1 \geq \kappa_2 > 0$ define the length of each window frame as κ_1 and the window overlap as $\kappa_2 - \kappa_1$. Furthermore, the transformation vector $\bar{\varphi}$ selects which transform to apply to each dimension of the control data. Generally, we select the Fourier PSD for *state trajectories*, and the Walsh PSD for *command trajectories*, since these trajectories tend to be nonsmooth and, in part, discontinuous (see Appendix A).

4.2.6 Vector quantization

In the previous section, we define the transformation from the data sets X^n to the feature matrices V^n. Let, $V = \{\bar{v}^n_t\}$, $t \in \{1, 2, \ldots, T_n\}$, $n \in \{1, 2, \ldots, N\}$, denote the set of all feature vectors, where $\bar{v}^n_t$ is the tth row of the nth feature matrix. In order to apply discrete-output HMMs, we now need to convert the feature vectors V to L discrete symbols, where L is the number of output observables in our HMM models. In other words, we want to replace the many $\bar{v}^n_t$ with L prototype vectors $Q_L = \{\bar{q}_l\}$, $l \in \{1, 2, \ldots, L\}$, known as the *codebook*, such that we minimize the total distortion $D(V, Q_L)$,

$$D(V, Q_L) = \sum_{n,t} \min_l d(\bar{v}^n_t, \bar{q}_l), \text{ where } d(\bar{v}^n_t, \bar{q}_l) = (\bar{q}_l - \bar{v}^n_t) \cdot (\bar{q}_l - \bar{v}^n_t), \tag{4.67}$$

over all feature vectors. We choose the well-known LBG vector quantization (VQ) algorithm [142] to perform this quantization. Figure 4.10 illustrates the algorithm, which generates codebooks of size $L = 2^m$, $m \in \{0, 1, 2, \ldots\}$, and can be stopped at an appropriate level of discretization given the amount of available data and the complexity of the system trajectories. For our data,

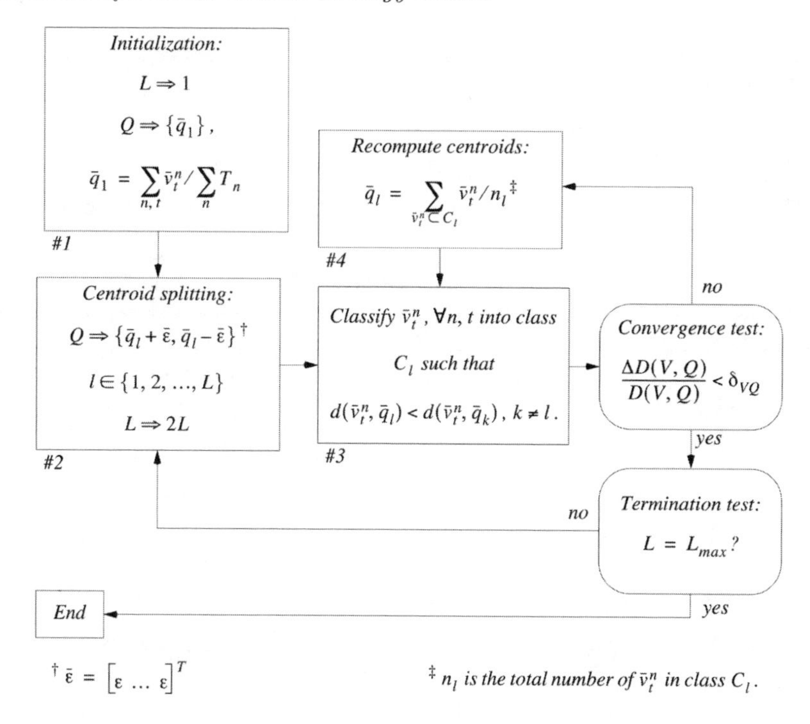

FIGURE 4.10: The LBG VQ algorithm generates VQ codebooks of increasing size.

we set the split offset $\varepsilon = 0.0001$ and the convergence criterion $\delta_{VQ} = 0.01$.[††] With these parameter settings, the centroids $\{\bar{q}_l\}$ usually converge within only a few iterations of the #3-#4 loop in Figure 4.10. As an example, Figure 4.11 illustrates the LBQ vector quantization for some random 2D data and $L \in \{1, 2, 4, 8, 16, 32\}$, while Figure 4.12 illustrates the quick convergence of the algorithm after centroid splitting for the same data and $L = 4$.

Given a trained VQ codebook Q_L, we convert the feature vectors V^n to a sequence of discrete symbols $O_n = \{o_1^n, o_2^n, \ldots, o_{T_n}^n\}$,

$$O_n = T_{VQ}^m(V^n, Q_L) = \{T_{VQ}^v(\bar{v}_1^n, Q_L), T_{VQ}^v(\bar{v}_2^n, Q_L), \ldots, T_{VQ}^v(\bar{v}_{T_n}^n, Q_L), \}, \tag{4.68}$$

$$\text{where, } o_t^n = T_{VQ}^v(\bar{v}_t^n, Q_L) = \text{index}\left[\min_l d(\bar{v}_t^n, \bar{q}_l)\right] \tag{4.69}$$

This completes the conversion from the multi-dimensional, real-valued data sets X^n to the discrete observation sequences $O_n = \{o_1^n, o_2^n, \ldots, o_{T_n}^n\}$. Com-

[††]These parameters are selected to achieve low distortion levels, while minimizing the number of iterations of the VQ algorithm. After the last split of the codebook $L = L_{max}$, we set $\delta_{VQ} \rightarrow \delta_{VQ}/10$.

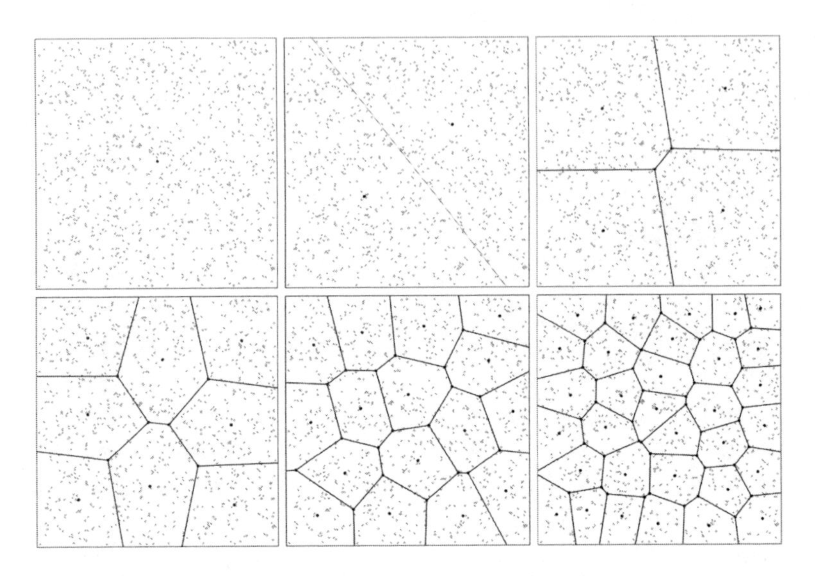

FIGURE 4.11: The LBG vector quantization for some random 2D data, as L equals 1, 2, 4, 8, 16 and 32.

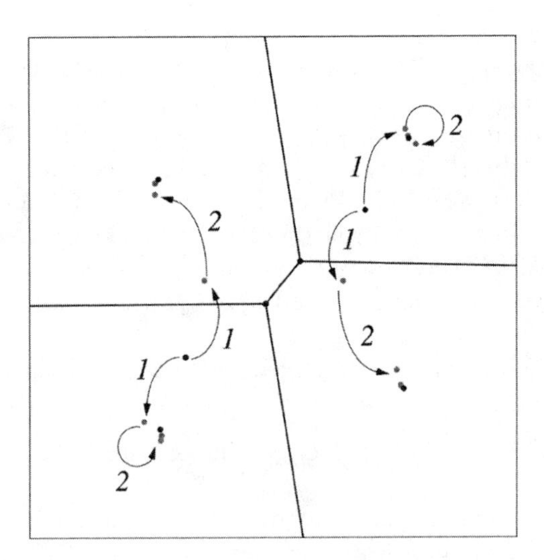

FIGURE 4.12: For a given codebook size $L = 4$, the LBG algorithm converges in only a few iterations after centroid splitting.

bining equations (4.40), (4.65), and (4.68), we can summarize the signal-to-symbol conversion of the data sets X^n as,

$$O_n = T_{VQ}^m \left\{ T_{(\bar{\varphi},[\kappa_1,\kappa_2])}^{mm} \left[N^m(X^n, \bar{s}) \right], Q_L \right\} = T_{all}\left(X^n, \bar{s}, \bar{\varphi}, [\kappa_1, \kappa_2], Q_L \right). \tag{4.70}$$

4.2.7 Discretization compensation

We have stated previously that we choose to use discrete-output HMMs in our similarity analysis because they involve *significantly* less computation in training than either continuous-output or semicontinuous-output HMMs. While computationally efficient, discretization of the output space can have some negative consequences when analyzing real-valued data. Consider the following example.

Assume that we want to determine the similarity between two control data sets, X^1 and X^2. We follow the signal-to-symbol conversion procedure described in the previous two sections, and convert the data sets to discrete observation sequences O_1 and O_2,

$$O_k = \{o_t^k\}, t \in \{1, 2, \ldots, T_k\}, k \in \{1, 2\}. \tag{4.71}$$

We also train corresponding n-state HMMs, λ_1 and λ_2, where

$$A_k = \begin{bmatrix} a_{11}^k & a_{12}^k & \cdots & a_{1,n}^k \\ a_{21}^k & a_{22}^k & \cdots & a_{2,n}^k \\ \vdots & \vdots & \ddots & \vdots \\ a_{n1}^k & a_{n2}^k & \cdots & a_{nn}^k \end{bmatrix}, B_k = \begin{bmatrix} b_1^k(1) & b_2^k(1) & \cdots & b_n^k(1) \\ b_1^k(2) & b_2^k(2) & \cdots & b_n^k(2) \\ \vdots & \vdots & \ddots & \vdots \\ b_1^k(L) & b_2^k(L) & \cdots & b_n^k(L) \end{bmatrix}, \tag{4.72}$$

$$\pi_k = \begin{bmatrix} \pi_1^k \\ \pi_2^k \\ \vdots \\ \pi_n^k \end{bmatrix}, k \in \{1, 2\}. \tag{4.73}$$

Now suppose that symbol l appears in the O_1 observation sequence (say at $t = \tau$), but does not appear in the O_2 observation sequence. This will force,

$$b_j^2(l) = 0, j \in \{1, 2, \ldots, n\}, \tag{4.74}$$

during the training of λ_2. Consequently, when we try to evaluate $P(O_1|\lambda_2)$ using the forward algorithm, we get,

$$\alpha_\tau(j) = \left[\sum_{i=1}^n \hat{\alpha}_{\tau-1}(i) a_{ij}^2 \right] b_j^2(o_\tau) = \left[\sum_{i=1}^n \hat{\alpha}_{\tau-1}(i) a_{ij}^2 \right] b_j^2(l) = 0, j \in \{1, 2, \ldots, n\} \tag{4.75}$$

$$c_T = 1/\left(\sum_{i=1}^{n} \alpha_T(i)\right) \to \infty, P(O_1|\lambda_2) = 1/\left(\prod_{t=1}^{T} c_t\right) \to 0, P_{12} \to 0 \quad (4.76)$$

$$\therefore \sigma(O_1, O_2) = 0. \quad (4.77)$$

Thus, the presence of a single observable l present in one observation sequence but not the other will force the similarity measure to be 0, even if the two observation sequences are identical in every other respect. This is not desirable, since the rogue observable l might be an event that is observed only rarely, or might even be the result of a measurement error or unintended control action on the part of the process that generated control trajectory X^1. Below, we consider two parameterized post-training solutions to this singularity problem within the context of discrete-output HMMs: (1) flooring and (2) semicontinuous evaluation.

Flooring [192] defines the common practice of replacing nonzero elements in the trained HMMs by some small value $\rho > 0$ and then renormalizing the rows of A and the columns of B to satisfy the probabilistic constraints in equations (4.88) and (4.89). If there are m zero elements in a probability vector (i.e., a row of A or a column of B), this method redistributes $(\rho m)/(1 + \rho m)$ of the total probability mass to the zero elements.

Semicontinuous evaluation [101] redefines the forward algorithm. Let $O = \{o_t\}$ denote a discrete observation sequence that has been vector quantized from a sequence of real vectors $V = \{\bar{v}_t\}, t \in \{1, 2, \ldots, T\}$, and a VQ codebook $Q = \{\bar{q}_l\}, l \in \{1, 2, \ldots, L\}$. For discrete evaluation on a Hidden Markov Model λ, $P(O|\lambda)$ is computed using the forward algorithm,

$$\alpha_1(i) = \pi_i b_i(o_1), i \in \{1, 2, \ldots, n\} \quad (4.78)$$

$$\alpha_{t+1}(j) = \left[\sum_{i=1}^{n} \alpha_t(i) a_{ij}\right] b_j(o_{t+1}), t \in \{1, 2, \ldots, T-1\} \quad (4.79)$$

$$P(O|\lambda) = \sum_{i=1}^{n} \alpha_T(i) \quad (4.80)$$

Semicontinuous evaluation proceeds almost identically, except that the $b_j(o_t)$ terms are replaced by $\tilde{b}_j(\bar{v}_t)$ terms,

$$\tilde{b}_j(\bar{v}_t) = \sum_{l=1}^{L} p\left(\bar{v}_t|C_l\right) b_j(l), \quad (4.81)$$

$$p\left(\bar{v}_t|C_l\right) = \exp\left[-\frac{1}{\sigma^2}(\bar{q}_l - \bar{v}_t) \cdot (\bar{q}_l - \bar{v}_t)\right], \quad (4.82)$$

where $p\left(\bar{v}_t|C_l\right)$ represents the estimated conditional probability density function that vector $\bar{v}_t$ belongs to class C_l, corresponding to the codebook vector $\bar{q}_l$, and σ is a user-defined smoothing parameter. Thus, in semicontinuous

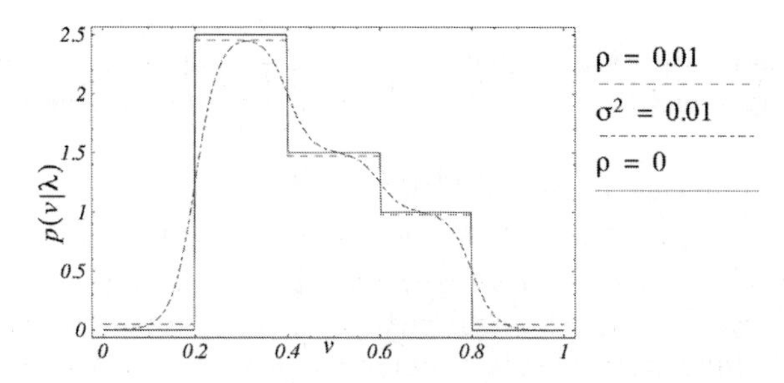

FIGURE 4.13: To avoid the singularity problem, the initial discrete pdf ($\rho = 0$) can be modified either through flooring ($\rho = 0.01$) or semicontinuous evaluation ($\sigma^2 = 0.01$).

evaluation, we view the codebook vectors as the peaks of Gaussian distributions with uniform variances σ^2. The complete forward algorithm (without scaling) is given by,

$$\alpha_1(i) = \pi_i \tilde{b}_i(\bar{v}_1), i \in \{1, 2, \ldots, n\} \tag{4.83}$$

$$\alpha_{t+1}(j) = \left[\sum_{i=1}^{n} \alpha_t(i) a_{ij} \right] \tilde{b}_j(\bar{v}_{t+1}), t \in \{1, 2, \ldots, T-1\}, j \in \{1, 2, \ldots, n\} \tag{4.84}$$

$$P(V|\lambda) = \sum_{i=1}^{n} \alpha_T(i), \text{ where } \lim_{\sigma \to 0} P(V|\lambda) = P(O|\lambda) \tag{4.85}$$

As an example, consider a single-state HMM λ with the discrete-output probability matrix B,

$$B = \left[0.0 \; 0.5 \; 0.3 \; 0.2 \; 0.0 \right]^T. \tag{4.86}$$

Also, let the discrete symbols l correspond to real-valued numbers v,

$$\frac{l-1}{5} \leq v < \frac{l}{5}, l \in \{1, 2, 3, 4, 5\}. \tag{4.87}$$

Figure 4.13 illustrates how the discrete probability density function (pdf) $p(v|\lambda)$ encoded by B is modified through flooring and semicontinuous evaluation, for specific values $\rho = 0.01$ and $\sigma^2 = 0.01$, respectively. Flooring of course maintains the discrete structure of the HMM, while semicontinuous evaluation smoothes between output classes.

We note that the smoothing of the output pdf achieved by semicontinuous evaluation is done so at a significant computational cost, in comparison to discrete evaluation. Assuming L output classes (i.e., symbols) and K dimensions for the $\bar{v}_t$ vectors, the computation of $\tilde{b}_j(\bar{v}_t)$ is $\mathcal{O}(LK)$, while $b_j(o_t)$ requires only one table lookup. Consequently, for typical values of L and K, the evaluation of $P(V|\lambda)$ will be orders of magnitude slower than the evaluation of $P(O|\lambda)$.

For the experiments in the next chapter, the similarity measure achieves roughly equivalent discrimination results with semicontinuous evaluation as with flooring. Therefore, because of the substantial computational burden of semicontinuous evaluation, unless otherwise noted, we choose flooring rather than semicontinuous evaluation to avoid the singularity problem, with $\rho = 0.0001$. For the HMMs in this book, this value of ρ redistributes less than 0.1% of the probability mass in the state transition matrices A, and less than 0.5% probability mass in the output probability matrices B.

4.2.8 HMM training

The last step of the similarity analysis involves training Hidden Markov Models λ corresponding to each observation sequence O. To do this we use the iterative Baum-Welch algorithm. Throughout this book, we initialize the Baum-Welch algorithm by setting the Hidden Markov Model parameters to random, nonzero values, subject to the necessary probabilistic constraints,

$$\sum_{j=1}^{n_s} a_{ij} = 1, i \in \{1, 2, \ldots, n_s\}, \tag{4.88}$$

$$\sum_{l=1}^{L} b_j(l) = 1, j \in \{1, 2, \ldots, n_s\}, \tag{4.89}$$

where a_{ij} is the probability of transiting from state i to state j, $b_j(l)$ is the probability of observing symbol l in state j, and n_s is the number of HMM states that we choose for the models λ. Let $\lambda^{(k)}$ denote the HMM λ after k iterations of the Baum-Welch algorithm, and let $\lambda^{(m)}$ denote the current iteration of Baum-Welch. Then, we stop training if,

$$\frac{P\left(O|\lambda^{(k)}\right) - P\left(O|\lambda^{(k-1)}\right)}{P\left(O|\lambda^{(k)}\right)} < \delta_{HMM}, k \in \{m, m-1, \ldots, m-4\}, \tag{4.90}$$

where $\delta_{HMM} = 0.000001$. This type of stringent convergence test is required, because in practice, the Baum-Welch algorithm frequently stalls over consecutive iterations. Figure 4.14, for example, plots $-\log P(O|\lambda^{(k)})/T$ for some typical human data. We must be careful that we do not stop training on these types of plateaus if further improvements can be achieved. Otherwise, the assumption that the P_{ii}, defined in (4.14), represent near-optimal global maxima would be violated, along with properties #2 and #3 in equations (4.21) and (4.22), respectively.

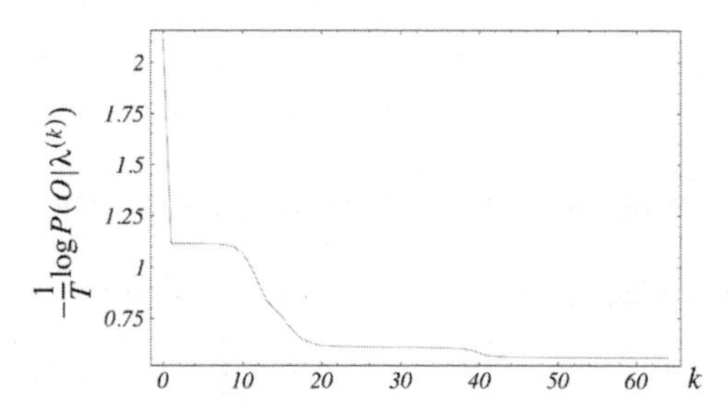

FIGURE 4.14: The Baum-Welch algorithm can stall over several iterations, until further improvements are realized.

4.3 Human-to-model comparisons

Now, we will apply the proposed similarity measure as a means of validating the control trajectories of the different HCS models presented in the previous chapter. In other words we will quantify our previous qualitative observations about the similarity between the learned models and their respective human control training data.

We select the following parameters for our similarity analysis:

$$\bar{\varphi} = [HF\ HF\ HF\ W\ W]^T, \kappa_1 = 16, \kappa_2 = \kappa_1/2, L = 128, n_s = 8, \quad (4.91)$$

so that for a control trajectory $X^* = \left[\bar{v}_\xi\ \bar{v}_\eta\ \bar{\omega}\ \bar{\delta}\ \bar{\phi}\right]$, the corresponding observation sequence O^* will be given by,

$$O^* = T_{all}\left(X^*, \bar{s}, \bar{\varphi}, [\kappa_1, \kappa_2], Q_L\right) \quad (4.92)$$

where the scale vector $\bar{s}$ is chosen and the VQ codebook Q_L is trained over all control trajectories in the similarity analysis.

Table 4.3 and Figure 4.15 report the resulting human-to-model similarity measures σ for Larry, Moe, Groucho, and Harpo, and their respective *Cq*, *Ck*, and hybrid HCS models. In addition, we provide similarity results for the hybrid HCS models, where we let the statistical models λ_i be uniform, such that equation (3.69),

$$P(A_i|S_l) = b_i(l)P(A_i) / \sum_{i=1}^{N} b_i(l)P(A_i) \quad (4.93)$$

reduces to,

$$P(A_i|S_l) = P(A_i). \quad (4.94)$$

In other words, the choice of control action at each time step depends strictly on the priors $P(A_i)$, when the λ_i are uniform.

Table 4.3: Human-to-model similarity[a]

Individual	Dimensions	Cq	Ck	Just $P(A_i)^b$	Hybrid
Larry	$\{\bar{x}, \bar{u}\}^c$	0.100	0.161	0^d	0.450
	$\{\bar{u}\}^e$	0.128	0.432	0^d	0.657
Moe	$\{\bar{x}, \bar{u}\}$	0.087	0.088	0.046	0.555
	$\{\bar{u}\}$	0.117	0.146	0.187	0.594
Groucho	$\{\bar{x}, \bar{u}\}$	0.101	0.096	0.014	0.457
	$\{\bar{u}\}$	0.319	0.172	0.132	0.561
Harpo	$\{\bar{x}, \bar{u}\}$	0^d	0.003	0.012	0.578
	$\{\bar{u}\}$	0^d	0.003	0.024	0.609

a. Each individual's run is compared to the model trajectory over road #5.
b. Models λ_i are uniform over the entire input space. Condition (5-27) is enforced.
c. All state and control variables are included for the similarity analysis.
d. Model is unstable over road #5.
e. Only the control variables $\{\delta, \phi\}$ are included in the similarity analysis.

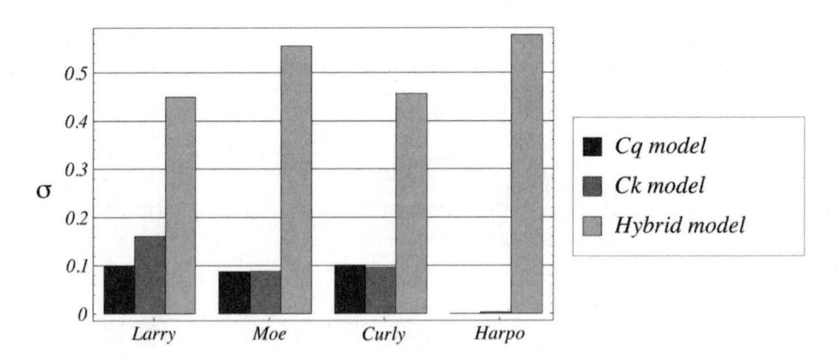

FIGURE 4.15: Human-to-model similarity for different modeling approaches and $\{\bar{x}, \bar{u}\}$.

Table 4.4: Hybrid model-to-human matching

σ	*Larry*	*Moe*	*Groucho*	*Harpo*
Larry's model	**0.450**	0.329	0.315	0.069
Moe's model	0.126	**0.555**	0.338	0.217
Groucho's model	0.152	0.377	**0.457**	0.206
Harpo's model	0.013	0.134	0.127	**0.578**

Table 4.5: *Ck* model-to-human matching

σ	*Larry*	*Moe*	*Groucho*	*Harpo*
Larry's model	0.161	**0.166**	0.118	0.157
Moe's model	0.056	**0.088**	0.063	0.041
Groucho's model	0.056	0.066	**0.096**	0.040
Harpo's model	0.006	0.008	**0.012**	0.003

From Table 4.3, the *Cq* and *Ck* models exhibit roughly the same similarity to each model's corresponding human control data. These similarity values are, however, rather low. Therefore, neither the *Cq* nor the *Ck* learning algorithm is able to model the driving control strategies with a high degree of fidelity to the source training data.

In comparison, we note that the hybrid controllers have much more in common with the source training data than do either the *Cq* or *Ck* models. Finally, we observe that without models λ_i, the hybrid control strategies degenerate, and are no longer representative of the human's control strategy. This confirms that the statistical models λ_i impart useful information to the hybrid control strategies, and that the improved fidelity of the hybrid controllers is not simply due to random thrashing about of the acceleration command ϕ.

As an additional validation check, we now show that a particular individual's hybrid model not only closely matches the control data for that individual, but also exhibits a lesser degree of similarity with other individual's control data. Table 4.4 reports the similarity of each individual's model with each individual's control strategy for the hybrid HCS models. We observe that the highest degree of similarity occurs between a specific individual and his model. In contrast, we observe from Table 4.5, that the *Ck* models do not necessarily exhibit the highest degree of similarity (however low) with their respective training data.

In summary, we have demonstrated that the hybrid models exhibit greater fidelity to the human training data than either of the cascade network-based modeling approaches. Which learning approach – continuous or hybrid – is preferred may ultimately depend on the specific application for the HCS model. If the model is being developed towards the eventual control of a real robot or vehicle, then the continuous modeling approach might be preferred as a good starting point. Continuous models extrapolate control strategies to a greater range of inputs, show greater inherent stability, and lend themselves more readily to theoretical performance analysis. If, on the other hand, the model is being developed in order to simulate different human behaviors in a virtual reality simulation or game, then the discontinuous control approach might be preferred, since fidelity to the human training data and random variations in behavior would be the desired qualities of the HCS model. Thus, depending on the application, we believe a need exists for both types of modeling approaches.

5

Evaluation of Human Control Strategy

5.1 Introduction

Since most HCS models are empirical, few if any guarantees exist about their theoretical performance. In the previous work, a stochastic similarity measure, which compares model-generated control trajectories to the original human training data, has been proposed for validating HCS models [168]. This similarity measure can then be applied towards validating a learned model's fidelity to the training data by comparing the model's dynamic trajectories in the feedback loop with the human's dynamic trajectories. While this similarity measure can ensure that a given HCS model adequately captures the characteristics of a human's driving skill, it does not measure a particular model's skill or performance. In other words, it does not (nor can it) tell us which model is better or worse. Thus, performance evaluation forms an integral part of HCS modeling research, without which it becomes impossible to rank or prefer one HCS controller over another. * Moreover, only when we have developed adequate performance criteria, can we hope to optimize HCS models with respect to those performance criteria.

In general, skill or performance can be defined through a number of subtask-dependent as well as subtask-independent criteria. Some of these criteria may conflict with one another, and which is the most appropriate for a given task depends in part on the specific goals of the task. Overall, there are two approaches for defining performance criteria: (1) subtask analysis and (2) intrinsic analysis.

In subtask analysis, we will examine performance within the context of a certain subtask, based on the task of human driving, for example. For the task of driving, many candidate performance criteria exist, such as driving stability and driving safety. Rather than select such specific criteria, however, we prefer to decompose the driving task into subtasks, and then define criteria that measure performance for the most meaningful subtasks. For a driving problem, for example, we can consider the driving task to be a combination of the following subtasks: (1) driving along a straight road, (2) turning through a curve in the road, and (3) avoiding obstacles. Of these subtasks, avoiding obstacles is perhaps the most useful, as it can measure important characteristics of a HCS model, including its ability to change speeds quickly while maintaining vehicle safety and stability. Thus, we define a number of performance criteria with respect to specific subtasks, such as a criterion based on an HCS model's ability to avoid sudden obstacles, a criterion based on a HCS model's ability to negotiate tight turns in a safe and stable manner, a criterion based on transient response, and a criterion based on the time delay of driving response. Each of these performance measures tests an HCS model's performance outside the range of its training data.

In intrinsic analysis, we will examine a given model's behavior on a more global scale. Once again, we consider the task of human driving. For a given HCS model, we might be interested in such measures as average speed, passenger comfort, driving smoothness, and fuel efficiency. These measures are not based on any single subtask, but rather are aggregate measures of performance. In other words, they measure the intrinsic characteristics of a particular HCS model.

For the problem of HCS model evaluation, we first record driving data from different individuals through a dynamic driving simulator. For each driver, we then train a HCS model with the flexible cascade neural network learning architecture. Because different drivers exhibit different styles or control strategies, their respective models will likewise differ. It is the goal of this work to define performance criteria by which the driving models' performances can be evaluated and ranked.

5.2 Obstacle avoidance

Safety is an important aspect of driving ability. Obstacle avoidance is one important measuring stick for gauging a model's safety during automated driving. Collision avoidance concerns the avoidance of static obstacles such as disabled vehicles and the large objects blocking the forward path, and avoidance of dynamical obstacles such as other vehicles with lower speeds in the same lane. Collision can be avoided either by deceleration to full stop or to the speed of the moving objects without hitting the objects, or by executing a timely lane change maneuver to avoid them. Both methods are constrained by the vehicle dynamics, human control strategy, and environmental conditions. For the static obstacle, the best way to avoid the collision is by using the lane change maneuver. Thus, lane change obstacle avoidance is an important performance criterion by which we can gauge a model's performance. In this section, we will develop and evaluate such criteria for different HCS models.

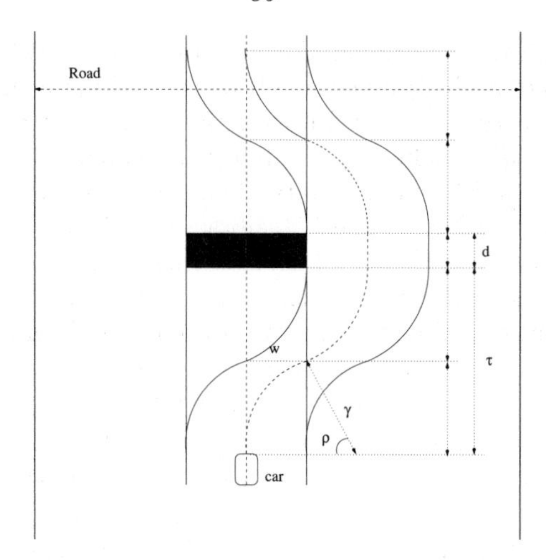

FIGURE 5.1: Virtual path for obstacle avoidance.

5.2.1 Virtual path equivalence

Since our HCS models receive only a description of the road ahead as input from the environment, we reformulate the task of obstacle avoidance as *virtual path following*. Assume that an obstacle appears τ meters ahead of the driver's current position. Furthermore, assume that this obstacle obstructs the width of the road ($2w$) and extends for d meters along the road. Then, rather than follow the path of the actual road, we wish the HCS model to follow a virtual path as illustrated in Figure 5.1 above. This virtual path consists of

- two arcs with radius of curvature γ, which offset the road median laterally by $2w$, followed by
- a straight-line segment of length d, and
- another two arcs with radius of curvature γ which return the road median to the original path.

By analyzing the geometry of the virtual path, we can calculate the required radius of curvature γ of the virtual path segments in terms of obstacle width $2w$ and obstacle distance τ:

$$\gamma^2 = (\frac{\tau}{2})^2 + (\gamma - w)^2 = \frac{\tau^2}{4} + \tau^2 - 2\gamma w + w^2 \tag{5.1}$$

$$\gamma = \frac{\tau^2}{8w} + \frac{w}{2} \tag{5.2}$$

The corresponding sweep angle ρ for the curves is given by,

$$\rho = \sin^{-1}(\frac{\tau/2}{\gamma}) = \sin^{-1}(\frac{\tau}{\frac{\tau^2}{4w} + w}) \tag{5.3}$$

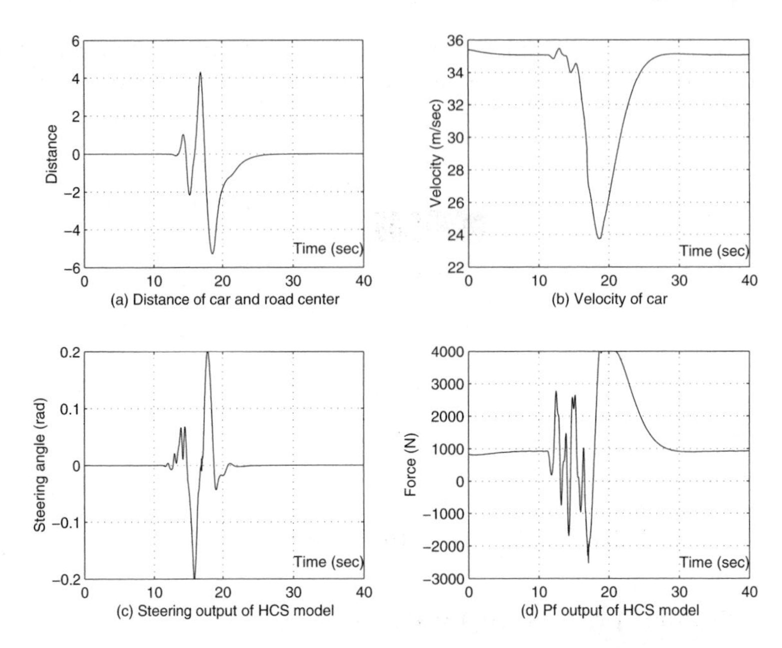

FIGURE 5.2: Example of obstacle avoidance by a HCS model.

Consider an obstacle located $\tau = 60$m ahead of the driver's current position. For this obstacle distance and $w = 5$m, γ evaluates to 92.5m. This is less than the minimum radius of curvature (100m) that we allow for the roads over which we collect our human control data. Therefore, the obstacle avoidance task in part tests each HCS model's ability to operate safely outside the range of its training data.

As an example, Figure 5.2 illustrates one HCS model's response to the virtual path created by an obstacle distance of 60m.

Figure 5.2(a) plots the vehicle's lateral distance from the road median through the virtual path. We observe that through the virtual path, the vehicle deviates sharply from the road median by over 4m. In addition, Figure 5.2(b) shows that the velocity of the car drops substantially from approximately 35m/sec to a low of about 23m/sec on the virtual path. The model's corresponding steering (δ) and force (P_f) outputs are plotted in Figures 5.2(c) and 5.2(d), respectively.

5.2.2 Lateral offset estimation

As we observed in Figure 5.2(a), a driving model may deviate significantly from the center of the road during the obstacle avoidance maneuver. Below, we derive the important relationship between the obstacle detection distance τ and a model's corresponding maximum lateral deviation ψ. First, we take

N measurements of ψ for different values of τ, where we denote the ith measurement as (τ_i, ψ_i). Next, we assume a polynomial relationship of the form,

$$\begin{aligned}\psi_i &= \alpha_p \tau_i^p + \alpha_{p-1}\tau_i^{p-1} + \cdots + \alpha_1 \tau_i + \alpha_0 + e_i \\ &= \Gamma_i^T \alpha + e_i \end{aligned} \tag{5.4}$$

where e_i is additive measurement error. Then, we can write,

$$\begin{aligned}\psi_1 &= \Gamma_1^T \alpha + e_1 \\ \psi_2 &= \Gamma_2^T \alpha + e_2 \\ &\cdots \\ \psi_N &= \Gamma_N^T \alpha + e_N \end{aligned} \tag{5.5}$$

or, in matrix notation,

$$\Psi = \Gamma \alpha + e \tag{5.6}$$

where, $\Psi = [\psi_1, \psi_2, \cdots, \psi_N]^T$ is the observation vector, $\Gamma = [\Gamma_1, \Gamma_2, \cdots, \Gamma_N]^T$ is the regression matrix, and $e = [e_1, e_2, \cdots, e_N]^T$ is the error vector.

Assuming white noise properties for e ($E\{e_i\} = 0$ and $E\{e_i e_j\} = \sigma_e^2 \delta_{ij}$ for all i, j), we can minimize the least-squares error criterion,

$$V(\hat{\alpha}) = \frac{1}{2}\varepsilon^T \varepsilon = \frac{1}{2}\sum_{k=1}^{N} \varepsilon_k^2 = \frac{1}{2}(\Psi - \Gamma\hat{\alpha})^T(\Psi - \Gamma\hat{\alpha}) \tag{5.7}$$

with the optimal, unbiased estimate $\bar{\alpha}$,

$$\bar{\alpha} = (\Gamma^T \Gamma)^{-1} \Gamma^T \Psi \tag{5.8}$$

assuming that $(\Gamma^T \Gamma)$ is invertible.

For example, consider the HCS model in Figure 5.2. We plot its measured ψ values for τ ranging from 20 to 100 meters in Figure 5.3. Superimposed on top of the measured data is the estimated fifth-order relationship ($p = 5$) between ψ and τ. We observe that the polynomial model fits the data closely and appears to be sufficient to express the relationship between ψ and τ.

5.2.3 Obstacle avoidance threshold

We observe from Figure 5.3 that as the obstacle detection distance decreases, the maximum lateral offset increases. Thus, for a given model and initial velocity $v_{original}$, there exists a value τ_{min} below which the maximum offset error will exceed the lane width w_l. We define the driving control for obstacle distances above τ_{min} to be stable; likewise, we define the driving control to be unstable for obstacle distances below τ_{min}.

Now, define an obstacle avoidance performance criterion J_1 to be,

$$J_1 = \frac{\tau_{min}}{v_{original}} \tag{5.9}$$

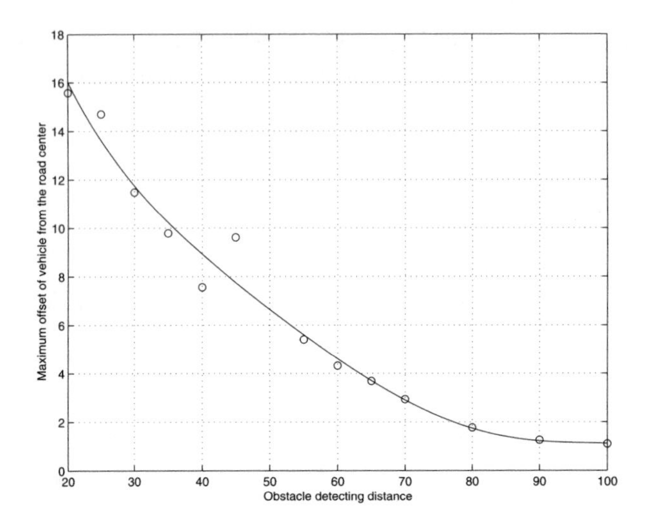

FIGURE 5.3: Maximum lateral offset as a function of obstacle detection distance.

where $v_{original}$ is the velocity of the vehicle when the obstacle is first detected. J_1 measures to what extent a given HCS model can avoid an obstacle while still controlling the vehicle in a stable manner. The normalization by $v_{original}$ is required, because slower speeds increase the amount of time a driver has to react and therefore avoiding obstacles becomes that much easier.

Below, we calculate the J_1 from three HCS models, trained on real driving data from Tom, Dick, and Harry, respectively. Figure 5.4 plots ψ as a fifth-order function of τ for three different models. The solid line in the figure describes curve J_1 of Harry's HCS model, while dash and dash-dot lines show the result of Dick's and Tom's HCS model, respectively.

From Figure 5.4, it is easy to approximate τ_{min} for each HCS model, thus, the corresponding β performance criteria for each model is,

$$J_1^{Tom} = \frac{45}{35} = 1.3 \tag{5.10}$$

$$J_1^{Dick} = \frac{35}{35} = 1.0 \tag{5.11}$$

$$J_1^{Harry} = \frac{18}{35} = 0.51 \tag{5.12}$$

Thus, as an obstacle avoider, Harry's model clearly outperforms Tom's and Dick's models, since J_1^{Harry} is the lowest performance measure among the three models.

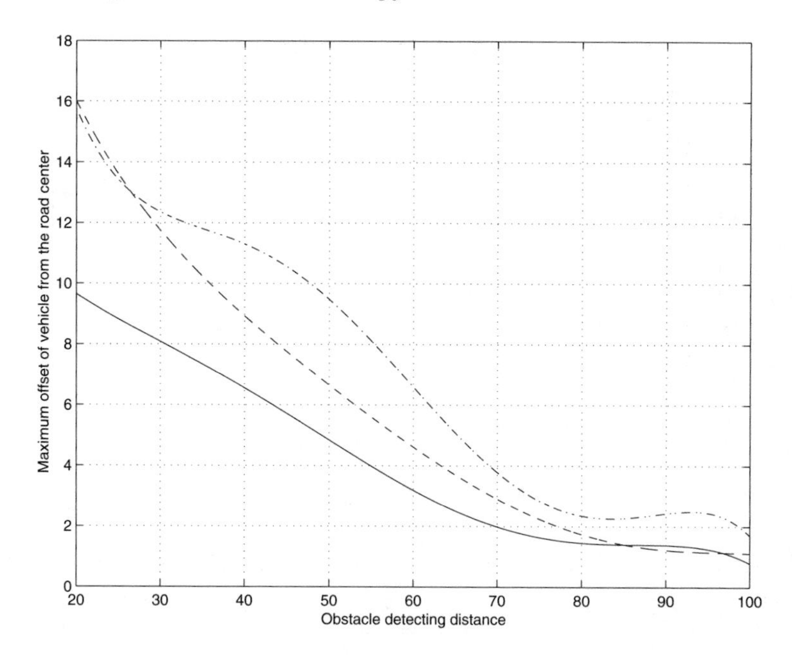

FIGURE 5.4: Maximum lateral offset for Dick's, Tom's, and Harry's model.

5.2.4 Obstacle avoidance velocity loss

The performance criterion J_1 measures the stability of a particular HCS model in avoiding an obstacle. It does not, however, directly measure how skillfully the model avoids the obstacle. Consider, for example, Figure 5.2(b). During the obstacle avoidance maneuver, the velocity of the vehicle drops sharply so that the model can adequately deal with the tight maneuvers required. Below, we define a performance $IAVD$, which is defined as *integral of the absolutes velocity difference* between the velocity before obstacle detection and time-dependent velocity during the obstacle avoidance maneuver:

$$IAVD = \int_{t_0}^{t_f} |v_{original} - v_{virtual}| dt \tag{5.13}$$

where $v_{original}$ is the velocity before obstacle detection, and $v_{virtual}$ is the time-dependent velocity during the obstacle avoidance maneuver.

Consider once again three HCS models for Tom, Dick, and Harry. Each model can successfully avoid the obstacle when τ ranges from 50 to 100 meters. For any given τ, we obtain three values of $IAVD$ corresponding to the three HCS models. Figure 5.5 shows three curves describing the relationships between $IAVD$ value and the obstacle distance τ.

In Figure 5.5, solid line describes curve of $IAVD$ of Harry's HCS model, while dash and dash-dot lines show same result of Dick's and Tom's HCS

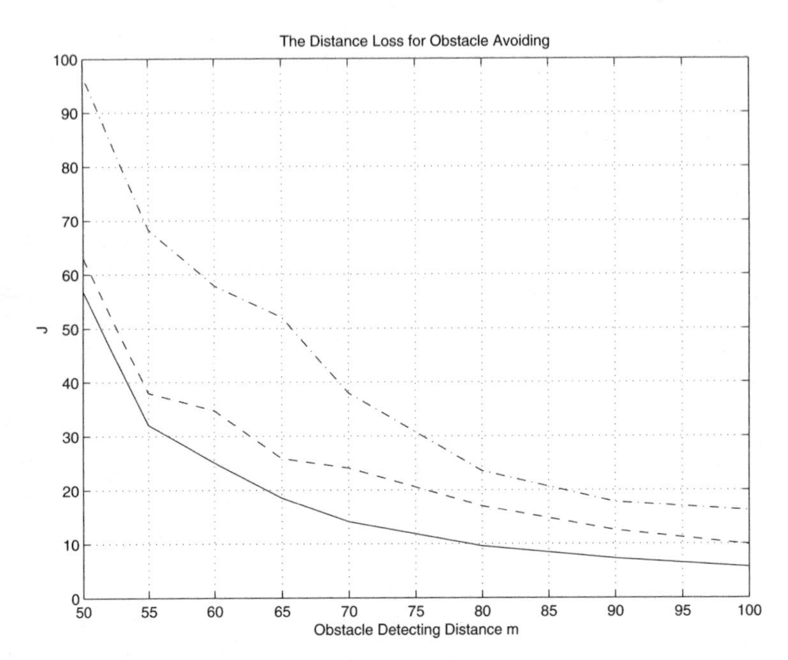

FIGURE 5.5: IAVD as a function of obstacle detection distance with Harry (solid), Dick (dash), and Tom (dash-dot).

model, respectively. We observe that $IAVD$ will increase with the decrease of obstacle detection distance τ. This pattern is preserved for all HCS models. When an obstacle is detected, the driver should change his path from a straight line to a set of curve-segments to avoid the obstacle. When the detected distance τ is small, the driver should have a rapid response. This will dramatically alter the vehicle's velocity. As we know, a smaller $IAVD$ indicates a better performance. Once again, we observe that Harry's model performs the best when evaluated with the $IAVD$ index, since its $IAVD$ is smaller for each τ than either Tom's or Dick's model.

5.3 Tight turning

In this section, we analyze performance as a function of how well a particular HCS model is able to navigate tight turns. First, we define a special road segment consisting of two straight-line segments connected directly (without a transition arc segment) at an angle ζ. For small values of ζ, each HCS model is able to successfully drive through the tight turn. For larger values

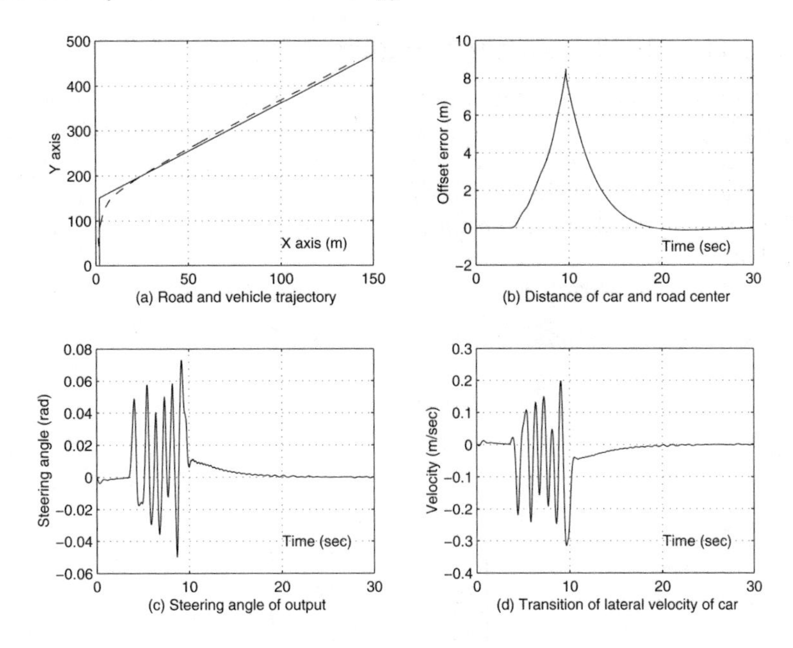

FIGURE 5.6: Example model driving behavior through a tight turn.

of ζ, however, some models fail to execute the turn properly by temporarily running off the road or even losing complete sight of the road.

5.3.1 Tight turning connections

Figure 5.6 illustrates, for example, how Harry's model transitions through a tight turn for $\zeta = 5\pi/36rad$.

Figure 5.6(a) plots the two straight-line segments connected at an angle ζ. The solid line describes the road median, while the dashed line describes the actual trajectory executed by Harry's HCS model. The length of the initial straight-line segment is chosen to be long enough (150m) to eliminate transients by allowing the model to settle into a stable state. This is equivalent to allowing the vehicle to drive on a straight road for a long period of time before the tight turn appears in the road. Figure 5.6(b) plots the lateral offset from the road median during the tight-turn maneuver. Here, Harry's model maximally deviates about 8m from the road center. Both before and after the turn, the lateral offset converges to zero. Figure 5.6(c) plots the commanded steering angle for Harry's HCS model, and Figure 5.6(d) plots the corresponding changes in velocity.

5.3.2 Threshold with tight angle

Now, define the maximum lateral offset error corresponding to a tight turn with angle ζ to be ρ. We want to determine a functional relationship between ρ and ζ for a given HCS model. First, we take N measurements of ρ for different values of ζ, where we denote the ith measurement as (ζ_i, ρ_i). Then, similar to Section 3.2, we assume a polynomial relationship between ρ and ζ such that,

$$\rho_i = \alpha_p \zeta_i^p + \alpha_{p-1} \zeta_i^{p-1} + \cdots + \alpha_1 \zeta_i + \alpha_0 + e_i \tag{5.14}$$

The least-squares estimate of the model $\hat{\alpha}$ is given by,

$$\hat{\alpha} = (\hat{\zeta}^T \hat{\zeta})^{-1} \cdot \hat{\zeta}^T \cdot \hat{\rho} \tag{5.15}$$

where

$$\hat{\rho} = [\rho_1, \rho_2, \cdots \rho_N]^T$$

$$\hat{\zeta} = \begin{bmatrix} \zeta_1^p & \zeta_1^{p-1} & \cdots & \zeta_1 & 1 \\ \zeta_2^p & \zeta_2^{p-1} & \cdots & \zeta_2 & 1 \\ \vdots & \vdots & & \vdots & \vdots \\ \zeta_N^p & \zeta_N^{p-1} & \cdots & \zeta_N & 1 \end{bmatrix}$$

$$\hat{\alpha} = [\alpha_p, \alpha_{p-1}, \cdots, \alpha_0]^T$$

Again, we consider the HCS models of Harry, Dick, and Tom. Because Tom's HCS model cannot handle the tight turning in the given ζ range, it will run-off the road when the corner-angle is greater than $\pi/9$. So we consider the tight-turning performance of Tom to be poor compared to Harry, and Dick's HCS models.

For ζ ranging from $-4\pi/9$rad to $4\pi/9$rad and assuming a fifth-order model ($p = 5$), we arrive at the following estimate for Harry's model,

$$\rho_{Harry} = 2.787\zeta^5 - 0.584\zeta^4 - 0.599\zeta^3 - 4.286\zeta^2 + 11.68\zeta - 0.330 \tag{5.16}$$

and the following estimate for Dick's model,

$$\rho_{Dick} = -1.734\zeta^5 + 1.076\zeta^4 + 2.258\zeta^3 - 0.243\zeta^2 + 21.29\zeta - 0.679 \tag{5.17}$$

Equations 5.16 and 5.17 are plotted in Figure 5.7, where the solid line corresponds to Harry's model and the dashed line corresponds to Dick's model. For a given road width, we can determine the values of ζ for which each model stays on the road. For example, assume a road width of 20m. Then, the maximum allowable lateral offset is $\pm 10m$. From Figure 5.8, where the boundaries are explicitly drawn, we observe that Harry's model can execute tight turns from -0.65rad to 1.05rad, while Dick's model can only execute tight turns from -0.45rad to 0.48rad. Thus, Harry's model generates stable driving for a wider range of conditions than does Dick's model.

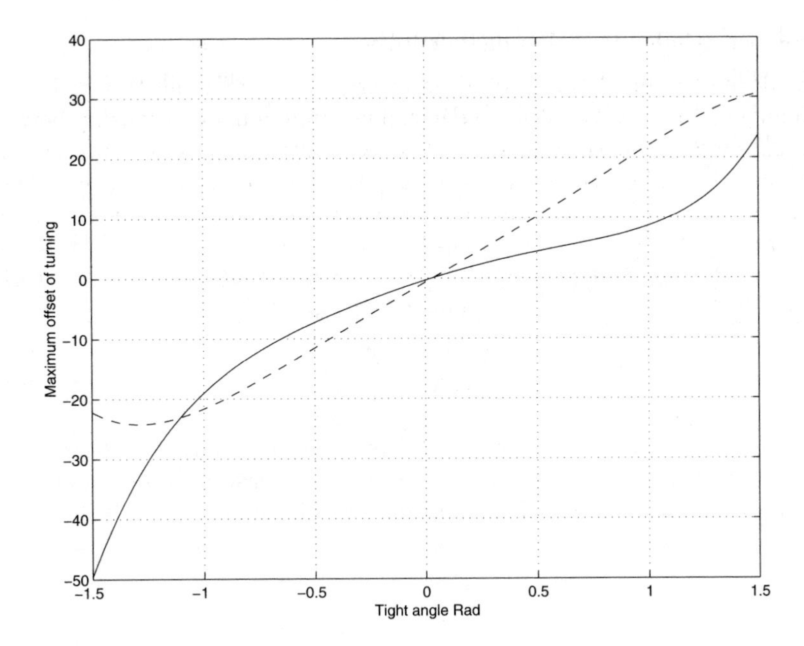

FIGURE 5.7: Maximum lateral offset in tight turns from Dick's and Harry's model.

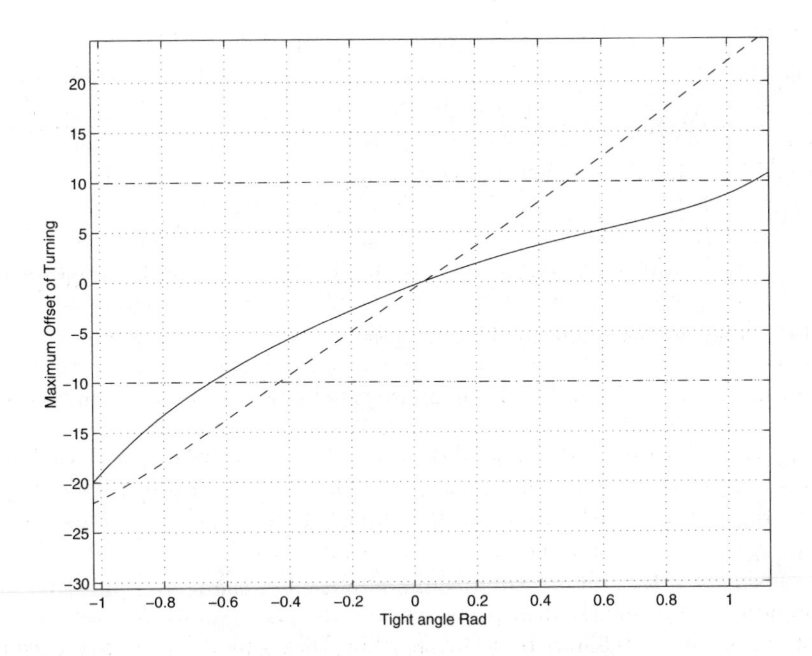

FIGURE 5.8: Harry's model stays on the road for a greater range of tight turns.

Our performance criterion of tight turning can be described by the tight angle range. We note that the first-order coefficient in the approximation function dominates this angle range, so we define the tight-turning performance criterion J_2 to be,

$$J_2 = \alpha_1 \tag{5.18}$$

where α_1 is the linear coefficient in the fifth order model of equations 5.16 and 5.17. A smaller value of J_2 indicates better performance of tight turns, because it gives a large stable handling range ζ. In this case,

$$J_2^{Harry} = 11.68 \tag{5.19}$$

$$J_2^{Dick} = 21.29 \tag{5.20}$$

Thus, in tight turning, Harry's model shows better performance than Dick's model because J_2^{Harry} is the lowest performance measure. Tom's model cannot finish the tight turning because its range for handling for tight turns is too small.

5.4 Transient response

In traditional control theory, the performance of feedback control systems is studied by the time-domain performance specifications. Because HCS models are inherently dynamical controllers, their performance may be specified in terms of transient response and steady-state response as well. We would like to consider the model's transient response while the vehicle turns. The *transient response* is defined as the response that disappears with time. The *steady-state response* is that which exists a long time following any input signal initiation.

The specifications for HCS models normally include several time-response indices for specified input commands as well as a desired steady-state accuracy. The time-domain performance specifications are important because human driving control systems are inherently time-domain systems. Standard performance measures in time-domain are usually defined in terms of the step response of a system. The swiftness of the response is measured by rise time T_r, and the peak time T_p. The closeness of the response to the desired response, as represented by the overshoot M_p and settling time, T_s, is defined as the time required for the system to settle to within a certain percentage of the input amplitude.

The standard test input signal commonly used is the step input. A tight turn during driving gives a start time for time-domain performance analysis. In our research, we have defined a special road connection consisting of two straight-line segments connected directly at an angle ζ. In the driving simulation, when the vehicle drives between these two road segments, the orientation

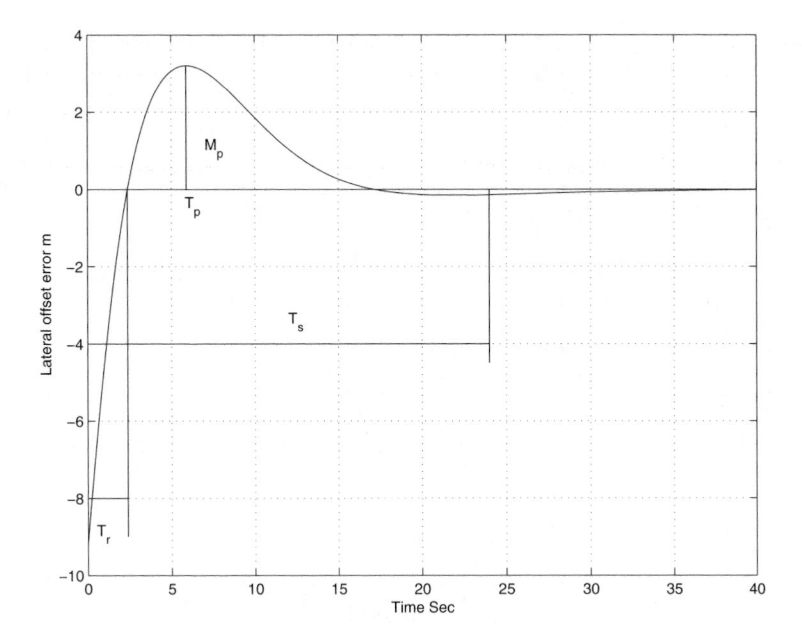

FIGURE 5.9: Transient response of lateral offset for a given tight turning.

of the vehicle changes by an angle ζ. As we described in Section 2.5, the road description can be viewed as the input of the whole system. Thus we use this turn angle ζ as the step input. The resulting time-domain performances are then studied. For a given step angle input, we have two choices for studying time-domain response:

- the transient response of the lateral offset between the vehicle center and the road center, and
- the transient response of the vehicle's orientation angle.

Here, we start the work from transient response of the lateral offset. From subplot (b) in Figure 5.6, we notice that there is a "peak" point which displaces the maximum lateral offset. We also know this point occurs at the road connection point. So, we define this time as "zero" to analyze the response specification.

Figure 5.9 shows an example of transient response of lateral offset for a given tight turning.

At time "zero," the lateral offset has the maximum absolute value. After this time, the lateral offset tends to a stable value while the transient finishes. Here, we define some time-domain measurement for the transient response in Figure 5.9.

- **Rise time, T_r,** is defined as the time required for the system to cross the stable point.

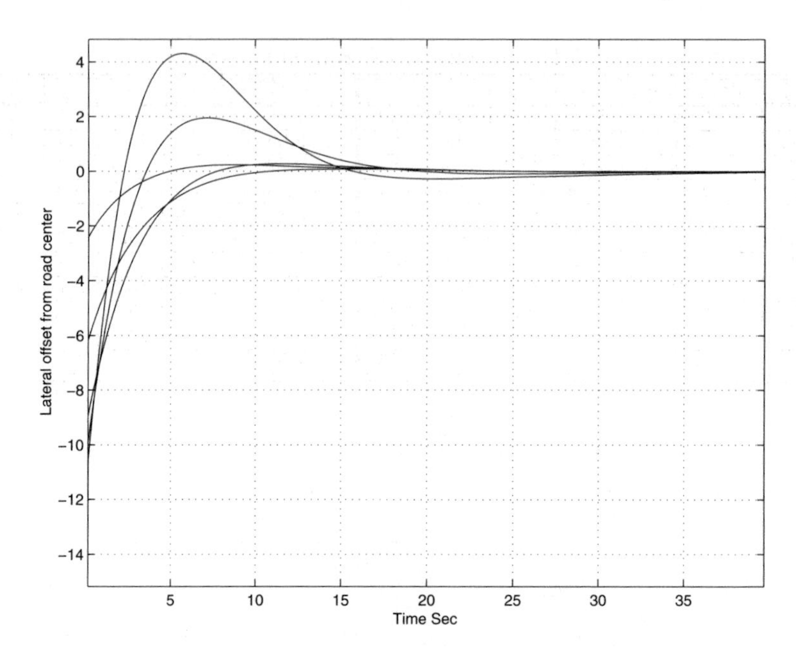

FIGURE 5.10: Transient response of lateral offset for different turn angles.

- **Peak time,** T_p, is defined as the time required for the system to obtain the maximum overshoot.
- **Setting time,** T_s, is defined as the time required for the system to settle within a certain percentage of the input amplitude.
- **Overshoot,** M_p, is defined as the peak value of the time response.

For the vehicle driving problem, we do not use the time response of unit step input, but study the relationship between the lateral offset response and turn angle ζ. Because this response has relationship with the vehicle's velocity, we assume all response in a given velocity, 20m/s. Figure 5.10 shows lateral offset response with the different turn angle $\zeta \in \{10°, 20°, 30°, 40°, 50°\}$.

From Figure 5.10, we observe that responses are quite different with different turn angles ζ. For a smaller turn angle ζ, the vehicle shows a smaller offset and small peak value. We put the related response time in Table 5.1.

This table shows the result of transient response for HCS models of Harry and Dick. The left half in the table gives the results of Dick's model and the right half for Harry's. For any given turn angle, Harry's model gives a smaller value of maximum offset, rising time, and peak time than Dick's HCS model. That means Harry's model has a smaller error with respect to the road center. We use the square of the error to describe this performance. Here "error" means the difference between the vehicle center and road center.

Table 5.1: Parameters of transient response

ζ	Max. Offset	T_r	M_p	T_p	T_s	Max. Offset	T_r	M_p	T_p	T_s
10	2.42	11.0	1.50	14.0	9.2	1.88	1.3	4.36	3.5	20
15	3.21	9.4	1.90	13.0	8.2	2.06	1.3	4.34	3.4	22
20	3.92	8.0	3.00	12.0	18.0	2.39	1.1	5.83	3.2	25
25	4.59	7.8	3.00	11.5	15.0	2.63	1.0	6.36	3.0	25
30	5.23	5.1	8.80	8.8	15.1	3.10	0.9	7.84	2.8	26
35	5.87	4.9	9.50	8.5	16.3	3.54	0.8	8.42	2.5	28
40	6.55	3.5	15.7	7.2	16.8	4.33	0.8	9.23	2.3	29
45	7.32	2.7	24.0	6.3	15.4	4.95	0.7	13.2	2.1	30
50	8.25	2.2	30.0	5.7	20.5	5.60	0.7	19.7	2.0	30
55	9.43	2.0	35.5	5.2	24.8	6.27	0.6	24.4	1.8	30
60	10.97	1.8	46.0	5.0	27.0	7.54	0.4	31.9	1.8	31

This performance index is calculated by *integral of the square of the error*, ISE, which is defined as

$$ISE = \int_{o}^{T} e^2(t)dt \tag{5.21}$$

For two given models, we calculate ISE with some given turn angles in Figure 5.11. This is the relationship between ISE and the turn angle.

In Figure 5.11, solid line is ISE result of Harry's HCS model while dash line is ISE result of Dick's. We notice that Harry has a better performance than Dick.

5.5 Time delay

For vehicle driving, time delay is an important criterion for describing the rapidity of driver response. This criterion indicates the sensitivity of an HCS model. If a system has a small delay time, it will respond better during an accident. Here, we study the time delay based on the tight turning problem.

Consider again the transient response of the vehicle orientation angle. Assume the vehicle has a zero orientation on the first road segment, then the orientation will change to another value when vehicle changes to another road segment. By recording the orientation value of the transient, we obtain another transient figure, such as Figure 5.12.

The time "zero" here is defined as the time when the HCS model responds to the bend in the road. That means the road connection point is involved in the road description $\bar{z}$. The first part (about 5 seconds) of the orientation angle transient plot has a vibration, and the remaining part has a smooth curve. We define the delay time T_d as the time required from "zero" to the cross point between the extend line of smooth curve and the "zero" angle

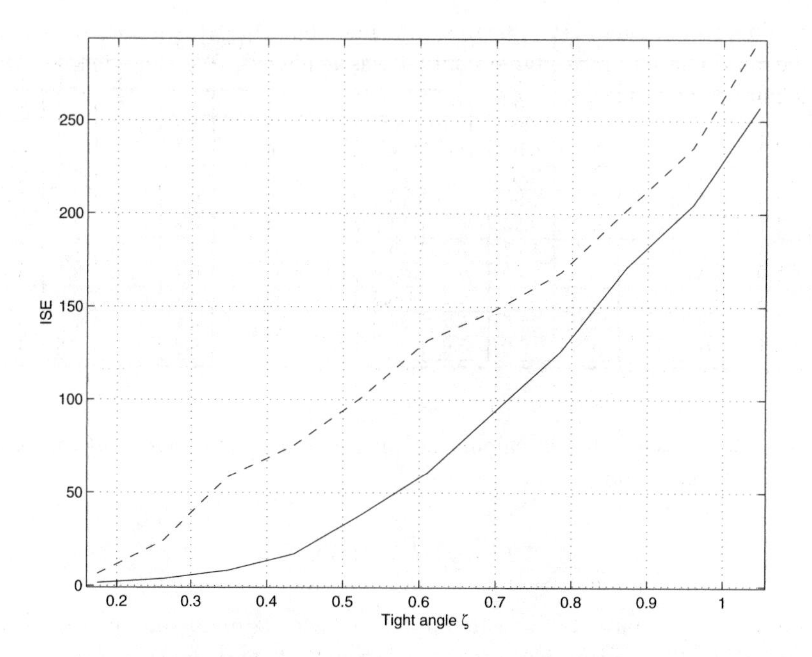

FIGURE 5.11: Integral of square of the offset with road center.

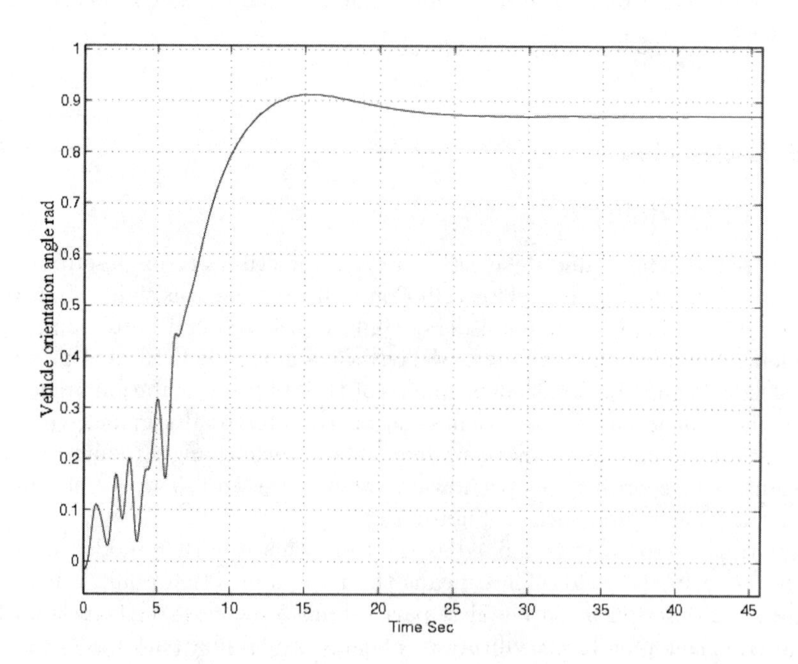

FIGURE 5.12: Transient response of vehicle's orientation angle.

axis. This time delay also indicates the time that the HCS model responded to the road change until stable control was achieved. We thus define a new performance criterion as,

$$J_3 = T_d \tag{5.22}$$

Consider Harry's and Dick's model again; two transient orientation angle curves are shown in Figure 5.13.

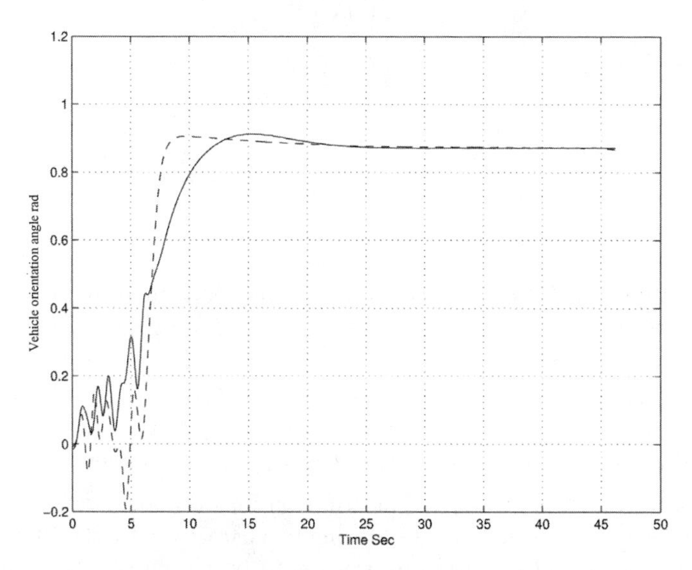

FIGURE 5.13: Transient orientation angle of Harry (solid) and Dick (dash).

Then we obtain,

$$J_3^{Harry} = 3.5 \text{ sec} \tag{5.23}$$

$$J_3^{Dick} = 6.0 \text{ sec} \tag{5.24}$$

The HCS model of Harry gives a better performance for time delay.

5.6 Passenger comfort

Until now, we have introduced performance criteria based on specific external subtasks. For the real task of driving, many other candidate performance criteria, such as average speed, passenger comfort, driving smoothness, and fuel efficiency, exist. These measures are not based on any single subtask, but

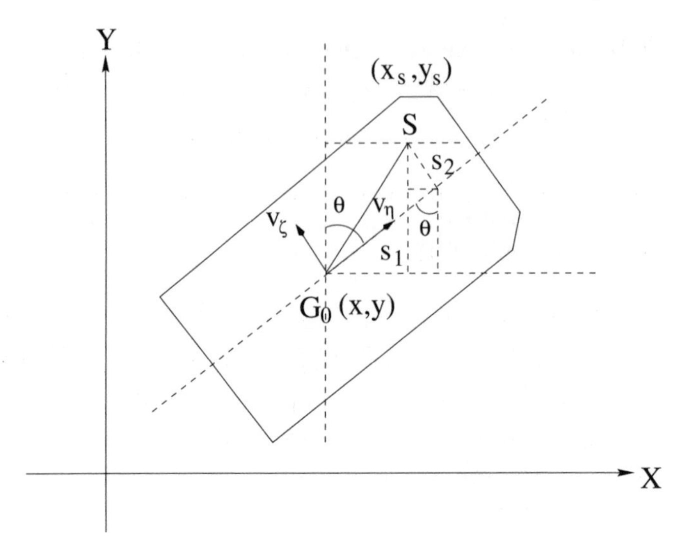

FIGURE 5.14: Diagram of vehicle and human system.

rather are aggregate measures of performance. In other words, they measure the intrinsic characteristics of a particular HCS model. That motivates us to develop other performance criteria based on the model's intrinsic analysis.

Passenger comfort is an important criterion for evaluating human driving control strategies. Consider a person sitting in a car driven by the HCS model. His/her comfort would be primarily influenced by the forces that passenger experiences in the car. Whenever the HCS model changes the applied force P_f on the car, the passenger will feel a longitudinal force. Similarly, whenever the HCS model changes steering δ, the passenger will experience a lateral force. Below we quantify passenger comfort as a function of the applied forces on the vehicle under HCS model control.

Consider a vehicle shown in Figure 5.14. Let the configuration of the system be described by the vehicle's center of mass (x, y), the angle θ between the positive Y axis and the axis of symmetry of the car, and the location of the passenger S, (x_s, y_s). Furthermore, define the distance from S to the axis of symmetry as s_2 and define the distance from S to the center of mass along the axis of symmetry as s_1.

The coordinates of the point S at the vehicle are

$$x_s = x + s_1 \sin\theta - s_2 \cos\theta \tag{5.25}$$

$$y_s = y + s_1 \cos\theta + s_2 \sin\theta \tag{5.26}$$

From equations 5.25 and 5.26, we can determine the coordinate velocity of the point S as

$$\dot{x}_s = \dot{x} + s_1\dot{\theta}\cos\theta + s_2\dot{\theta}\sin\theta \tag{5.27}$$
$$\dot{y}_s = \dot{y} - s_1\dot{\theta}\sin\theta + s_2\dot{\theta}\cos\theta \tag{5.28}$$

The actual velocity of the point S as a function of the coordinate velocities is given by,

$$\begin{aligned} v_s^2 &= \dot{x}_s^2 + \dot{y}_s^2 \\ &= (\dot{x} + s_1\dot{\theta}\cos\theta + s_2\dot{\theta}\sin\theta)^2 + (\dot{y} - s_1\dot{\theta}\sin\theta + s_2\dot{\theta}\cos\theta)^2 \\ &= \dot{x}^2 + s_1^2\dot{\theta}^2\cos^2\theta + s_2^2\dot{\theta}^2\sin^2\theta + 2\dot{x}s_1\dot{\theta}\cos\theta + 2s_2\dot{x}\dot{\theta}\sin\theta \qquad (5.29) \\ &\quad +2s_1s_2\dot{\theta}^2\cos\theta\sin\theta + \dot{y}^2 + s_1^2\dot{\theta}^2\sin^2\theta + s_2^2\dot{\theta}^2\cos^2\theta \\ &\quad -2s_1\dot{y}\dot{\theta}\sin\theta + 2s_2\dot{y}\dot{\theta}\cos\theta - 2s_1s_2\dot{\theta}^2\sin\theta\cos\theta \\ &= \dot{x}^2 + \dot{y}^2 + s_1^2\dot{\theta}^2 + s_2^2\dot{\theta}^2 + 2s_1\dot{\theta}(\dot{x}\cos\theta - \dot{y}\sin\theta) \qquad (5.30) \\ &\quad +2s_2\dot{\theta}(\dot{x}\sin\theta + \dot{y}\cos\theta) \end{aligned}$$

As we described in Chapter 2, the longitudinal acceleration of the vehicle is given by,

$$\dot{v}_\eta = (P_f + P_r - F_{\xi f}\delta)/m + v_\xi\dot{\theta} - (\text{sgn } v_\eta)c_d v_\eta^2 \tag{5.31}$$

and the lateral acceleration of the vehicle is given by,

$$\dot{v}_\xi = (P_f\delta + F_{\xi f} + F_{\xi r})/m - v_\eta\dot{\theta} - (\text{sgn } v_\xi)c_d v_\xi^2 \tag{5.32}$$

Now, the accelerations experienced by the passenger include not only the vehicle's acceleration, but also the centrifugal force, given by,

$$\frac{v_s^2}{R} = \frac{1}{R}[\dot{x}^2 + \dot{y}^2 + s_1^2\dot{\theta}^2 + s_2^2\dot{\theta}^2 + 2s_1\dot{\theta}(\dot{x}\cos\theta - \dot{y}\sin\theta) + 2s_2\dot{\theta}(\dot{x}\sin\theta + \dot{y}\cos\theta)] \tag{5.33}$$

The centrifugal force generally points in the direction of the negative lateral acceleration of the vehicle. By combining the vehicle and centrifugal accelerations, we then arrive at the following expression for the total acceleration at point S:

$$a = \sqrt{\dot{v}_\eta^2 + (\dot{v}_\xi - \frac{v_s^2}{R})^2} \tag{5.34}$$

A part of test road which is about 20km long is shown in Figure 5.15. The road is a series of connected road segments, including straight lines, right turn curves, and left turn curves.

In our definition of the "comfort" performance criterion J_4, we normalize acceleration felt by the passenger by the speed of the vehicle, since higher speeds generate higher accelerations through a given curve:

$$J_4 = \frac{a_{mean}}{v_{mean}} \tag{5.35}$$

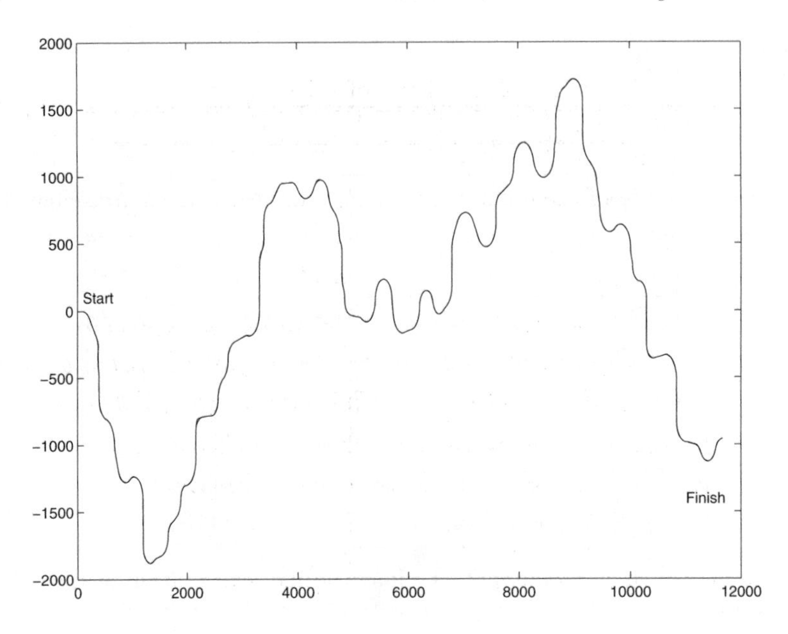

FIGURE 5.15: A 20km road example used to collect data.

Thus, J_4 is defined as the ratio of average acceleration over average speed for a given road.

Let us now look at how different HCS models perform with respect to this performance criterion. First, we collect driving data from three human operators - Tom, Dick, and Harry. After training an HCS model for each individual, we then run that person's model over three different roads (1, 2, and 3). Furthermore, we split each of those runs into two groups, A and B, where group A represents the first half of each run, and group B represents the second half of each run. Thus, for example, Tom1-A represents the first half of Tom's HCS model's run over road 1.

Table 5.2 shows the velocity statistics for different HCS models – Tom, Dick, and Harry, on the given roads (1, 2, and 3). Data of both group A and group B are given in Table 5.2.

Table 5.3 shows the acceleration statistics for different HCS models – Tom, Dick, and Harry, on the given roads (1, 2, and 3). Data of both group A and group B are given in Table 5.3.

Table 5.4 gives some aggregate statistics for each of these model-generated data sets. Specifically, it lists the percentage of occurrences for which the acceleration in a particular data set is larger than one g, $2g$, and $3g$, respectively. These percentages give us a rough idea about the comfort level of each model driver. If we average the percentages for each HCS model, we find that Tom's model generates accelerations above one g 22.36% of the time, accelerations above $2g$ 1.09% of the time, and accelerations above $3g$ 0.065% of the

Table 5.2: Statistics of velocity data

Data Name	V_{max}	V_{min}	V_{mean}	V_{median}	V_{SD}
Tom1 - A	47.48	21.13	31.39	30.74	3.947
Tom1 - B	44.18	23.13	31.81	31.77	3.360
Dick1 - A	43.97	15.26	31.44	31.59	4.617
Dick1 - B	46.08	24.80	34.06	34.23	3.448
Harry1 - A	33.47	1.001	24.57	25.54	4.869
Harry2 - B	31.22	1.000	24.42	25.18	4.531
Tom2 - A	40.23	23.36	30.51	30.26	3.139
Tom2 - B	45.98	22.66	30.33	30.68	3.849
Dick2 - A	41.54	33.84	31.83	31.29	3.784
Dick2 - B	43.27	16.46	32.75	32.81	4.254
Harry2 - A	26.15	17.00	21.82	21.83	2.059
Harry2 - B	27.53	21.03	23.74	23.46	1.597
Tom3 - A	45.96	22.92	31.79	31.41	3.092
Tom3 - B	44.48	20.05	32.10	31.47	3.779
Dick3 - A	47.43	15.06	33.35	33.49	4.481
Dick3 - B	43.28	22.95	33.32	33.30	4.003
Harry3 - A	28.48	23.45	25.63	25.49	1.125
Harry3 - B	31.44	18.16	25.38	25.56	2.421

Table 5.3: Statistics of acceleration data

Data Name	a_{max}	a_{min}	a_{mean}	a_{median}	a_{SD}
Tom1 - A	31.58	0.022	7.569	6.434	5.587
Tom1 - B	24.16	0.139	6.684	6.217	4.987
Dick1 - A	44.53	0.020	9.236	7.344	7.554
Dick1 - B	45.73	0.020	8.940	7.400	6.689
Harry1 - A	19.98	0.000	4.083	3.169	3.846
Harry1 - B	21.46	0.020	4.747	3.572	4.220
Tom2 - A	28.13	0.021	6.536	5.990	4.065
Tom2 - B	28.28	0.153	6.267	5.516	3.940
Dick2 - A	42.30	0.020	9.409	7.600	7.649
Dick2 - B	40.34	0.050	8.912	7.344	6.393
Harry2 - A	12.27	0.010	2.840	2.629	2.200
Harry2 - B	15.55	0.013	3.009	2.340	2.730
Tom3 - A	22.52	0.004	6.729	6.314	3.899
Tom3 - B	25.78	0.063	7.049	6.673	4.124
Dick3 - A	53.20	0.049	11.23	8.426	9.753
Dick3 - B	48.93	0.034	11.02	8.600	9.137
Harry3 - A	16.48	0.030	3.977	3.577	3.030
Harry3 - B	23.69	0.011	4.459	4.369	4.285

Table 5.4: Aggregate statistics of acceleration data

Data Name	$a > g(9.8m/s^2)$	$a > 2g(19.6m/s^2)$	$a > 3g(29.4m/s^2)$	$\rho = \frac{a_{mean}}{V_{mean}}$
Tom1 - A	28.20%	3.85%	0.39%	0.2411
Tom1 - B	20.99%	0.79%	0	0.2101
Tom2 - A	19.11%	0.74%	0	0.2142
Tom2 - B	18.00%	0.63%	0	0.2185
Tom3 - A	22.15%	0.21%	0	0.2117
Tom3 - B	25.76%	0.31%	0	0.2196
Dick1 - A	36.43%	8.69%	1.63%	0.2938
Dick1 - B	36.85%	6.79%	0	0.2625
Dick2 - A	37.73%	8.87%	2.19%	0.2956
Dick2 - B	37.25%	7.31%	0.43%	0.2721
Dick3 - A	43.20%	15.17%	5.13%	0.3367
Dick3 - B	43.35%	14.97%	4.61%	0.3307
Harry1 - A	5.93%	0.48%	0	0.1662
Harry1 - B	8.13%	1.30%	0.013%	0.1941
Harry2 - A	1.16%	0	0	0.1302
Harry2 - B	2.69%	0	0	0.1267
Harry3 - A	4.93%	0	0	0.1552
Harry3 - B	11.49%	1.28%	0	0.1757

time. The same statistics for Dick's model are 39.14%, 10.30%, and 2.33%, respectively. Similarly, for Harry's model the statistics are 5.72%, 0.51%, and 0.00%, respectively.

From these results, we would expect that Harry's HCS model offers the smoothest ride of the three models, since it generates the smallest forces. Driving with Dick's model, on the other hand, would prove to be quite uncomfortable. Calculating performance criterion J_4 for each model confirms these qualitative observations. For Tom's model, J_4 varies from 0.2101 to 0.2411, and the average is given by,

$$J_4^{Tom} = 0.2192 \tag{5.36}$$

For Dick's model, J_4 varies from 0.2625 to 0.3367, and the average is given by,

$$J_4^{Dick} = 0.2986 \tag{5.37}$$

Finally, for Harry's model, J_4 varies from 0.1303 to 0.1941, and the average is given by,

$$J_4^{Harry} = 0.1508 \tag{5.38}$$

We observe that J_4 is the smallest for Harry's model, and that that value is much smaller than J_4 for Dick's model.

5.7 Driving smoothness

Another way to evaluate the smoothness of a given driver's control strategy is through frequency analysis of the instantaneous curvature of the road and the corresponding instantaneous curvature of the vehicle's path. As a HCS model steers the car along the road, the curvature of the vehicle's path will in general not be the same as that of the road. Below, we will use this difference between the two curvatures to evaluate the driving smoothness of a given model in the frequency domain. We will show that the resulting performance measure yields results consistent with the J_4 passenger comfort performance criterion defined in the previous section.

Let us define $u(k)$ as the instantaneous curvature of the road at time step k, and let $z(k)$ be the instantaneous curvature of the vehicle's path at time step k. We can view the road's curvature $u(k)$ as the input to HCS model, and $z(k)$ as the output of HCS model.

To calculate the frequency response from u to z, we use *power spectral density (PSD)* to find the domain frequency. We first partition the complete data into N groups, where each group is of length L. Hence, the kth element of group i is given by,

$$\begin{aligned} u_i(k) &= u[k+(i-1)L] \\ z_i(k) &= z[k+(i-1)L] \\ i &= 1, 2, \cdots N; \qquad 1 \leq k \leq L \end{aligned} \tag{5.39}$$

We also define the following convolutions for each group of data i:

$$\begin{aligned} I_{u_i,L}(w) &= \frac{1}{L} U_i(jw) U_i^*(jw) = \frac{1}{L} ||U_i(jw)||^2 \\ I_{u_i z_i,L}(jw) &= \frac{1}{L} U_i(jw) Z_i^*(jw) \\ i &= 1, 2, \cdots, N \end{aligned} \tag{5.40}$$

where,

$$U_i(jw) = \sum_{k=1}^{L} u_i(k) H_k e^{-jwk} \tag{5.41}$$

$$Z_i(jw) = \sum_{k=1}^{L} z_i(k) H_k e^{-jwk} \tag{5.42}$$

Define the discrete Fourier transform and,

$$H_k = 0.54 - 0.46 \cos[\frac{2\pi(k-1)}{L-1}], \; k \in \{1, 2, \cdots, L\} \tag{5.43}$$

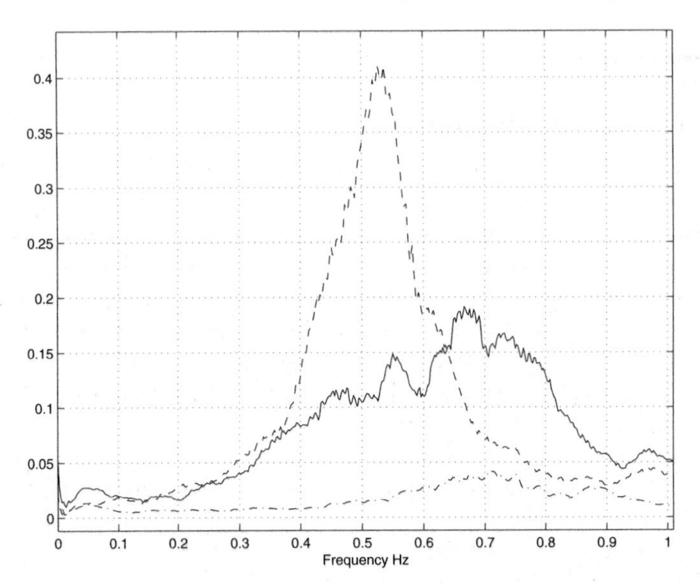

FIGURE 5.16: Power spectral density analysis.

defines the Hamming coefficients, which we include to minimize the spectral leakage effects of data windowing.

By summing up the terms in Equation 5.41,

$$S_{u,L}(w) = \frac{1}{N}\sum_{i=1}^{N} I_{u_i,L}(w) \tag{5.44}$$

$$S_{uz,L}(jw) = \frac{1}{N}\sum_{i=1}^{N} I_{u_i z_i,L}(jw) \tag{5.45}$$

We define the frequency response $G(jw)$ for a given HCS model as,

$$G(jw) = S_{uz,L}(jw)/S_{u,L}(w) \tag{5.46}$$

Figure 5.16 plots $|G(jw)|$ for HCS models corresponding to Tom, Dick, and Harry, Each group of data corresponds to 40 seconds ($L = 2000$ at 50Hz), and the data for each model is collected over road 1. In Figure 5.16 the solid line corresponds to Tom, the dash-dotted line corresponds to Dick, and the dashed line corresponds to Harry.

Given the plots of $|G(jw)|$, we now define the following smoothness performance criterion:

$$J_5 = f_{domain} \tag{5.47}$$

where f_{domain} corresponds to the domain frequency of each $|G(jw)|$ curve.

We obtain the following smoothness results for the three models:

$$J_5^{Harry} = 0.52Hz \tag{5.48}$$

$$J_5^{Tom} = 0.66Hz \tag{5.49}$$

$$J_5^{Dick} = 0.72Hz \tag{5.50}$$

5.8 Summary

Modeling human control strategy analytically is difficult at best. Therefore, an increasing number of researchers have resorted to empirical modeling of human control strategy as a viable alternative. This in turn requires that performance criteria be developed, since few if any theoretical guarantees exist for these models. In this chapter, we have developed several such criteria based on the task of human driving. These criteria are based on subtask analysis and intrinsic analysis. Based on subtask analysis, we developed performance criteria, including obstacle avoidance, tight-turning, transient response, and time delay. Based on intrinsic analysis, we developed two performance criteria as well. One criterion is based on the "comfort" level of the passengers. This performance criterion normalizes the acceleration felt of the passenger by the speed of the vehicle. The other performance criterion is domain frequency, which is calculated by the PSD analysis. This criterion describes the smoothness of the driving. These two criteria examine a given model's behavior on a more global scale. These measures are not based on any single subtask, but rather are aggregate measures of performance. In other words, they measure the intrinsic characteristics of a particular HCS model.

We model human driving using the cascade neural network architecture, and evaluate the performance of driving models derived from different individuals using the developed performance criteria. Then we compare the HCS models. We summarize the developed series of performance criteria for the three driving HCS models in Table 5.5. We conclude that Harry's HCS model gives the best performance of these three models.

Table 5.5: Summary of performance criteria

HCS model	J_1	$IAVD$	J_2	ISE	J_3	J_4	J_5
Tom	1.30	High	X	X	X	0.2192	0.66
Dick	1.00	Media	21.29	High	6.0	0.2986	0.72
Harry	0.51	Lower	11.68	Lower	3.5	0.1508	0.52

6

Performance Optimization of Human Control Strategy

6.1 Introduction

Performance evaluation is only a part of the solution for effectively applying models of human control strategy. Once the skills of an HCS and corresponding models have been evaluated, we would like to optimize performance of the HCS models based on specific criteria. A HCS model's performance is usually only as good as the data with which it is trained. While humans are in general very capable of demonstrating intelligent behaviors, they are far less capable of demonstrating those behaviors without occasional errors and random (noise) deviations from some nominal trajectory. An empirical learning algorithm will necessarily incorporate those problems in the learned model, and will consequently be less than optimal. Furthermore, control requirements may differ between humans and robots, where stringent power or force requirements often have to be met. A given individual's performance level, therefore, may or may not be sufficient for a particular application.

In a previous chapter, we introduced performance measures for evaluating the performance of our driving models. Below, we will develop an algorithm for optimizing a learned control strategy model with respect to one of those performance criteria. This optimization is difficult in principle because

- we have no explicit gradient information

$$G(\omega) = \frac{\partial}{\partial \omega} J(\omega) \tag{6.1}$$

- each experimental measurement of $J(\omega)$ requires a significant amount of computation.

We lack explicit gradient information, since we can only compute our performance measures empirically. Hence, gradient-based optimization techniques, such as steepest descent and Newton-Raphson, are not suitable. And because each performance measurement is potentially computationally expensive, genetic optimization, which can require many iterations to converge, also does not offer a good alternative. Hence, in this chapter, we propose an iterative optimization algorithm, based on simultaneous perturbed stochastic approximation (SPSA) for improving the performance of learned HCS models. This algorithm leaves the learned model's structure intact, but tunes the parameters of an HCS model in order to improve performance. It requires no analytic formulation of performance, and only two experimental measurements of a user-defined performance criterion per iteration. The initial HCS model serves as a good starting point for the algorithm, since it already generates stable control commands.

6.2 Simultaneously perturbed stochastic approximation

Stochastic approximation (SA) is a well-known iterative algorithm for finding roots of equations in the presence of noisy measurements through stochastic gradient approximation. SA has potential applications in a number of areas relevant to statistical modeling and control, such as parameter estimation, adaptive control, stochastic optimization, and neural network weight-estimation. The SA algorithm can solve problems where the relationship between the objective function and the solution is very complex. For the reason that we have little *a priori* knowledge about the structure of human control strategy, no *a priori* mathematics model can be used for HCS models. Thus to optimize HCS models, some methods such as steepest descent and Newton-Raphson are not suitable. *Simultaneously perturbed stochastic approximation* (SPSA) is a particular multivariate SA technique which requires as few as two measurements per iteration and shows fast convergence in practice.

Consider a problem of finding a root of the gradient equation as

$$\frac{\partial L(\theta)}{\partial \theta} = 0 \tag{6.2}$$

When $L(\theta)$ is observed directly, many methods can be used to find the optimal solution θ^*. In the case where $L(\theta)$ is not distinct, the SA algorithm is appropriate. Denote θ_k as our estimate for θ^* at the kth iteration of the SA algorithm. Let θ_k be defined by the recursive relationship

$$\theta_{k+1} = \theta_k - \alpha_k G_k \tag{6.3}$$

where G_k is the simultaneously perturbed gradient approximation at the kth iteration

$$G_k = \frac{L_k^{(+)} - L_k^{(-)}}{2c_k} \begin{bmatrix} 1/\triangle_{k1} \\ 1/\triangle_{k2} \\ \cdots \\ 1/\triangle kp \end{bmatrix} \tag{6.4}$$

and

$$L_k^{(+)} = L(\theta_k + c_k \triangle_k) \tag{6.5}$$
$$L_k^{(-)} = L(\theta_k - c_k \triangle_k) \tag{6.6}$$
$$\triangle_k = [\triangle_{k1} \triangle_{k2} \cdots \triangle_{kp}]^T \tag{6.7}$$

where $\triangle_k$ is a vector of mutually independent, zero-mean random variables, the sequence $\{\triangle_k\}$ is independent and identically distributed, and the $\{\alpha_k\}$, $\{c_k\}$ are positive scalar sequences restricted to

$$\alpha_k \rightarrow 0, \quad c_k \rightarrow 0 \; as \; k \rightarrow \infty, \tag{6.8}$$
$$\sum_{k=0}^{\infty} \alpha_k = \infty, \quad \sum_{k=0}^{\infty} (\frac{\alpha_k}{c_k})^2 < \infty \tag{6.9}$$

Thus, at iteration k, we evaluate the performance by the evaluation criterion for two values $L^{(+)}$ and $L^{(-)}$ from θ_k in order to arrive at a new estimate for θ_{k+1}.

Example: Consider a lower order system. We use SPSA to solve the parameter estimation problem. Given a continuous function

$$f(x) = x^2 + 5.4x + 8 = a_2 x^2 + a_1 x + a_0 \tag{6.10}$$

the coefficient $\theta = [a_2 \; a_1 \; a_0]^T$ should be estimated. Let the initial value of the coefficients start at some distance from the real value, such as $\theta = [1.5 \; 4.6 \; 0]$. Using the SPSA algorithm, we find that the parameter estimation $\theta = [a_2 \; a_1 \; a_0]^T$ converges to the real parameter at the conditions of

$$\alpha_k = 0.001/k \tag{6.11}$$
$$c_k = 1/k^0.25 \tag{6.12}$$
$$\theta_0 = [1.5 \; 4.6 \; 0]^T \tag{6.13}$$

The simulation result can be seen in Figure 6.1.

It shows transition of the summation of the square errors. After 1000 times iteration, the estimated parameter of θ is

$$\theta_{1000} = [1.0812 \; 5.3864 \; 7.9868]^T \tag{6.14}$$

which approximates to the real value of the parameters $\theta_{real} = [1.0 \; 5.4 \; 8.0]^T$. In this example, the objective function is defined as the summation of the square errors. The value of the objective function is changed from 380 to 5. This algorithm is shown to converge in Figure 6.1.

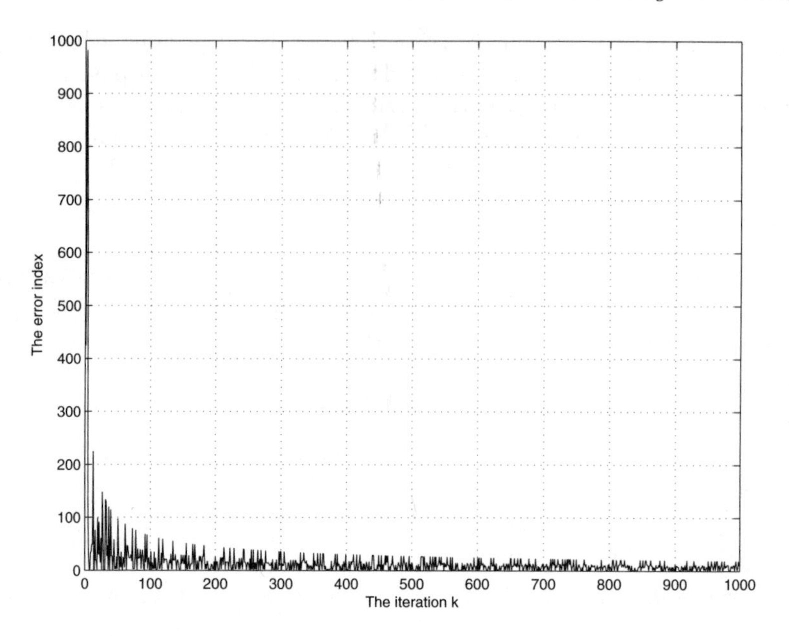

FIGURE 6.1: Example of parameter estimation problem.

6.3 Iterative optimization algorithm

We have introduced some performance measures for evaluating the performance of our driving models in Chapter 3. Below, we will develop an algorithm for optimizing a learned control strategy model with respect to one of those performance criteria.

Since a HCS model offers an initially stable model, it represents a good starting point from which to further optimize performance. Let

$$\omega = [w_1 \quad w_2 \quad \cdots \quad w_n]^T \tag{6.15}$$

denote a vector consisting of all the weights in the trained HCS model $\Gamma(\omega)$, and let $J(\omega)$ denote any one performance criterion. We would now like to determine the weight vector ω^* which optimizes the performance criterion $J(\omega)$. This optimization is difficult in principle because (1) we have no explicit gradient information,

$$G(\omega) = \frac{\partial}{\partial \omega} J(\omega) \tag{6.16}$$

and (2) each experimental measurement of $J(\omega)$ requires a significant amount of computation. We lack explicit gradient information, since we can only compute our performance measures empirically. Hence, gradient-based optimization techniques, such as steepest descent and Newton-Raphson, are not suitable. And because each performance measure evaluation is potentially computationally expensive, genetic optimization, which can require many iterations to converge, also does not offer a good alternative. Therefore, we turn to simultaneously perturbed stochastic approximation (SPSA) to carry out the performance optimization.

Denote ω_k as our estimate of ω^* at the kth iteration of the SA algorithm, and let ω_k be defined by the following recursive relationship:

$$\omega_{k+1} = \omega_k - \alpha_k \bar{G}_k \tag{6.17}$$

where $\bar{G}_k$ is the simultaneously perturbed gradient approximation at the kth iteration,

$$\bar{G}_k = \frac{1}{p}\sum_{i=1}^{p} G_k^i \approx \frac{\partial}{\partial \omega} J(\omega) \tag{6.18}$$

$$G_k^i = \frac{J_k^{(+)} - J_k^{(-)}}{2c_k} \begin{bmatrix} 1/\triangle_{kw_1} \\ 1/\triangle_{kw_2} \\ \cdots \\ 1/\triangle_{kw_n} \end{bmatrix} \tag{6.19}$$

Equation 6.18 averages p stochastic two-point measurements G_k^i for a better overall gradient approximation, where,

$$J_k^{(+)} = J(\omega_k + c_k \triangle_k) \tag{6.20}$$

$$J_k^{(-)} = J(\omega_k - c_k \triangle_k) \tag{6.21}$$

$$\triangle_k = [\triangle_{kw_1} \triangle_{kw_2} \cdots \triangle_{kw_n}]^T \tag{6.22}$$

where $\triangle_k$ is a vector of mutually independent, zero-mean random variables (e.g., symmetric Bernoulli distributed), the sequence $\{\triangle_k\}$ is independent and identically distributed, and $\{\alpha_k\}$ and $\{c_k\}$ are positive scalar sequences satisfying the following properties.

$$\alpha_k \to 0, \quad c_k \to 0 \; as \; k \to \infty, \tag{6.23}$$

$$\sum_{k=0}^{\infty} \alpha_k = \infty, \; \sum_{k=0}^{\infty} (\frac{\alpha_k}{c_k})^2 < \infty \tag{6.24}$$

The weight vector ω_0 is of course the weight representation of the initially stable learned cascade model. Large values of p in equation 6.18 will give more accurate approximations of the gradient. Figure 6.2 illustrates the overall performance-optimization algorithm.

From Figure 6.2, we can summarize the iterative algorithm as follows.

- **Step 1:** Using the current HCS model, put all node weights in a vector to obtain the current model parameters ω_k.

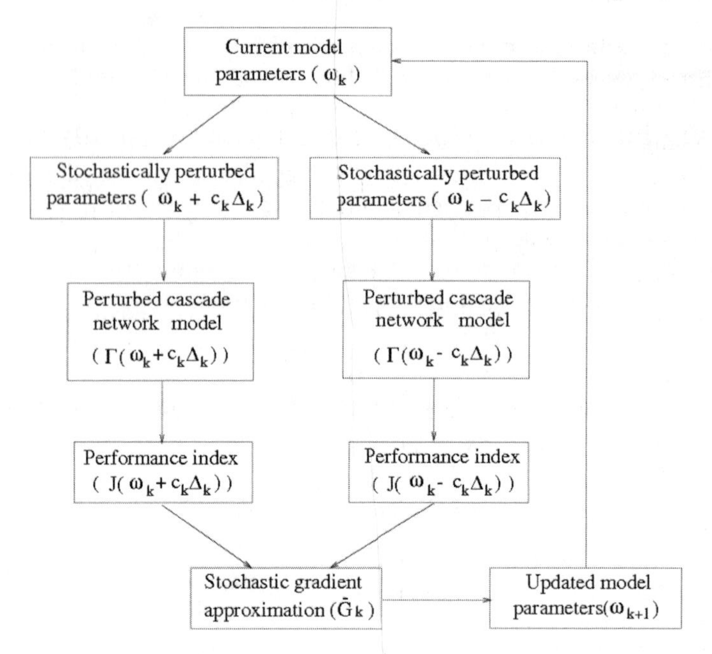

FIGURE 6.2: Overall approach to stochastic optimization structure.

- **Step 2:** Stochastically perturb the parameters $c_k\triangle_k$ in both positive and negative directions. The perturbed vector $c_k\triangle_k$ has the same structure with ω_k.

- **Step 3:** Using the two perturbed parameters, obtain two perturbed cascade network HCS models $\Gamma(\omega_k + c_k\triangle_k)$ and $\Gamma(\omega_k - c_k\triangle_k)$.

- **Step 4:** Using the two perturbed HCS models, drive the simulator vehicle to obtain driving data.

- **Step 5:** Based on the collected driving data, calculate two performance indices $J(\omega_k + c_k\triangle_k)$ and $J(\omega_k - c_k\triangle_k)$.

- **Step 6:** Use equation 6.18 to obtain the stochastic gradient approximation G_k.

- **Step 7:** Update the parameters and go back to Step 1.

6.4 Model optimization and performance analysis

Here, we test the performance optimization algorithm on control data collected from two individuals, Harry and Dick. In order to simplify the problem somewhat, we keep the applied force constant at $P_f = 300N$. Hence, the user is asked to control only the steering δ.

For each person, we train a two-hidden-unit HCS model with $n_s = n_c = 3$, and $n_r = 15$. Because we are keeping P_f constant, the total number of inputs for the neural network models is therefore $n_i = 42$.

Now, we would like to improve the tight-turning performance criterion J_2 defined in Chapter 3 for each of the trained models. In the SPSA algorithm, we empirically determine the following values for the scaling sequences $\{c_k\}$, $\{\alpha_k\}$:

$$\alpha_k = 0.000001/k, \quad k > 0 \tag{6.25}$$

$$c_k = 0.001/k^{0.25}, \quad k > 0 \tag{6.26}$$

We also set the number of measurements per gradient approximation in equation 6.18 to $p = 1$. Finally, denote J_2^k as the criterion J_2 after iteration k of the optimization algorithm; hence, J_2^0 denotes the performance measure prior to any optimization.

Figure 6.3 plots $100 \times J_2^k / J_2^0$, $0 \leq k \leq 60$, for HCS models corresponding to Dick and Harry. We note that for Dick, the performance index J_2 improves from $J_2^0 = 25.5$ to $J_2^{60} = 12.5$. For Harry, the improvement is less dramatic; his model's performance index improves from $J_2^0 = 17.7$ to $J_2^{60} = 16.1$. Thus, the performance optimization algorithm is able to improve the performance of Dick's model by about 55% and Harry's model by about 9% over their respective initial models. In other words, the optimized models are better able to negotiate tight turns without running off the road.

From Figure 6.3, we observe that most of the improvement in the optimization algorithm occurs in the first few iterations. Then, as $k \to \infty$, J_2^k converges to a stable value since $\alpha_k, c_k \to 0$. Clearly, the extent to which we can improve the performance in the trained HCS models depends on the characteristics of the original models. Dick's initial performance index of $J_2^0 = 25.5$ is much worse than Harry's initial performance index of $J_2^0 = 17.7$. Therefore, we would expect that Dick's initial model lies further away from the nearest local minimum, while Harry's model lies closer to that local minimum. As a result, Harry's model can be improved only a little, while Dick's model has much larger room for improvement.

Below we discuss some further issues related to performance optimization, including

- the effect of performance optimization on other performance criteria, and

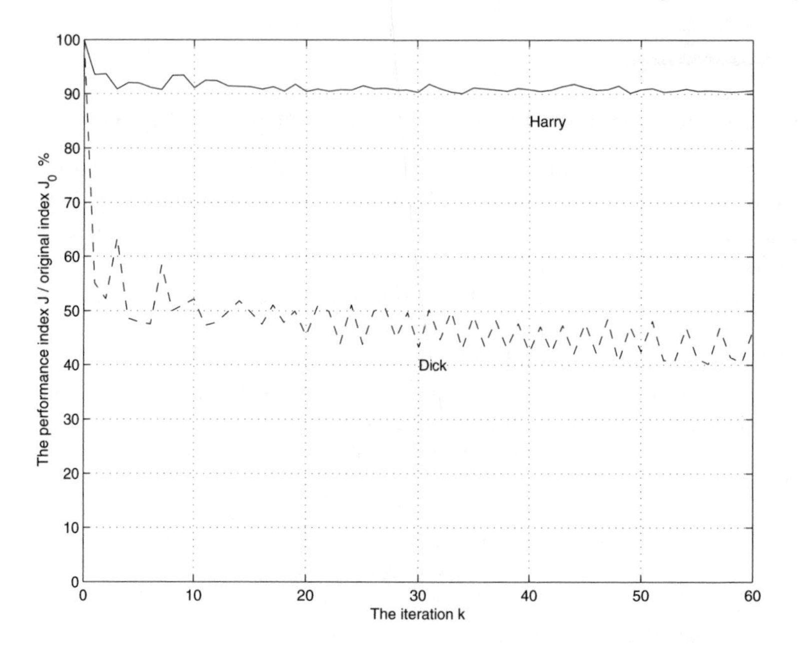

FIGURE 6.3: $100 \times J_2^k/J_2^0$, $0 \leq k \leq 60$, for HCS models corresponding to Dick and Harry.

- the similarity of control strategies before and after performance optimization.

First, we show how performance improvement with respect to one criterion can potentially affect performance improvement with respect to a different criterion. Consider Dick's HCS model once again. As we have already observed, his tight turning performance criterion improves from $J_2^0 = 25.5$ to $J_2^{60} = 12.5$. Now, let J_1^0 denote the obstacle avoidance performance criterion for Dick's initial HCS model, and let J_1^{60} denote the obstacle avoidance performance criterion for Dick's HCS model, optimized with respect to J_2. Figure 6.4 plots the maximum offset from the road median as a function of the obstacle detection distance τ for Dick's initial model (solid line) and Dick's optimized model (dashed line), where $v_{initial} = 25$.

From Figure 6.4, we can calculate J_1^0 and J_1^{60}:

$$J_1^0 = \frac{42}{35} = 1.20 \tag{6.27}$$

$$J_1^{60} = \frac{36}{35} = 1.03 \tag{6.28}$$

Thus, Dick's optimized HCS model not only improves tight turning performance, but obstacle-avoidance performance as well. This should not be too surprising, since the tight turning and obstacle avoidance behaviors are in

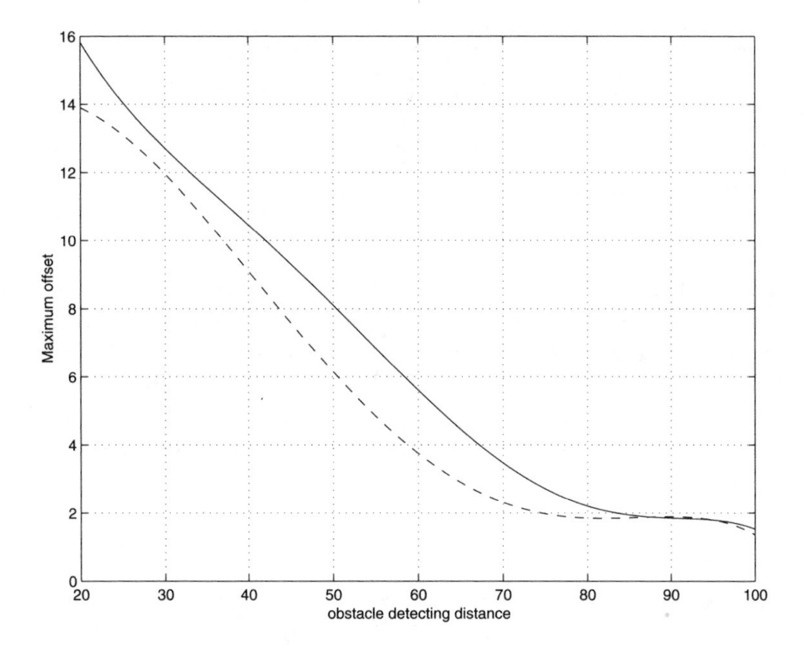

FIGURE 6.4: Maximum lateral offset for original and final HCS models.

fact closely related. During the obstacle avoidance maneuver, tight turns are precisely what is required for successful execution of the maneuver.

Second, we would like to see how much performance optimization changes the model's control strategy away from the original human control approach. To do this we turn to a hidden Markov model-based similarity measure developed for comparing human-based control strategies. Let H_x denote the human control trajectory for individual x, let M_x denote control trajectories for the unoptimized model corresponding to individual x, and let O_x denote control trajectories for the optimized model (with respect to J_2) corresponding to individual x. Also let $0 \leq \sigma(A, B) \leq 1$ denote the similarity measure for two different control trajectories A and B, where larger values indicate greater similarity, while smaller values indicate greater dissimilarity between A and B.

For each individual, we can calculate the following three similarity measures:

$$\sigma(H_x, M_x) \tag{6.29}$$

$$\sigma(H_x, O_x) \tag{6.30}$$

$$\sigma(M_x, O_x) \tag{6.31}$$

Table 6.1 lists these similarities for Dick and Harry.

From our experience with this similarity measure, we note that all the values in Table 6.1 indicate significant similarity. Specifically, the similarities for

Table 6.1: Similarity of control strategies

	$x = Dick$	$x = Harry$
$\sigma(H_x, M_x)$	0.762	0.573
$\sigma(H_x, O_x)$	0.434	0.469
$\sigma(M_x, O_x)$	0.544	0.823

$\sigma(H_x, O_x)$ (0.434 and 0.469) suggest that even after performance optimization, a substantial part of the original human control strategy is preserved. Furthermore, the other similarity measures are consistent with the degree of performance improvement in each case. For Dick, where a substantial performance improvement of 55% is achieved, the similarity between the initial and optimized models is far less than for Harry, where the performance improvement is more incremental.

6.5 Summary

In this chapter, we have proposed an iterative optimization algorithm, based on simultaneously perturbed stochastic approximation (SPSA), for improving the performance of learning models of human control strategy. The algorithm keeps the overall structure of the learned models intact, but tunes the parameters (i.e., weights) in the model to achieve better performance. It requires no analytic formulation of performance, only two experimental measurements of a defined performance criterion per iteration. We have demonstrated the viability of the approach for the task of human driving, where we model the human control strategy through cascade neural networks. While performance improvements vary between HCS models, the optimization algorithm always settles to stable, improved performance after only a few iterations. Furthermore, the optimized models retain important characteristics of the original human control strategy.

7

Transfer of Human Control Strategy

7.1 Introduction

In this chapter, we will focus on another important application of modeling human control strategy – namely HCS model transfer. * Transferring human control strategy has a lot of applications in areas from virtual reality and robotics to intelligent highway systems. In the previous chapters, a preliminary transfer of human control strategies from human experts to human apprentices has been studied. On the other hand, transferring human control strategy from one HCS model to another HCS model is still an open issue. The HCS model transferring is important because it can be widely used in machine learning, pilot training, games, and virtual reality.

To develop an algorithm to accomplish the transfer of the control strategies discussed above, we need a number of related results which have been discussed in the previous chapters.

- First, we need to be able to successfully abstract the human control strategy to a reliable computational model.
- Second, we need to be able to evaluate our resulting models both with respect to how well they approximate the human control data, and how well they meet certain performance criteria.

The previous work on these two issues forms the basis of the algorithm proposed herein. The algorithm requires that we first collect control data from different expert drivers, such as *expert A* and *expert B*. From driving data, we can then abstract two HCS models, one for the expert A and one for the expert B, using the previously developed techniques. We define them as *HCS-Expert-A* and *HCS-Expert-B*, respectively. Although these two models show acceptable performance in general, *HCS-Expert-B* still wants improvement based on one or two performance criteria. On the other hand, *HCS-Expert-A* performs better based on the same criteria. Our main idea here is to improve the *HCS-Expert-B*'s performance while still retaining important characteristics of his own. In other words, *HCS-Expert-A*'s knowledge will transfer to *HCS-Expert-B* based on the given transferring algorithm.

In this chapter, we will focus our attention on method of HCS transferring. All HCS models are cascade neural network models which are trained from the individual driver data. Based on the model validation, we guarantee HCS models have the same characteristic as the individual drivers. Here, the HCS transfer process is in fact a learning process. We use one expert HCS model (teacher) to guide the other HCS model (apprentice). By changing the weights or the structure, the latter HCS model will have a similar "manner" as the teacher HCS model.

We will introduce two methods of transferring in this chapter:

- One method is based on the stochastic similarity measure. In this transfer learning algorithm, we propose to raise the similarity between teacher HCS model and apprentice HCS model. The overall algorithm consists of two steps. In the first step, we let the teacher model influence the eventual structure of the HCS model. Once an appropriate model structure has been chosen, we then tune the parameters of the apprentice model through simultaneously perturbed stochastic approximation, an optimization algorithm that requires no analytic formulation, only two empirical measurements of a user-defined objective function.

- The other method is based on the model compensation. In this transferring algorithm, we introduce another HCS model to compensate the difference between the teacher HCS model and the apprentice HCS model. The new compensated apprentice model is a combination of original apprentice model and the neural network compensation.

7.2 Model transfer based on similarity measure

We are now interested in using one HCS model to teach a less-skilled individual HCS model to control the system for better performance. Any HCS model based on cascade neural network is defined by the structure and weighting parameters. Using the trained HCS model, we can also obtain HCS driving data by applying a HCS model on the driving simulator. In Section 5.2, we have introduced HMM similarity of two HCS models with observation sequences which obtain the similarity among different models. In this section, we will introduce a learning algorithm for transferring skill from a teacher HCS model to an apprentice HCS model. Rather than simply discard the apprentice HCS model, we attempt to preserve important aspects of the apprentice model, while at the same time improving the apprentice model's performance. In the proposed algorithm, the teacher model serves as the guide or teacher to the apprentice model, and influences both the eventual structure and parametric representation of the apprentice model. As we will demonstrate shortly, the similarity measure defined in the previous section will play a crucial role in this transfer learning algorithm.

7.2.1 Structure learning

The main goal of HCS transferring is that the apprentice HCS model should have the same characteristics as the teacher HCS model. The transferring process is also a learning process in the view of the apprentice. Here, we focus our attention on the structure learning.

Recall that in cascade neural network learning, the structure of the neural network is adjusted during training, as hidden units are added one at a time until satisfactory error convergence is reached. Suppose that we train a HCS model from human control data provided by a teacher using cascade learning. The final trained expert model will then consist of a given structure, which, in cascade learning, is completely defined by the number of hidden units in the final model.

Now, suppose that we have collected training data from an apprentice – that is, from an individual less skilled than the teacher. What final structure should his learned model assume? One answer is that we let the model converge to the "best" structure as was the case for the teacher model. Since we already have a teacher model at hand, however, we can let the teacher model inform the choice of structure for the apprentice model.

Figure 7.1 illustrates our approach for structure learning in apprentice HCS models, as guided by a teacher HCS model. We first train the teacher HC-S model in the usual fashion. Then, we train the apprentice HCS model, but impose an additional constraint during learning – namely, that the final structure (i.e., the number of hidden units) of the apprentice model and the teacher model be the same.

7.2.2 Parameter learning

Even though we impose the same structure on the teacher and apprentice HCS models, they will clearly converge to different parametric representations. The teacher model will be similar to the teacher's control strategy, while the apprentice model will be similar to the apprentice's control strategy. We would now like to tune the apprentice model so as to retain part of the control strategy encoded within, while at the same time improving the performance in the apprentice model. Once again we will use the teacher model as a guide in this learning, by examining the similarity (as defined previously) between the teacher and apprentice models.

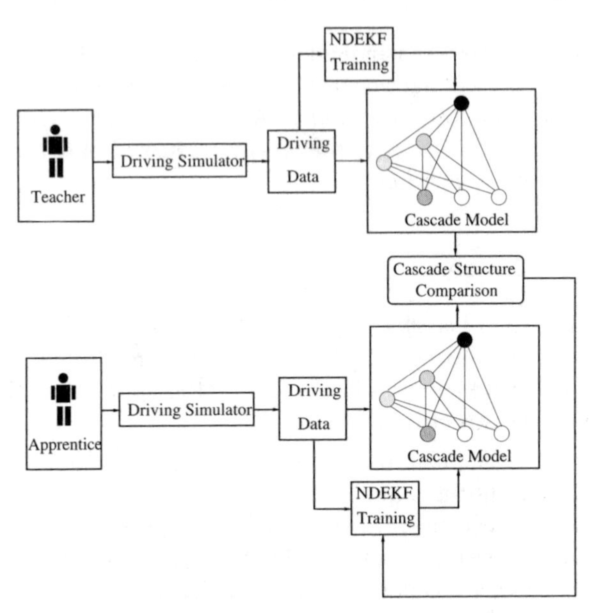

FIGURE 7.1: Structure learning in apprentice HCS model.

Let,

$$\omega = [w_1 \quad w_2 \quad \cdots \quad w_n] \tag{7.1}$$

denote a vector consisting of all the weights in the apprentice HCS model $\Gamma(\omega)$. Also let $\sigma(O_t, O_p(\omega))$ denote the HMM similarity between the apprentice HCS model and the teacher HCS model. Here O_t is the observation sequence of the teacher's model, and $O_p(\omega)$ is the observation sequence of the apprentice model. We would now like to determine a weight vector ω^* which raises the teacher/apprentice similarity $\sigma(O_t, O_p(\omega^*))$, while at the same time retaining part of the apprentice control strategy.

Determining a suitable ω^* is difficult in principle because (1) we have no explicit gradient information

$$G(\omega) = \frac{\partial}{\partial \omega}\sigma(O_t, O_p(\omega)) \tag{7.2}$$

and (2) each experimental measurement of $\sigma(O_t, O_p(\omega))$ requires a significant amount of computation. We lack explicit gradient information, since we can only evaluate the similarity measure empirically.

Again, we use *simultaneously perturbed stochastic approximation (SPSA)* to adjust ω. Denote ω_k as our estimate of ω^* at the kth iteration of the SA algorithm, and let ω_k be defined by the following recursive relationship:

$$\omega_{k+1} = \omega_k - \alpha_k \bar{G}_k \tag{7.3}$$

where $\bar{G}_k$ is the simultaneously perturbed gradient approximation at the kth iteration,

$$\bar{G}_k = \frac{1}{q}\sum_{i=1}^{q} G_k^i \approx \frac{\partial}{\partial \omega}\bar{\sigma}(O_t, O_p(\omega)) \tag{7.4}$$

$$G_k^i = \frac{\bar{\sigma}_k^{(+)} - \bar{\sigma}_k^{(-)}}{2c_k}\begin{bmatrix} 1/\triangle_{kw_1} \\ 1/\triangle_{kw_2} \\ \cdots \\ 1/\triangle_{kw_n} \end{bmatrix} \tag{7.5}$$

Equation (7.4) averages q stochastic two-point measurements G_k^i for a better overall gradient approximation, where,

$$\bar{\sigma}_k^{(+)} = \sigma(O_t, O_p(\omega_k + c_k \triangle_k)) \tag{7.6}$$

$$\bar{\sigma}_k^{(-)} = \sigma(O_t, O_p(\omega_k - c_k \triangle_k)) \tag{7.7}$$

$$\triangle_k = [\triangle_{kw_1} \triangle_{kw_2} \cdots \triangle_{kw_n}]^T \tag{7.8}$$

The SPSA algorithm is used to minimize the objective function and we use $\bar{\sigma} = 1 - \sigma(O_t, O_p(\omega))$ as our objective function, and where $\triangle_k$ is a vector of mutually independent, mean-zero random variables (e.g., symmetric Bernoulli distributed), the sequence $\{\triangle_k\}$ is independent and identically distributed, and the $\{\alpha_k\}$, $\{c_k\}$ are positive scalar sequences satisfying the following properties:

$$\alpha_k \to 0, \quad c_k \to 0 \ as \ k \to \infty, \tag{7.9}$$

$$\sum_{k=0}^{\infty} \alpha_k = \infty, \quad \sum_{k=0}^{\infty} (\frac{\alpha_k}{c_k})^2 < \infty \tag{7.10}$$

The weight vector ω_0 is of course the weight representation in the initially stable learned cascade apprentice model. Larger values of q in equation (7.4) will give more accurate approximations of the gradient. Figure 7.2 illustrates the overall parameter learning algorithm.

7.2.3 Experimental study

Here, we test the transfer learning algorithm on control data collected from three individuals: Tom, Dick, and Harry. In the previous work, we have observed that Harry's driving model performs better with respect to certain important performance measures. Therefore, we view Harry as the teacher, and Dick and Tom as the apprentices. Furthermore, in order to simplify the problem somewhat, we keep the applied force constant at $P_f = 300N$. In other words, we ask each driver to control the steering δ only.

In the experiment, we first train the teacher model on Harry's control data. The final trained model consists of two hidden units with $n_s = n_c = 3$, and

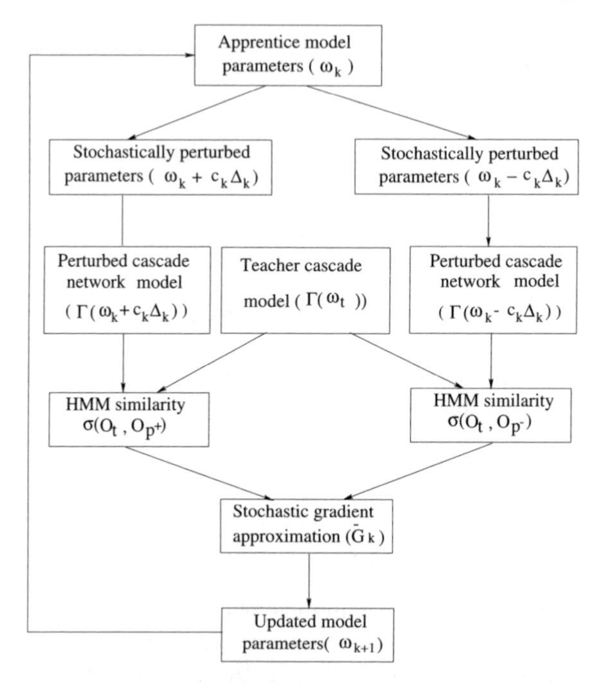

FIGURE 7.2: Stochastic HCS model parameter learning algorithm.

$n_r = 15$; because we are keeping P_f constant, the total number of inputs for the cascade network model therefore is $n_i = 42$. Keeping in mind that we want the final structure of the apprentice models to be the same as the teacher's structure, we also train Dick's and Tom's models to two hidden units each, with the same number of inputs ($n_i = 42$).

We now would like to improve the performance of the apprentice models, while still retaining some aspects of the apprentice control strategies. In other words, we would like to improve the similarity $\sigma(O_t, O_p)$ between Harry's expert model and each of the apprentice models using SPSA algorithm discussed in the previous section.

With SPSA algorithm, we empirically determine the following values for the scaling sequences $\{\alpha_k\}$, $\{c_k\}$:

$$\alpha_k = 0.00001/k^2, \quad k > 0 \tag{7.11}$$

$$c_k = 0.001/k^{1.25}, \quad k > 0 \tag{7.12}$$

Furthermore, we set the number of measurements per gradient estimation in equation (7.4) to $q = 1$. Finally, we denote $\sigma(O_t, O_p^k)$ as the similarity $\sigma(O_t, O_p)$ after iteration k of the learning algorithm; hence, $\sigma(O_t, O_p^0)$ denotes the similarity prior to any weight adjustment in the apprentice models.

Figure 7.3 plots the similarity measure between Tom's and Dick's apprentice models and Harry's expert HCS model, respectively, as a function of iteration

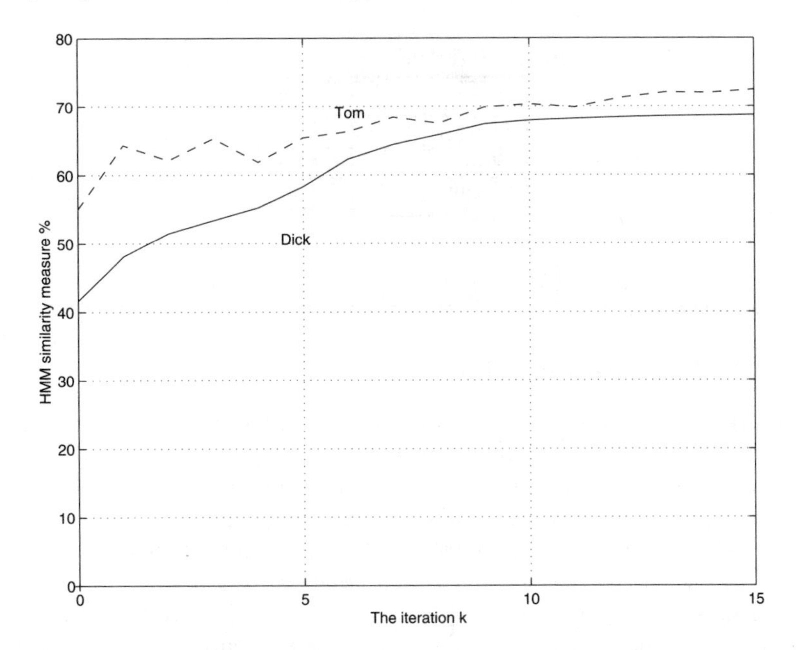

FIGURE 7.3: Similarity between apprentice/teacher HCS models during transfer learning.

k in the transfer learning algorithm. We observe that for Dick's model, the similarity to Harry's model improves from $\sigma(O_t, O_p^0) = 41.5\%$ to $\sigma(O_t, O_p^{15}) = 68.5\%$. Although for Tom's model the change is less dramatic, his model's similarity nevertheless rises from $\sigma(O_t, O_p^0) = 55.0\%$ to $\sigma(O_t, O_p^{15}) = 72.2\%$. Thus, the transfer learning algorithm improves the similarity of Dick's model by approximately 37% and Tom's model by about 17.2% over their respective initial models.

Since the similarity in control strategies improves, we would expect that apprentice performance improves as well. In order to test this, we examine model performance as measured by the obstacle avoidance performance criterion J defined in Section 5.2. Let J_t denote the performance criterion value for Harry's teacher model. Also, let J_p^0 denote apprentice performance before learning, and let J_p^{15} denote apprentice performance after transfer learning. We note that the performance criterion is defined so that smaller values indicate better obstacle avoidance performance.

Table 7.1 lists these performance values for Dick and Tom. From Table 7.1, we note that Harry's performance does not change, of course, since we keep his model fixed. The models for Dick and Tom do, however, improve with respect to the obstacle avoidance performance measure. The adjusted apprentice models can be viewed as hybrid models, which combine the apprentice control strategy with some of the improved techniques of the expert model.

Table 7.1: Obstacle avoidance performance measure

	J_p^0	J_p^{15}
Harry	0.51	0.51
Tom	1.23	0.87
Dick	1.37	1.13

7.3 Model compensation

In the previous section, we introduced a transferring method based on the similarity measure. That algorithm keeps the apprentice HCS model to the same structure as the teacher HCS model. This method raises the similarity by changing the weights of HCS model. This algorithm lets the apprentice HCS model develop characteristics similar to the teacher HCS model. On the other hand, when the teacher and the apprentice HCS model already exist, we need another way to achieve the proposed transfer. At this time, two HCS models are not assumed to have the same structure. The algorithm we developed is called model compensation.

In HCS model transferring problems, we propose to replace an actual human teacher instructor with a virtual instructor. An apprentice can obtain advice through the teacher HCS model. Suppose an apprentice driver is learning to control a dynamic system, such as driving a vehicle. We first collect data from an expert driver to train a teacher HCS model, defined as $M_{teacher}$. Based on the model validation method, we ensure that $M_{teacher}$ has a high similarity with the expert driving data. That means the $M_{teacher}$ shows good performance in some cases. We then utilize this expert HCS model to train a human operator, or apprentice HCS model, defined as $M_{apprentice}$.

We ask $M_{teacher}$ driving the simulator vehicle on a given road. And $M_{apprentice}$, like $M_{teacher}$, drives the vehicle on the same road. We assume two models have the same input. Then, we collect two models' output. Figure 7.4 shows the two control strategies of the steering angle δ.

In Figure 7.4, the upper curve shows $M_{apprentice}$ steering angle output, and the lower curve shows the steering angle output of $M_{teacher}$. We notice that control strategies are quite different between $M_{apprentice}$ and $M_{teacher}$. For any driving state, although we keep the same environment and same input vector, $M_{apprentice}$ only gives a small output steering angle. For $M_{teacher}$, the control style is quite different. The output of the HCS model is much larger than that of $M_{apprentice}$. Although two HCS models achieve stable driving, $M_{teacher}$ gives a better performance. For each state, there is a "gap" between the output of $M_{teacher}$ and $M_{apprentice}$. This output difference reflects the different skill level of the "teacher" and "apprentice." The idea of our model compensation algorithm is to find another HCS model to compensate the

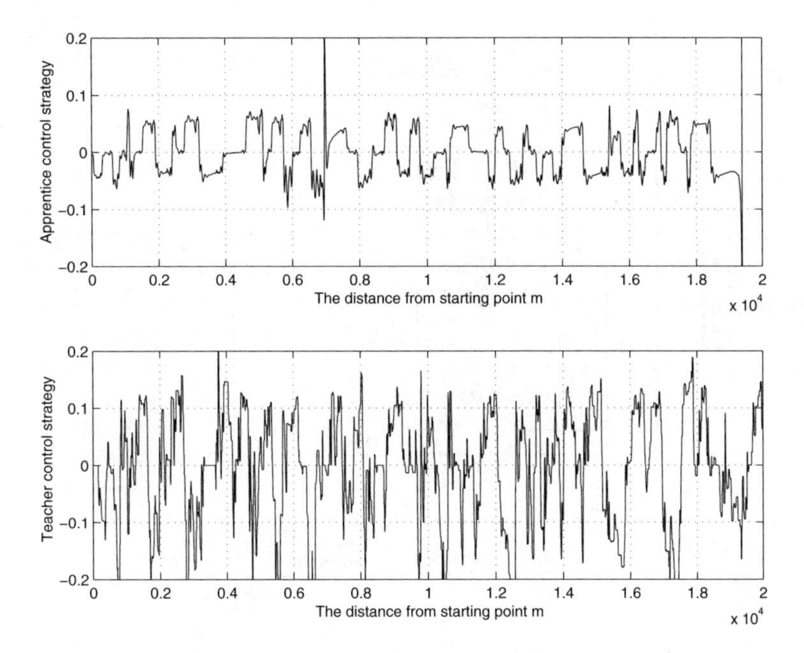

FIGURE 7.4: Comparison control strategy between teacher and apprentice HCS models.

"gap" between the apprentice and teacher.

We define the model compensation algorithm in the structure shown in Figure 7.5.

In Figure 7.5, both $M_{teacher}$ and $M_{apprentice}$ have the same input and output structure. Thus, both HCS models can drive a vehicle using the same input vector. Because cascade neural networks have a flexible structure, the two models are quite different not only due to their weight values but also due to the number of hidden nodes. This difference in structure results in the different "characteristics" of the individual HCS models.

For any time state, a "HCS input generator" creates an input vector based on the state of the vehicle, state of the control strategy, and the description of the road,

$$\{\nu_\xi(k-n_s), \cdots, \nu_\xi(k-1), \nu_x i(k), \nu_\eta(k-n_s), \cdots, \nu_\eta(k-1), \nu_\eta(k),$$
$$\dot{\theta}(k-n_s), \cdots, \dot{\theta}(k-1), \dot{\theta}(k)\} \tag{7.13}$$

$$\{\delta(k-n_c), \cdots, \delta(k-1), \delta(k), P_f(k-n_c), \cdots, P_f(k-1), P_f(k)\} \tag{7.14}$$

$$\{x(1), x(2), \cdots, x(n_r), y(1), y(2), \cdots, y(n_r)\} \tag{7.15}$$

This input vector is introduced either to "$M_{teacher}$" or to "$M_{apprentice}$." The "teacher/apprentice switch" chooses either one model's output to "driving simulator." When the given road driving finished, "HCS model comparison" records the difference of the two models' control strategy. Using the difference set and the recorded corresponding input vectors, we again use NDEKF

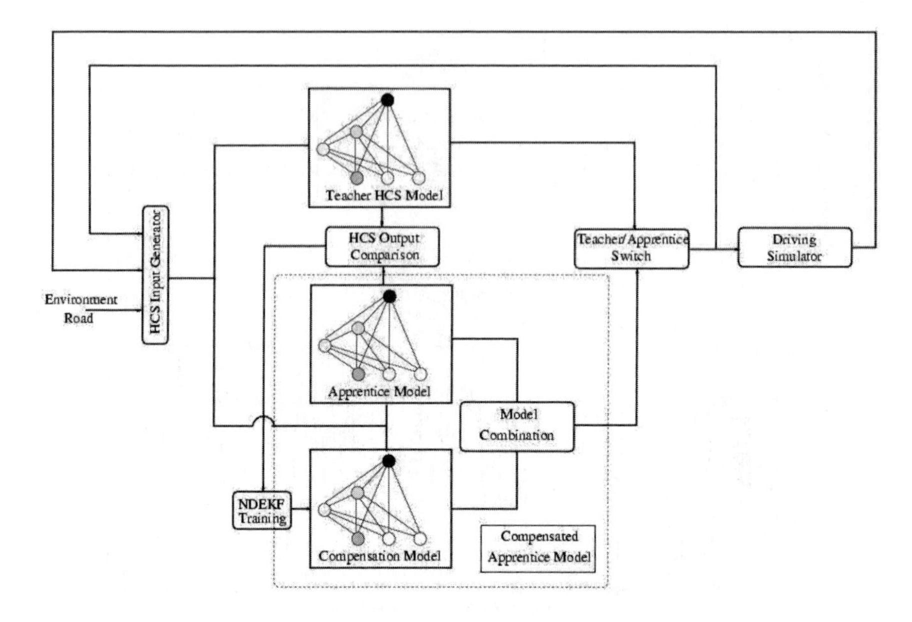

FIGURE 7.5: Diagram of model compensation.

training algorithm to create a new cascade neural network model, which is called the compensation model, defined as $M_{compensation}$. Because $M_{teacher}$, $M_{apprentice}$, and $M_{compensation}$ have the same input structure and output structure, we then use "model combination" to create a compensated apprentice model. The dash line box in Figure 7.5 shows the new apprentice HCS model which is a combination $M_{apprentice}$ and $M_{compensation}$. This new model, defined as M_{new}, is used to drive the same simulator vehicle on the same road.

An example result is shown in Figure 7.6.

In Figure 7.6, "apprentice" shows the steering control strategy of apprentice HCS model $M_{apprentice}$, "expert" shows the steering control strategy of teacher HCS model $M_{teacher}$, and "apprentice+compensation" shows the steering control strategy of compensated apprentice HCS model M_{new}. We find that the last two curves show a high similarity. For a more detailed comparison, we illustrate the three HCS model's control strategies driving on the road segment from 1200m to 2800m, in Figure 7.7.

In Figure 7.7, the solid line shows the output of $M_{teacher}$, the dashed line gives the output of $M_{apprentice}$, and the dash-dot line shows the output of M_{new}. We find that there is a big "gap" between $M_{teacher}$ and $M_{apprentice}$. But the "gap" between $M_{teacher}$ and M_{new} is much smaller. This result shows the new compensated apprentice HCS model nearly has the same "behavior" as the teacher model.

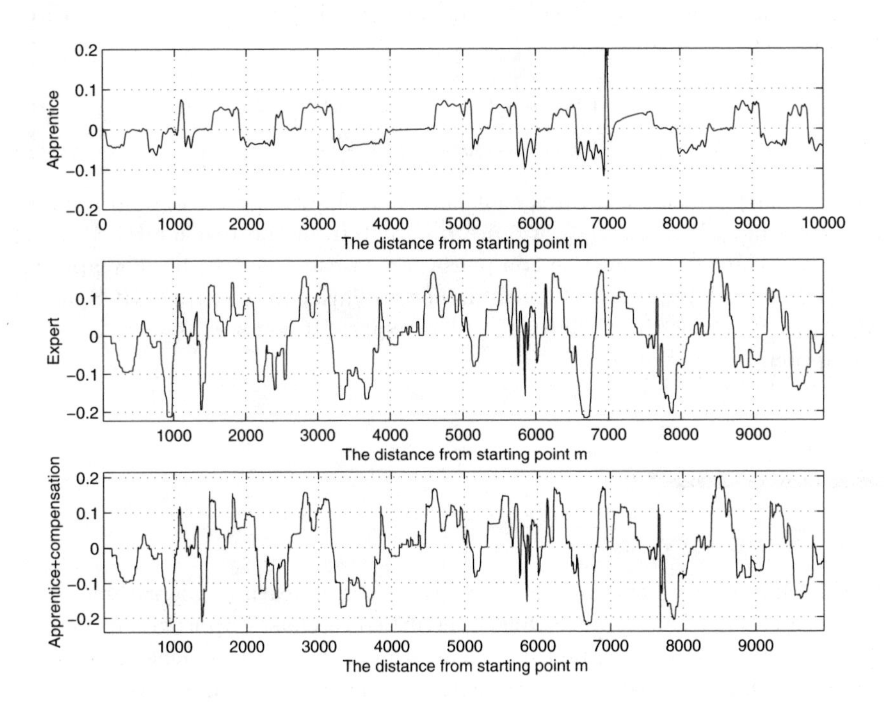

FIGURE 7.6: Control strategies of $M_{apprentice}$, $M_{teacher}$, and M_{new}.

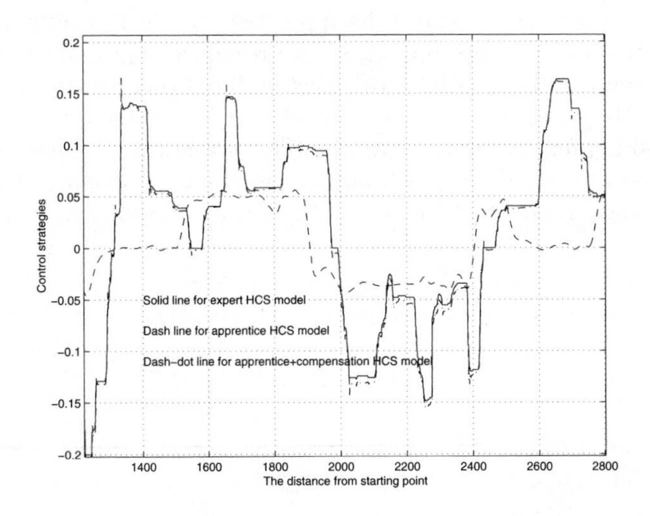

FIGURE 7.7: Comparison of models' steering control strategies.

Using the stochastic similarity to calculate the similarity values for the three HCS models, the result is

$$\sigma(O_{teacher}, O_{apprentice}) = 0.1822 \tag{7.16}$$

$$\sigma(O_{teacher}, O_{new}) = 0.8436 \tag{7.17}$$

This similarity result indicates that the control strategy of original apprentice HCS model $M_{apprentice}$ is quite different from the teacher HCS model $M_{teacher}$. But after using the compensation cascade model, the new apprentice HCS model M_{new} gives a much higher similarity to the expert HCS model $M_{teacher}$. It means the "behavior" of the expert HCS model is transferred to the apprentice HCS model.

7.4 Summary

In this chapter, we have proposed two transferring learning algorithms. One algorithm is based on simultaneously perturbed stochastic approximation (SPSA), for improving the similarity between apprentice and teacher models of human control strategy. This transferring algorithm includes two aspects. One is a structure learning method which keeps the apprentice's structure the same as its teacher. The other is parameter learning which tunes the parameters (i.e., weights) in the model to achieve better similarity with the teacher. It requires no analytic formulation, but only two experimental HMM similarity measurements between the apprentice and the teacher per iteration. We have demonstrated the apprentice's performance improved while the similarity improved with the help of the teacher. The other transferring learning algorithm is based on the model compensation. This method need not assume teacher and apprentice models have the same structure. The new apprentice HCS model, which is the combination of the original apprentice HCS model and the compensation HCS model, obtains the "behavior" of teacher HCS model.

8

Transferring Human Navigational Skills to Smart Wheelchair

In the previous chapter, we discussed how we could transfer complicated human control strategy from one model to another. In this chapter, we will present a case study in which human navigational skills are transferred to a wheelchair.

In practice, the environments in which mobile robots operate are usually modeled in highly complex forms, and as a result autonomous navigation and localization can be difficult. The difficulties are exacerbated for practical robots with limited on-board computational resources and complex planning algorithms, since this paradigm of environmental modeling requires enormous computational power. A novel navigation/localization learning methodology is presented to abstract and transfer the human sequential navigational skill to a robotic wheelchair by showing the platform how to respond in different local environments along a demonstrated, designated route using a lookup-table representation. This method utilizes limited on-board range sensing information to concisely model local unstructured environments, with respect to the robot, for navigation or localization along the learned route in order to achieve good performance with low on-line computational demand and low-cost hardware requirements. Experimental study demonstrates the feasibility of this method and some interesting characteristics of navigation, localization, and environmental modeling problems. Analysis is also conducted to investigate performance evaluation, advantages of the approach, choices of lookup-table inputs and outputs, and potential generalization of this study.

8.1 Introduction

8.1.1 Related work

In short, the problem of navigation is related to three main questions under a variety of technical scopes and limitations: (1) Where am I? (2) Where is the goal location relative to me? And (3) How do I get to the goal from the current location? The first question is actually a localization problem. Extensive research has been conducted in order to solve these problems with different practical and theoretical considerations [203].

The paradigm of learning-by-demonstration has attracted some attention in recent years for its ability to solve navigation problems. Route programs, which are constituted by step-by-step codes, are the traditional paradigm for instructing a machine. Although this paradigm is explicit and rigorous for human understanding and machine implementation, in some situations it is hard to describe the desired, exact instructions in specific and proper code statements. Examples include instructing a robot to follow human stochastic actions. Learning-by-demonstration is a good paradigm for tackling this difficulty. With an appropriate problem formulation and experimental design, the instructing tasks for the robot can be demonstrated by a human operator. Subsequently, the demonstrated data can be abstracted and learned by the robot, by constructing computational models with existing machine/statistical learning techniques [12], [166], such as neural network and support vector machine.

In the navigation learning area, a research group led by Inoue at the University of Tokyo has investigated a sequence of images obtained during human-guided movement [217]. A model of the mobile robot's route was described in a way that simplified the comparison of stored information and current visual information. Image information about the route was required for representing the environment. As they themselves noted, their approach was more suitable for routes in surroundings such as corridors. Another research group led by Miura at Osaka University has investigated a method using human involvement whereby an operator guided a mobile robot to a destination by remote control [116]. During a single demonstration, the robot developed a map by observing the surrounding environment in stereovision, after which it localized, computed, and followed the shortest path to a pre-determined destination. The objective of the human-guided movement in that case was to build a map with less effort. In another variation, Kraiss and Kuttelwesch have simulated vehicle navigation through a maze with five evenly spaced horizontal bars interspersed with randomly placed cut-outs. The simulated vehicle was taught by emulating a human teacher who drove a point robot.

However, in practice, the environments in which mobile robots operate are usually modeled in highly complex forms, such as geometric or image representations, and as a result autonomous navigation and localization can be difficult. The difficulties are exacerbated for practical robots with limited onboard computational resources and complex planning algorithms, since this paradigm of environmental modeling requires enormous computational power. In fact, environment modeling and localization can be considered as dual problems, because in order to localize a mobile robot many localization systems require a world model to match observed environment characteristics with the modeled ones. This is also true because in order to build a world model, most systems require precise localization of the robot [95], [137]. The most significant types of environment representations are cell-decomposition models, geometrical models, and topological models. Typically, modeling consists of several location sensing techniques, such as scene analysis, triangulation, proximity, and dead reckoning. The first technique, scene analysis, refers to the detection of scene features for inferring the objection location [49], [10]. The second, triangulation, refers to the use of the geometric properties of triangles to compute object location. The third, proximity, refers to the detection of an object when it is near to a known location by taking advantage of the limited range of a physical phenomenon. The fourth, dead reckoning, refers to incremental positioning methods such as odometry and inertial navigation systems.

With the above taxonomy, this study belongs to the scene analysis framework and learning-by-demonstration paradigm. This chapter presents a novel environment modeling approach by explicit local range sensing.

We will highlight three main parts in this chapter. First, we will illustrate our methodology and how it can be used to approach the navigation and localization problems. Second, an experimental study is presented to show the feasibility of the approach and the corresponding results. Finally, we will evaluate and discuss some characteristics observed from the experimental results.

8.2 Methodology

8.2.1 Problem formulation

The goal of this study is to abstract and transfer the human dynamic (sequential) navigational skill to a mobile robot. We will do so by showing the robot how to respond (output) in different local environments along a demonstrated, designated route, instead of achieving some general or optimal motion planning objectives (e.g., a "versatile" motion planner for tackling in various operational situations) [84], which demands complex algorithms and expensive platform hardware.

The objective of our study is to utilize limited on-board range sensing information to concisely model local unstructured (cluttered) environments, with respect to the robot, for navigation or localization along a learned route in order to achieve good performance with low on-line computational demand and low-cost hardware requirements. With this objective in mind, we formulate the approach into two consecutive sub-tasks: (1) we model the local environments in polar coordinate representations with on-board range sensing, and (2) we construct a mapping between sensor patterns input (polar) and control command output (for navigation) or absolute Cartesian configuration output (for localization) for real-time pattern recognition and output estimation. The feasibility and performance of the approach on the navigation and localization applications are the foci in this study.

8.2.2 Theoretical foundation

A novel navigation/localization learning approach is presented in this study. Local environments, in which the robot is situated, are represented in the form of range sensor signals for pattern recognition during autonomous navigation or localization. Range sensor signals are chosen as the learning inputs since they are a compact route representation describing the surrounding environmental information (which cannot be described by the robot's Cartesian x-y trajectory) in terms of "polar distance" with respect to the robot. Human dynamic (sequential) navigational skill is learnt and transferred to a mobile robot by showing the robot how to navigate/localize in different local environments throughout a demonstrated route by the use of a lookup-table in real-time. This is adopted for abstracting the human skill in the form of reactive mapping between raw sensor pattern inputs and human demonstrated control command (for navigation) or assigned/estimated robot configuration (for localization) outputs, acquired during the demonstrations. The learned lookup-table is the reactive [10] sensor-control/configuration mapping that explicitly stores the relations between different local environmental features and the corresponding demonstrated control commands (for navigation), or assigned configuration (for localization). For navigation, the mapping is served as the classifier for the distribution of features in sensor space and the corresponding control commands. On-line control command is calculated (mapped) from the learned lookup-table with the on-line sampled new sensor signals.

The proposed approach for navigation/localization consists of five steps. (1) Model local environments with sensor patterns, which are sampled via several range sensors at various robot configurations along the demonstrated route. (2) Manually assign the corresponding output (demonstrated control command or estimated configuration) to each sampled sensor pattern. (3) Form a lookup-table for the sensor-control/configuration mapping. (4) Search on-line for the "closest" sensor pattern in the constructed lookup-table for the real-time sampled sensor signal. (5) Adopt the corresponding output of the "closest" sensor pattern in the table as the on-line output (respond).

In this approach, it is not necessary for local environments to form perfect reflections with the ultrasonic sensing. No noise filtering is adopted to preserve all the original information in the sampled signals. Therefore, the raw sensing data is used directly as lookup-table data without any pre-processing. Furthermore, the sensor signals considered are discrete-time, discrete-valued and have limited sensing range. The signal of each sensor constitutes a single dimension in the sensor space, which is high-dimensional and fix-sized. In general, each particular route is unique in that its series of local environments are different from that of other routes. Therefore, each specific route is characterized by a specific distribution of all sampled sensor patterns. Since different local environments have different features in general (depending on the resolution of sensing information), a longer demonstrated route has more sampled sensor patterns (vectors) in the sensor space with denser and more complex distribution. Since each sensor vector is associated with a corresponding demonstrated output (control command for navigation and robot configuration for localization), unstructured environments (such as a common household setting) are particularly suitable for the proposed approach. The reasons for this is that they have more distinctive local features for recognizing the corresponding mapping outputs along the demonstrated route.

Some interpretations can be drawn for illustrating the characteristics of the approach for navigation application. First, each pattern of sensor signals (vector) is used for environmental pattern recognition. Second, each associated control command is used for on-line incremental motion. Third, the complete sequence of sensor vectors (trajectory) sampled along the route is actually an acquisition of local environmental (spatial) information, which is the complete global environmental model. Fourth, the complete sequence of control commands (trajectory) applied along the route is actually an acquisition of a sequence of "sub-goal" information, which gives the complete and sufficient destination information.

Instead of performing function approximation with machine/statistical learning techniques (e.g., neural network and support vector machine) [156], [153] as in our previous preliminary study [43], the constructed mapping in this study is explicitly listed as a lookup-table for searching output in real-time with the on-line sampled sensor input. Doing so avoids approximation or simplification of the actual complex mapping relationship. Therefore, the interpolation (generalization) aspect of the approach is not considered since we assume the resolution of our environmental model is good enough for the proposed applications, as shown in our experimental study outlined in the latter part of this chapter.

8.3 Experimental study

8.3.1 Settings

The mobile robotic platform we adopted is modified from a commercial robotic wheelchair, TAO-6, manufactured by Applied AI, Inc. TAO-6 has two motors in a mid wheel drive configuration providing good power, acceleration, and stability (Figure 8.1). We used seven wide-angle (around 60 degrees of directivity) ultrasonic range sensors and a keypad (with three discrete motion control states: move straight, turn left, and turn right), which are available as standard components on the original TAO-6 platform. The sensors were re-located to achieve greater coverage on the front local environment with respect to the robot. Each sensor returns the "distance to obstacle" value at a resolution of about 35mm per step (discrete value). The sensors have a limited range but accurately detect obstacles up to around 1.5m. They are op-

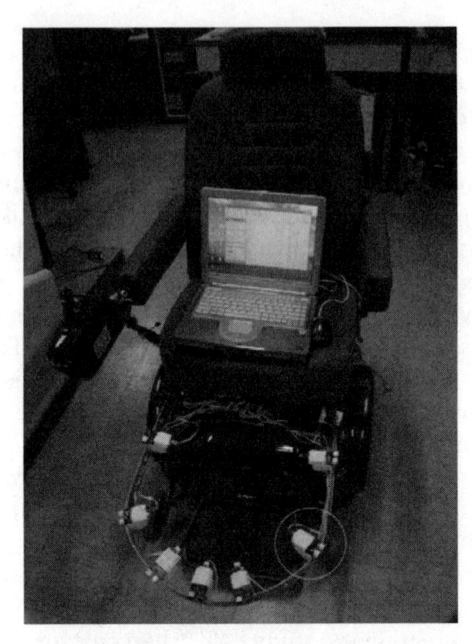

FIGURE 8.1: The wheelchair platform with seven ultrasonic sensors in the front (circled).

erated at different carrier frequencies, sampling simultaneously at a constant frequency of about 5Hz.

8.3.2 Experiment 1: Navigation

Procedure

The robot is controlled by a human operator via the use of keypad to navigate a designated route one time only in a static indoor unstructured environment (Figure 8.2), moving at a speed of about 0.8m/s with sensor sampling for collecting raw local range-sensor patterns (about 420 samples). At the time of each sampling, the control command, determined by the demonstrator and corresponding to the sampled sensor pattern, is also recorded (Figure 8.3). A lookup-table is constructed from all the raw sensor patterns (input) sampled along the route and their corresponding demonstrated control commands (output). In Figures 8.3 to 8.5, the commands 1, 2, and 3 refer to the discrete motion control commands "turn left," "move straight," and "turn right" respectively. To retain the simple input-output relationship, no previous input or output state is adopted.

FIGURE 8.2: A demonstrated route in an unstructured environment.

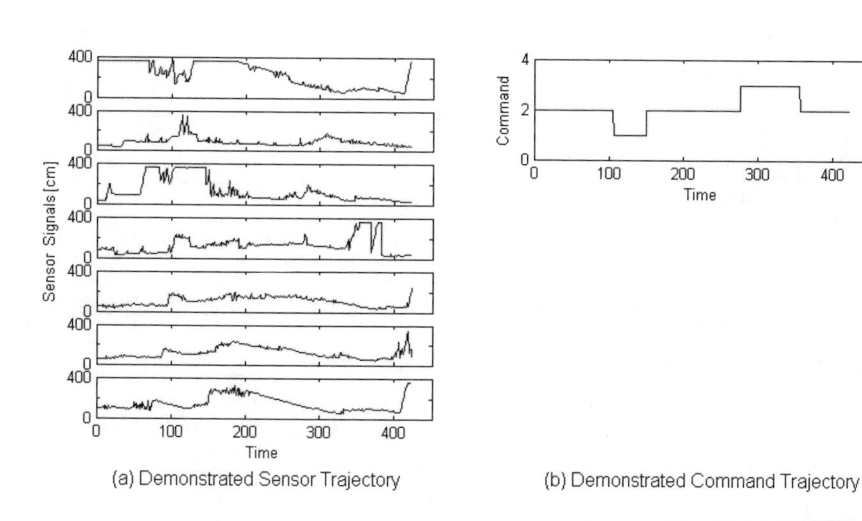

FIGURE 8.3: The (a) sensor trajectory and (b) corresponding command trajectory acquired in the demonstration.

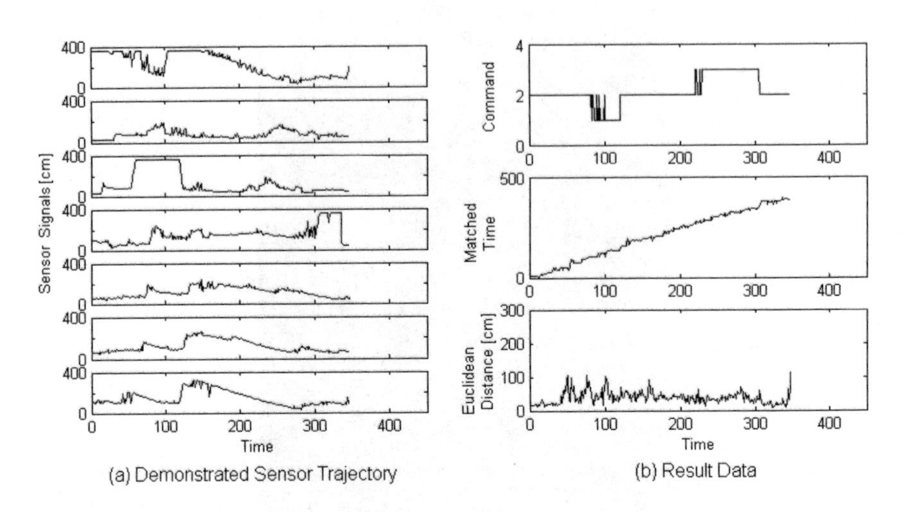

FIGURE 8.4: The (a) sensor trajectory and (b) result data acquired in the first trial of autonomous navigation.

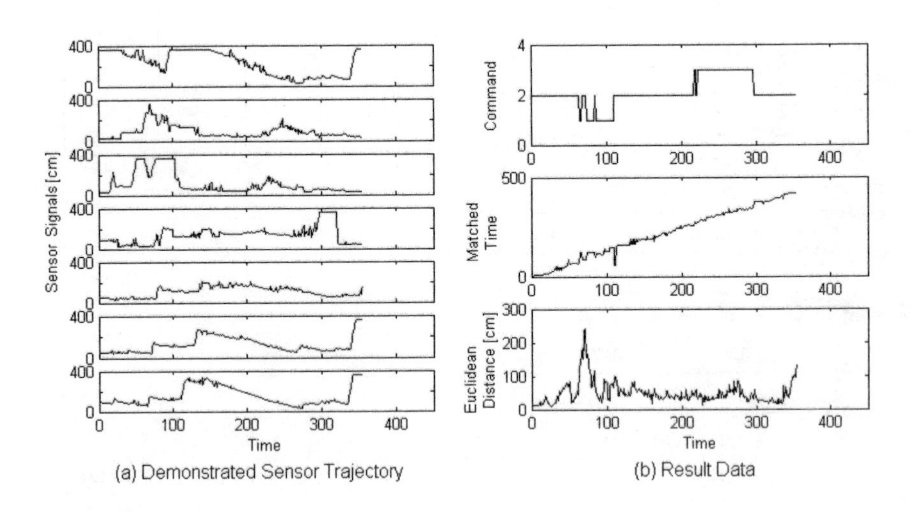

FIGURE 8.5: The (a) sensor trajectory and (b) result data acquired in the second trial of autonomous navigation.

Results

After one demonstration, the robot has the capacity to complete the navigation learning task with the constructed lookup-table at a speed of about 0.8m/s. Two trials of autonomous navigation are performed (Figures 8.4 and 8.5). In both autonomous navigations, the robot completed the task without collision with the surrounding obstacles. Each on-line sampled sensor pattern is matched with the "closest" sensor pattern stored in the lookup-table with a minimum Euclidean distance in the 7-D sensor space as the local environmental pattern recognition criteria. The corresponding control command output for the matched sensor pattern in the lookup-table is considered to be the corresponding control command of the on-line sampled sensor pattern.

From the Matched Time Result (Figures 8.4(b) and 8.5(b)), we observe that the on-line sampled sensor pattern does not match the "far-away" mismatched location, which does not always occur (a factor we analyze below) for our choice of demonstrated route. Therefore, the performance on local environmental recognition along the demonstrated route is good. The sampling rate is high enough to generate a lookup-table with fine resolution with the settings. It is possible to build a larger lookup-table with finer resolution for the local environment (but with a larger database and longer on-line searching time). This would include a number of sensor patterns, in which the matching Euclidean Distance is large (accounting for new configurations), and sampled in the two trials of autonomous navigation. The procedure for data collection is an application of active learning [240], [100] for the learning-by-demonstration paradigm; that is, by multi-phase human feedback demonstrations. This method is useful to strategically identify and collect critical training data, which is hard to sample for systems with different dynamic parameters in the demonstration and application stages.

We observed that the sensor patterns are very similar when sampling at the same configuration along all navigation routes (Figures 8.3(a), 8.4(a), and 8.5(a)). Hence, the navigation results are very similar for the two trials of autonomous navigation (Figures 8.4(b) and 8.5(b)), even under the stochastic, sequential effects on the dynamic parameters in the robot hardware such as friction. Since the skill model is reactive, the robot is capable of starting mid-way and continuing to navigate to the destination of the demonstrated route. Compared with the human demonstrated control command trajectory, the control command trajectories in the two trials of autonomous navigation have little noise (jerkiness). However, the robot acts as a low-pass filter and roughly performs similar motions as the human demonstration. Further study for a "smart filtering" technique is needed to investigate this problem.

FIGURE 8.6: A demonstrated route (A to B to A) in an unstructured environment.

8.3.3 Experiment 2: Localization

Procedure

The robot is controlled by human operation via the use of keypad to navigate a designated route five times in a static indoor unstructured environment, with the aim of obtaining the same motion profile (Figure 8.6). In each operation, the robot moves at a linear speed of about 0.8m/s or angular speed of about 0.8rad/s with sensor sampling for collecting raw local range-sensor patterns (about 260 samples) along the route. The demonstrated route in this experiment is different from that in Experiment 1, since the kinematic constraints of the wheelchair locomotion are not considered in this experiment.

Since the robot is operated at a roughly constant speed along the predefined route, the robot configuration (planar position and orientation) is assigned off-line by human estimation to the corresponding sensor pattern. A lookup-table is constructed from all the raw sensor patterns (input) sampled along the route and its labeled configuration (output). One (259 sensor patterns) and three (total 796 sensor patterns) navigational operations are used to construct two independent localization lookup-tables (Figure 8.7). Another two operations are used to evaluate off-line the localization performance of the two constructed lookup-tables (Figure 8.8). The accuracy of the assigned robot configurations, due to human estimation errors, remains an issue to be investigated in the latter part of this chapter.

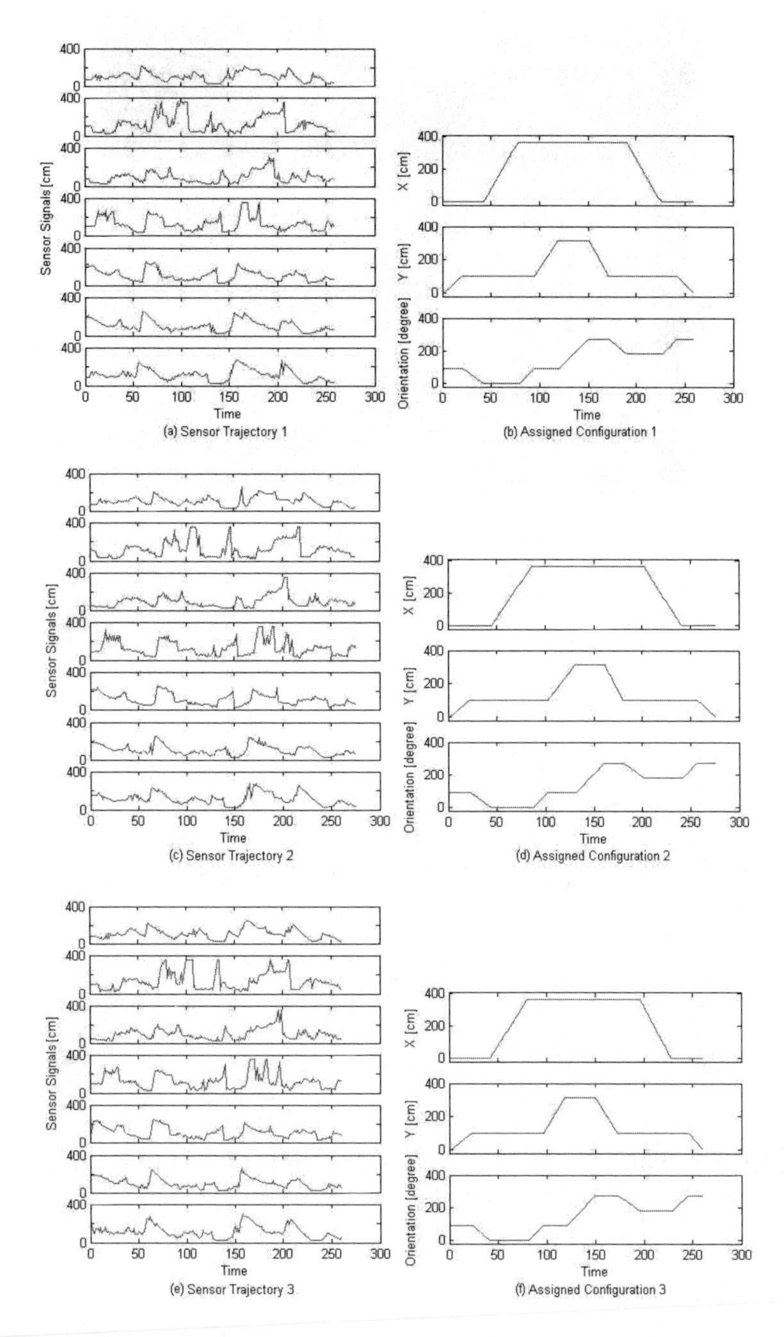

FIGURE 8.7: Three training demonstrations, (a & b), (c & d), and (e & f), for constructing two lookup-tables: (1) from Sensor Trajectory 1 and Assigned Configuration 1, and (2) from Sensor Trajectories 1, 2, & 3 and Assigned Configurations 1, 2, & 3.

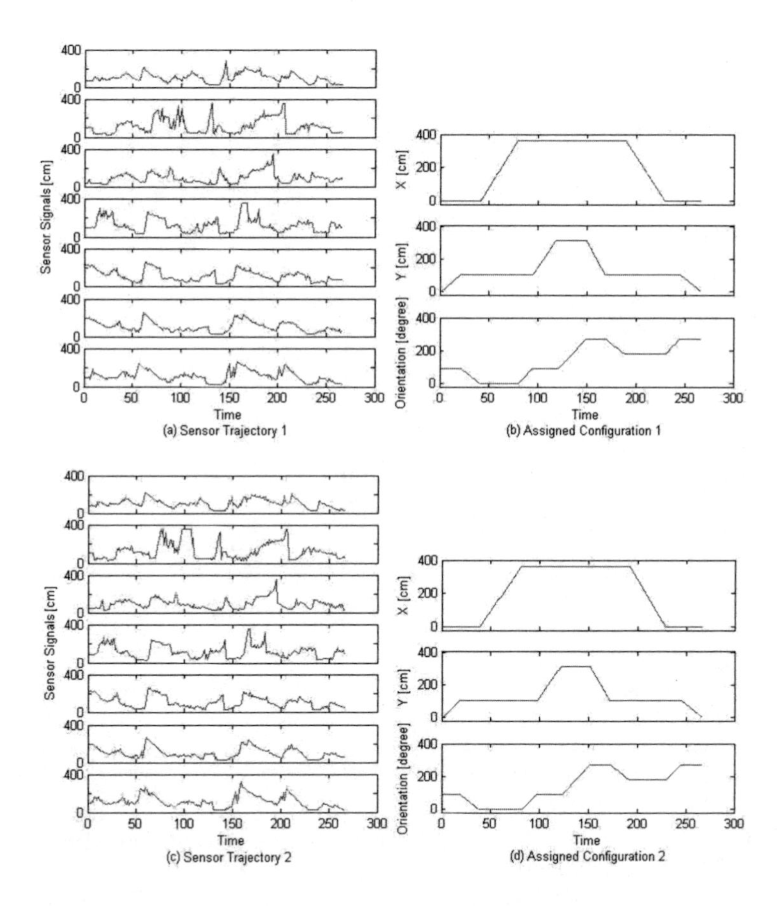

FIGURE 8.8: Two untrained demonstrations, (a & b), and (c & d), for evaluating the localization performance of the two lookup-tables.

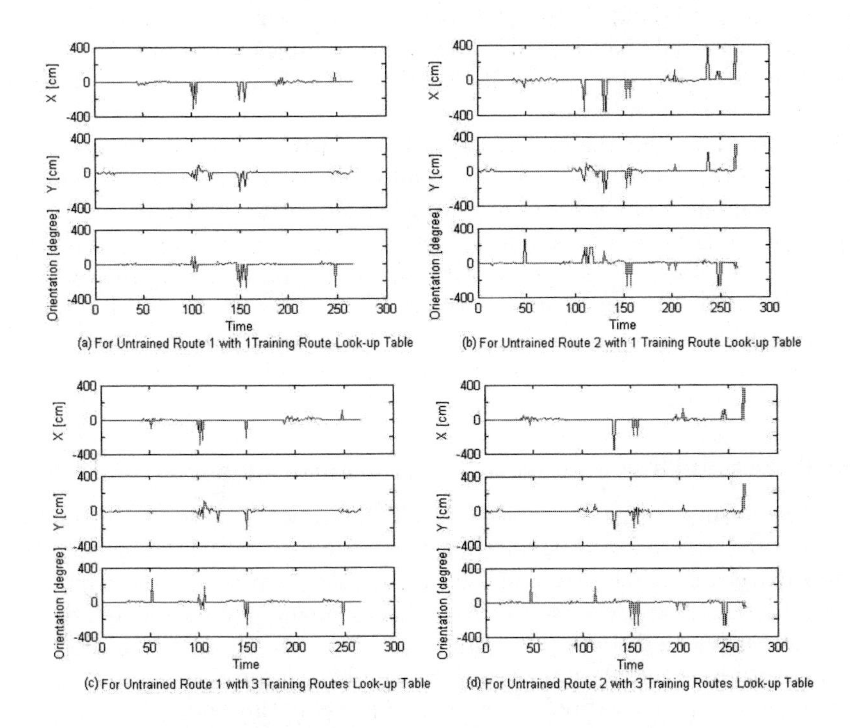

FIGURE 8.9: The errors in configuration exist in the two lookup-tables for the two untrained demonstrations.

Result 1: Localization performance

Each of the constructed localization lookup-tables is evaluated with two sets of new navigational operation data (Figure 8.8), in which each on-line sampled sensor pattern is matched with the "closest" sensor pattern stored in each lookup-table with minimum Euclidean distance in the 7-D sensor space as localization criteria (pattern recognition point of view). The corresponding configuration output for the matched sensor pattern in the table is considered to be the corresponding configuration of the on-line sampled sensor pattern. As demonstrated by the results, the errors in configuration output (x, y, and orientation) are bounded around ±25cm and ±20 degrees between the actual values in most instances, except for a few significant errors (Figure 8.9). Therefore, the proposed approach has an acceptable accuracy, even allowing for human assignment errors.

It is also clear that the two lookup-tables constructed from one and three operations obtain similar localization results for the two sets of new operation data. The number of constructing operations has little effect on the localization performance in our experimental settings. The sampling rate is thus high enough to generate a lookup-table with fine resolution by way of the settings. Moreover, we observe that the sensor patterns are very similar when sampling at the same configuration in the workspace in all operations (Figures 8.7 and 8.8). Hence, the localization results are very similar for the two untrained routes (Figure 8.9).

Result 2: Similar sensor patterns in various configurations

From the results in this experiment (Figure 8.10), we observe that at each of the locations where significant localization errors exist, the on-line sampled sensor pattern is matched to a "far-away," mismatched location at which the sensor pattern is still closest to the on-line sampled one on the 7-D sensor space in Euclidean distance. In Figure 8.10, the "Resultant Error" refers to the magnitude of the resultant vector formed by the three scalar errors (x, y, and orientation), and certainly represents the overall three output errors. The physical meaning here is that there are some locations at which sampled sensor patterns are very similar. This phenomenon is a reasonable approximation of our daily life experience (Figure 8.11).

Result 3: Small variations of major dimensions of environmental features along the route

We observe that for most of the time there are only minor variations in each sensor signal along the route (Figure 8.12). Also, at the times when significant variation exists in one sensor signal, the variations in the other sensor signals are relatively minor. In other words, significant variations do not occur simultaneously in the majority of sensors. Moreover, the average of the variations in seven sensor signals shows that the overall variations in all sensor

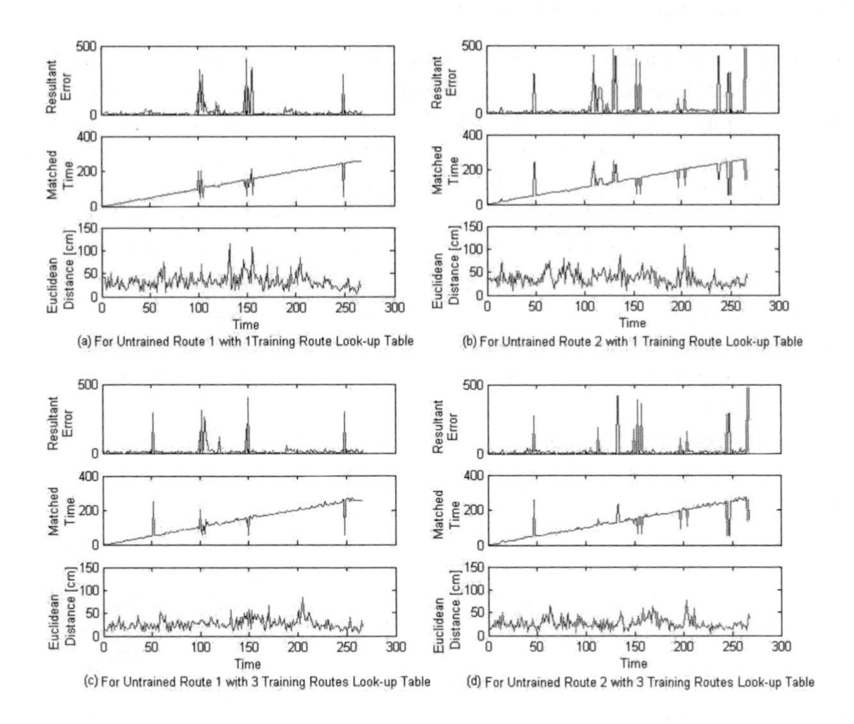

FIGURE 8.10: Significant errors appear when similar sensor patterns occur in various configurations.

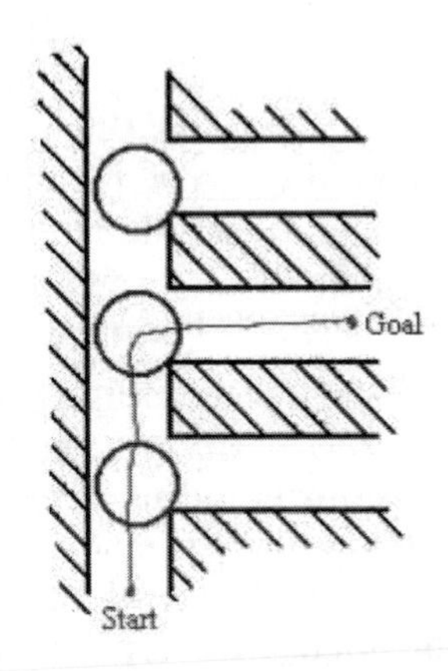

FIGURE 8.11: An example of the existence of similar local environments (circled).

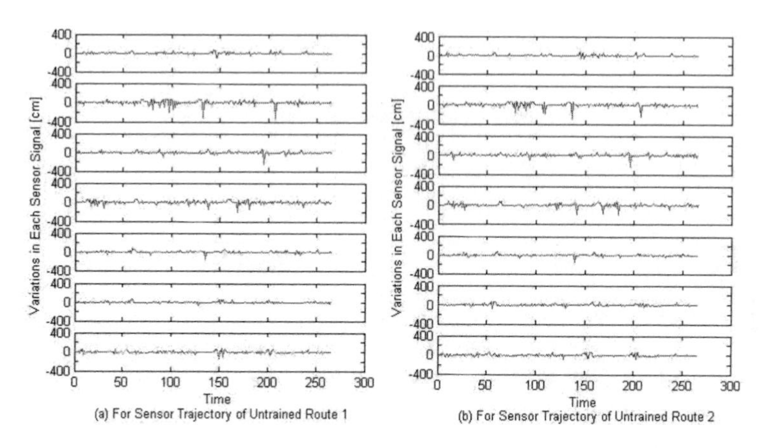

FIGURE 8.12: Variations in each sensor signal along the route.

signals alone the route is roughly bound around ±15cm (Figure 8.13), which is a small quantity compared with the recorded signal range in each sensor along the route. The physical meaning of this result is that nearby configurations have similar sensor patterns, which refers to similar local environmental features.

However, this result matches the one obtained in Result 2; that is, whenever large mean variations in sensor signals (relative to the adjacent untrained samples) occur in the route (Figure 8.13), large recognition distances (Euclidean) are measured (matching untrained samples to trained/lookup-table samples) at the corresponding nearby configurations (Figure 8.10). This phenomenon refers to the fact that the more divergence for the local environmental features obtained in nearby configurations, the larger the distance measure in recognition/matching is encountered.

Empirically these explanations match with experience in daily life where similar visual perception is received when a human being is in nearby locations/orientations in a static environment. Smaller differences in sensing configurations yield higher perceptional similarity. On the other hand, it is also reasonable to believe that smaller environmental changes should yield higher similarity between the original environment and changed environment at the same sensing configuration. Therefore, the robustness of the sensor-environment model for localization (pattern recognition) is likely to be proportional to the "weighting" of the environmental change with respect to the change of sensing configuration.

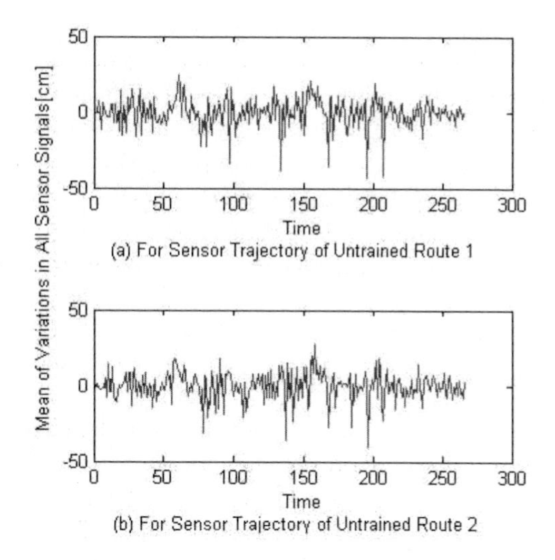

FIGURE 8.13: Mean of variations in all sensor signals along the route.

8.4 Analysis

8.4.1 Performance evaluation

The experimental study on the robot demonstrates the feasibility of this practical autonomous navigation approach for indoor route-specific application. As the learned model is able to lead the robot to accomplish the demonstrated route, the model is defined as being capable of effectively abstracting human navigational skill with limited environmental information sufficient for the navigational task. Yet, a source of stochastic nature of the two autonomous navigations, which is completed recursively (sequentially), is due to the little but acceptable localization (sensor pattern recognition) errors in local environment recognition observed in Experiment 2, incurring minor variations in the robot's motions during different trials of autonomous navigations. In our experience, the overall human skill model can typically be composed of a number of simpler sub-models such that each of which specializes a specific segment of the learned route depending on the complexity of the skill model. The transition between two consecutive skill models is dependent on the recognition of the ending patterns of the former model. A certain overlapping route segment for the skill models is needed for the transition.

For the localization task, there are two possible sources of error: (1) assigned mapping output (configuration), and (2) similar sensor patterns in different configurations. For the first error source, the acceptable error bound exists for the assigned mapping output, as shown in the experimental results (Figure 8.9). For the second error source, more sensors to achieve higher environmental sensing resolution are likely to reduce the similarity of the sensor patterns. It is likely that more sensors yield higher resolution of the environmental feature modeling. Another way to improve the accuracy of the system is to perform consecutive samplings in nearby positions or orientations. In this way the resultant lookup-table outputs are smoothed out by linear interpolation based on the actual position or orientation differences measured during samplings.

8.4.2 Advantages of the approach

Our approach possesses several advantages for practical applications. First, the environment is represented in an efficient and compact model, instead of the existing complex representations such as cell-decomposition, geometric, topological, or vision models. As a result, our model requires relatively less on-line computational demand, lower data storage memory, and cheaper hardware equipments for navigation and localization tasks. All of these benefit real-time planning. Second, concise local environmental information is sensed, extracted, and processed with the on-board computing in the self-contained robot in real-time, without any additional/extra positioning device. Third, the explicit lookup-table clearly represents the complex sensing-configuration mapping and is easier for analysis, compared with the use of a machine learning technique (cascade neural network with node-decoupled extended Kalman filter) for approximating the mapping relationship that was used in our previous preliminary study [43]. Fourth, with this learning-by-demonstration method, users with little technical knowledge are able to easily teach the robot in a simple and direct way. Fifth, since the trajectory planning is reactive and in real-time, the navigation planning has a certain fault-tolerant capability that momentary control-errors, ambiguous sensory data and small deficiencies in the lookup-table data are by-passed. Sixth, instead of simulating physical phenomena, the reactive mapping is obtained from human demonstrations. Hence, the arbitrary desired path shape for navigation is achieved as needed. Seventh, the robot's physical shape is abstracted in the skill model during the navigation learning process. Therefore, the approach is suitable for a mobile robot of any shape and size. Eighth, the localization approach can be used with existing simple positioning systems, such as a wheel encoder, to achieve low-cost error-compensation in order to increase the overall localization accuracy.

With the above advantages, this approach is useful, easy to implement, and suitable for simple, low-cost, and self-contained systems able to achieve reactive navigation and/or localization at designated routes/configurations in indoor, static, and unstructured environments (such as common household settings) with good performance. Moreover, using the methodology presented, the navigation problem and its associated environmental modeling issues are tackled for several specific mobility-assisting rehabilitation applications. For example, a wheelchair user can teach a robotic wheelchair to perform autonomously for learned navigation tasks that need to be performed frequently. For instance, a user might select a route from a list of learned routes so as to be transported to the destination with high-level machine autonomy. In other words, with this autonomous navigation ability, the robot can free the user from much of the low-level control now required in common electrically powered wheelchairs.

8.4.3 Choices of sensor-configuration mapping

The approach presented is suitable for the demonstrated route, as is the case for the route chosen in Experiment 1 (Figure 8.2), without involving too many robot configurations at which the sampled sensor patterns are very similar. In cases where many similar sensor patterns exist, the learned model is unable to classify the contradictions in the divergence of mapping outputs. To solve this problem, we can include the temporal dimension as an additional mapping input for better pattern recognition (control commands or configurations), counting from zero at the beginning of the route. Hence, the sampled sensor patterns collected along the route are "separated" by the time they are sampled. In other words, the local environmental features are now described by one more dimension: time. While real-time sensing information acts as the system's external feedback, the time-input serves as the internal information. This additional time-input adjustment is also tested for the route chosen in Experiment 1, yet the learned model is still able to complete the autonomous navigation task and without obvious difference in performance with the purely reactive case. Besides the time input, we can also adopt the previous sensor states as additional mapping inputs for better pattern recognition.

Although the above choices of lookup-table help to reduce similar patterns in difference control commands or configurations with similar sensor patterns, they have certain limitations for practical applications. For the time-input case, the robot is restricted to performing the same motion profile as that of demonstration in order to have the same temporal dimension or trajectory profile. For the previous sensor states case, the robot is constrained by having part of the motion profile the same as that of the demonstration(s) in order to have same previous states mapping input. In other words, both cases need to have kinematic constraints based on demonstrations. The curse of dimensionality referred to here has been widely addressed in the framework of dynamic programming in the literature [224], [54].

However, although not all the arbitrary indoor routes are free of the "similar sensor patterns" situation (for purely reactive mapping), the proposed method practically serves as a simple and powerful solution for the navigation learning tasks in certain applications. The addition of temporal dimension or previously state(s) can, in fact, solve this problem, even with the involvement of induced practical constraints such as the associated trade-outs.

8.4.4 Generalization of the study

Other than the contributions mentioned above, the study can be generalized in certain aspects. For the purposes of navigation, there are certain issues we may explore. First, how does the number and location of sensors on the robot affect navigation learning? Additionally, are there cases in which the limited number of sensors becomes problematic in general? Second, what is the effect on navigation performance when the lookup-table involves more than one demonstration? Third, what kinds of dynamic obstacles, and/or how many changes in the learned environment, can the trained model tolerate? In this study, since static obstacles cannot be identified from the surrounding environment, all static objects in the environment are considered as static obstacles for the navigation of the robot. The existence of new static obstacles can be considered as a change in the learned environment.

With regard to the localization task, the study can be generalized in some aspects. First, two choices of sensing information can be adopted for a localization system: color sensors (e.g., camera) and range sensors (e.g., ultrasonic or laser range finder). Both can be used individually for localization, while they also may form a hybrid/integration system for error-compensation purposes. Instead of comparing the localization performances between vision and range sensing [184], is it possible to achieve the synthesis of vision and range sensing? Second, it is possible to consider two kinds of sensing coverage: planar and spherical. Third, three types of Cartesian mapping output can be learned by demonstration navigation: route (1-D), area (2-D), and space (3-D) configurations. Therefore, this study focuses on one of the combinations of the above scopes.

8.5 Conclusion

In this study, a novel local environment modeling methodology for robotic wheelchair navigation and localization at a designated route with on-board ultrasonic range sensing is achieved by learning human navigational skill. For autonomous navigation along the demonstrated route, an explicit lookup-table is constructed with raw sensor patterns and human demonstrated control command outputs. Experimental results show the following: (1) After one demonstration, the robot has the capacity to complete the autonomous navigation task with the constructed lookup-table at a speed of about 0.8m/s without collision with surrounding obstacles. (2) The performance with regard to local environmental recognition along the demonstrated route is good. (3) The sampling rate is high enough to generate a lookup-table with fine resolution by way of the settings for the navigation learning task. (4) The navigation results are very similar for the two trials of autonomous navigation, even under the stochastic, sequential effects of the dynamic parameters in the robot hardware. (5) Compared with the human demonstrated control command trajectory, the control command trajectories in the autonomous navigation produce little noise. For localization at the demonstrated configurations (local environments), an explicit lookup-table is constructed with raw sensor patterns and human assigned position/orientation outputs. Experimental results demonstrate the following. (1) The errors in lookup-table output x, y, and orientation are bounded around ±25cm and ±20 degrees in most cases. (2) The sampling rate is high enough to generate a lookup-table with fine resolution describing the local features along the route. (3) The ultrasonic range sensing has high precision in sampling the same value in the same robot configuration in a workspace. (4) There are some locations at which the sampled sensor patterns are very similar in Euclidean distance. (5) Nearby configurations have similar sensor patterns that describe similar local environmental features.

9

Introduction to Human Action Skill Modeling

9.1 Learning action models from human demonstrations

This part of the book is a study of methods for learning action skills from human demonstrations. The goal of action learning is to characterize the s-tate space or action space explored during typical human performances of a given task. The action models that are learned from studying human performances typically represent some form of prototypical performance, and also characterize the ways that the human's performance of the action tend to vary over multiple performances due to external influences or stochastic variation. Action models can be used for gesture recognition, for creating realistic computer animations of human motion, for detailed study of an expert performer's motion, for building feed-forward signals for complex control systems around which custom feed-back controllers can be designed, and for evaluating the naturalness of the performances generated in real and simulated systems by reaction-skill models.

Although a great deal of work in robotics has gone into the study of methods for modeling human *reaction* skills, much less work has so far been done on the important related problem of modeling human *action* skills. Action learning is the characterization of open-loop control signals into a system, or the characterization of the output of either an open-loop or closed-loop system. While reaction learning generates a mapping from an input space or state space to a separate action/output space, action learning characterizes the state space or action space explored during a performance. Given data collected from multiple demonstrations of task performance by a human teacher, where the state of the performer or the task-state is sampled over time during each performance, the goal is to extract from the recorded data succinct models of those aspects of the recorded performances most responsible for the successful completion of the task. Reaction learning focuses on the muscle control a dancer uses to move his or her body, for example, while action learning studies the resulting dance. The methods presented here for action learning are based on techniques for reducing the dimensionality of data sets while preserving as much useful information as possible.

The original motivation for this work is for applications involving skill transfer from humans to robots by human demonstration. Instead of explicitly programming a robot to perform a given task, the object is to learn from example human performances a model of the human action skill which a robot may use to perform like its teacher. Consider a remote teleoperation scenario where a human expert on Earth is guiding the operation of a distant space robot performing a complex manipulation task. Instead of attempting closed-loop control over a communication link with a latency of several seconds, it would be more useful to use pre-built models of the human's low-level action skills. The models could be used locally at the teleoperation station to recognize and parameterize the operator's individual low-level actions as he or she controls a virtual-reality robot model in performance of the task. Concisely parameterized symbolic descriptions of the operator's actions can be sent to the remote robot as they are recognized. Provided that we have already transferred the skill models to the robot, it can then execute under local real-time supervision the actions desired by the operator.

Outside such a scenario, the methods presented in this book have a more general set of applications. Models of human action can be used for animation in video games, in movies, or for human-like agents for human-computer interaction (HCI); for analysis of what makes an expert's performances of a given task more successful than others; and for prescriptive analysis of how a novice might make their performances more like an expert's (e.g., how can I make my backhand more like that of Pete Sampras). Because this part of the book focuses on techniques for extracting human skill models and has not yet been extended to robotic performance of the tasks characterized by the models, the results here are more immediately useful for gesture recognition and these more analytical applications. The connection between the learned skills and robotic performance is the next logical step for study in this research effort.

What does action learning have to do with dimension reduction? Think about the process of reaching across a table to pick up a coffee mug. When we perform this task, it seems no more complicated than its description: reach toward the mug, grasp the handle, and then lift. We can accomplish this with hardly a conscious thought. To transfer this skill to a robot by the typical process of human to robot skill transfer, however, we need to record our motions in detail. When we instrument our hand and arms to record our motions as we perform the task, the resulting raw data are very different in quantity and quality from our simple description. To record the motion of the palm of our hand, for example, we need to sample three dimensions of position data and three dimensions of orientation information. Recording the motions of our finger joints and wrist typically requires about twenty more channels of information, and several additional channels could be collected for the other joints in our arm. To fully capture the motion of our arm and hand, we would normally sample all these variables many times a second, and write these data to a computer file. The result of recording the motion is a large quantity of high-dimensional data, where the overall structure of the task and the relative purpose and importance of each motion is obscured.

Once we have collected these data, we are left with the question of how to extract from them a useful model of the underlying human action skills. In this research we assume a two step approach: (a) high-level analysis of the overall structure of the task to break it into simpler component motions, and (b) learning a typical motion for each low-level component motion, as well as the most significant ways each such motion is most likely to vary from the prototypical motion.

The analysis step isolates those portions of performances which are similar in purpose and execution, and from which it should be easiest to build good models of typical motions and their variations. If we can build good models, we should be able to relate their basic features to their purpose within the structure of the overall task. We may also be able to learn from the low-level models something about the complexity of the skills underlying their performance.

A particular performance can be compared to a given model for gesture recognition, resulting in some measure of how likely it is that the performance belongs to the corresponding class of motions. If the performance does belong to that class, a good action model will let us concisely describe the most important ways that the performance differs from the prototypical motion. This concise description is a reduced-dimension representation of the performance.

The high-level task analysis, the first step of the action-learning process described above, is a simple process in most cases. Because we generally have a solid high-level understanding of how we perform tasks, we can use this knowledge to focus on the low-level action skills which we really want to model using the collected data. In our example, we know that our strategy for picking-up the mug is roughly reach, then grasp, then lift. Instead of building a single model of the entire task, we can instead write a high-level program which calls the low-level tasks "reach," "grasp," and "lift" as subroutines, and then focus on learning these low-level skills from the collected performance data.

Once we have performed the high-level analysis, we can focus on learning models for individual low-level action skills. Each low-level action will have an associated set of training data which is a subset of the overall performance data. The goal is to build a description of the action skill which is as simple and straightforward as the action itself, in a form which can adequately explain the associated high-dimensional training data. For example, although our raw data from the grasping of the mug handle has on the order of 30 data channels indicating the state over time of the many joints in the hand and arm, the hand configurations actually observed during the performances can be represented using only a few task-specific parameters. This is because some parts of the motion will be highly consistent for each grasping motion: we normally grasp the handle with our palm vertical, our thumb near the top of our hand, our wrist firm, and we usually curl all the joints in our fingers in unison as we grasp the handle. These consistencies in the motion of the hand can be modeled as linear or non-linear constraints between the different dimensions of the raw-performance data, and they effectively restrict the intrinsic dimensionality of our grasping motion. Once we have accounted for and represented these constraints, we can accurately describe the position of the hand with a simple parameterization such as the position of our fingertips as they approach the handle, how spread-apart they are, and how closed around the handle. Such a description is similar to our own understandings of the motion.

Given a library of low-level task descriptions of this type, parameterized representations of the typical configurations for each given motion, we can specify a performance of an action by supplying the name of a skill model and a set of parameters. There should be few parameters for adjusting a simple skill, and more parameters for adjusting a skill that is more complex. For the grasp, we might want to specify "use the grasp skill, and keep the fingers close together." In building a model of a low-level skill, then, we need to be able to do several things: (a) we need to be able to isolate the *intrinsic dimensionality* of the skill, meaning the number of parameters we need to adequately describe a given performance with respect to a particular skill model, (b) we need to be able to map a high-dimensional raw state description of a given performance to a description with this intrinsic dimensionality, and (c) we need to be able to map a lower-dimensional description back to a corresponding state description in the space of the raw data. The next section discusses how these tasks are formulated in this research. In this work, we address (b) and (c) for parametric and nonparametric models, and do some work on addressing (a) for parametric models. Although some potential approaches for addressing the determination of intrinsic dimensionality in nonparametric models will be discussed, the full investigation of this problem is still an issue for future research.

9.2 Formulation of the dimension reduction problem

This book deals with the problem of finding a low-dimensional representation for high-dimensional data sets collected from multiple human performances of a given task. Such performances consist both of some motions which are essential for successful task completion and some which are inessential. Because all the data correspond to successful performances of the task, the essential motions will always be present, and will probably be more consistent and less stochastic in nature than the inessential motions. In our example from the previous section, when we reach across a table to grasp the coffee mug, the essential aspects of the performance are the end point of our hand's trajectory toward the handle of the mug, the orientation and configuration of our fin-

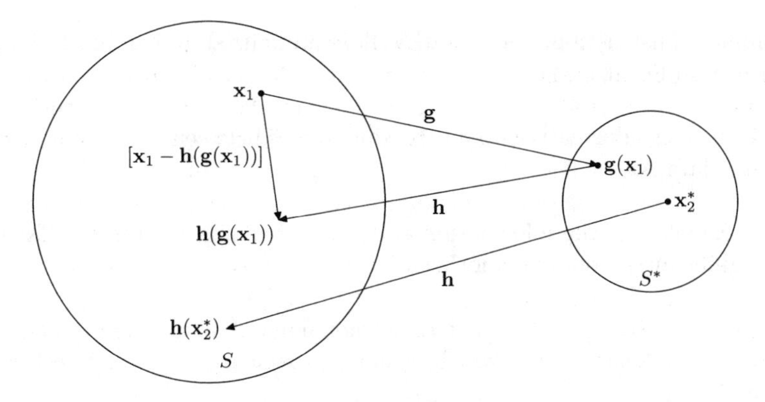

FIGURE 9.1: Mappings between raw data space S and feature space S^*.

gers when we grasp the handle, and the force closure of our final grasp. The inessential aspects are the particular path by which our hand moves to the handle, the orientation and configuration of our fingers in the earlier parts of our reach, the overall speed of the motion, and the small jitters in our hands and fingers over the course of the motion.

To assist in the extraction of the essential aspects of the performance, we will first need to isolate the low-level skills most suitable for learning, and we need to supply training data which we believe most directly shows the important aspects of the performance. In this case, the low-level skills may be a particular kind of reach and a specific kind of grasp, and the learning algorithm may be told to look at the motion of the hand relative to the mug rather than to a global frame of reference.

We formulate the extraction of the essential aspects of the performance as a dimension-reduction problem. Multiple human performances of a given task are recorded by sampling the performance state over time. Each resulting sample is an m-dimensional vector, where m may be a fairly high number (e.g., a Cyberglove records roughly 20 channels of data about the configuration of the finger joints in the hand, and tracking position and orientation of the hand requires six more channels). Given $\mathbf{X}_{(m \times n)}$, the training data matrix containing some or all of the samples from the example performances, the dimension reduction process generates a mapping $\mathbf{g} : \mathbb{R}^m \rightarrow \mathbb{R}^p$ $(p < m)$ to a p-dimensional space which preserves as much information as possible about each point, and a second mapping $\mathbf{h} : \mathbb{R}^p \rightarrow \mathbb{R}^m$ which maps feature points back to the original. We will call the input space $S \subseteq \mathbb{R}^m$ *raw data space*, and space $S^* \subseteq \mathbb{R}^p$ *feature space*. These spaces and the mappings between them are summarized in Figure 9.1. We optimize $\mathbf{g}$ and $\mathbf{h}$ such that mapping training data $\mathbf{X}$ into and back from the p-dimensional feature space results in a faithful reconstruction: $\|\mathbf{X} - \mathbf{h}(\mathbf{g}(\mathbf{X}))\|$ is minimized over $\mathbf{g}$ and $\mathbf{h}$.*

After this optimization,

- $\mathbf{g}$ encodes information about the similarities between the points in training data $\mathbf{X}$,

- $\mathbf{g}(\mathbf{X})$ encodes the information about how training vectors $\mathbf{x}_i \in \mathbf{X}$ differ significantly from one another,

- the image $\mathbf{h}(S^*) \subset S$ is the p-dimensional manifold in the m-dimensional raw data space within which or close to which the training vectors lie, and

- $[\mathbf{X} - \mathbf{h}(\mathbf{g}(\mathbf{X}))]$ is the set of residual vectors by which the low-level representation fails to model the high-level data.

If the norms of the residual (error) vectors are generally small, then these vectors likely encode inessential random motions in the task demonstrations. If the norm of the error vector for a given point is much higher than that of most, then it is not typical of the training data set. Moreover, if the norms of most error vectors are high, then either $\mathbf{g}$ and $\mathbf{h}$ are significantly sub-optimal, or the specified dimensionality for the feature space is lower than the intrinsic dimensionality (p) of the training set.

In addition to providing us with a great deal of information about the nature of the sub-task being analyzed, the method of dimension reduction provides a low-dimensional feature space in which to plan robotic performances of the sub-task. If we are able to generate mappings of sufficient quality, then performances planned within this feature space may closely resemble human performances.

An important special case of dimension reduction is modeling using a one-dimensional parameterization. We will discuss in Section 11.2 that the best one-dimensional parameterization of an action skill is a variable describing a temporal ordering of points. The process of building such a parameterization is *trajectory fitting*. If $\mathbf{g}$ and $\mathbf{h}$ are smooth mappings to and from a one-dimensional feature space, then the result of smoothly varying that single parameter over time from a starting value to a final value, and projecting that value into the raw performance space using mapping $\mathbf{h}$, can represent a "best-fit" model of the action. Another way to describe the fitted trajectory is "the one-dimensional model most likely to have generated training data $\mathbf{X}$."

The remainder of this chapter will review research related to this second part of the book.

*When $\mathbf{g}$ and $\mathbf{h}$ are written with matrix arguments, they operate on each individual column-vector of the matrix.

9.3 Related research

This section relates the work presented in the second part of the book to several areas of the robotics and machine learning literature. First, we describe the relationship between the high-level analysis described in Section 9.1 and current work on task-level recognition of human actions for learning assembly plans and for building symbolic descriptions of human motions. Next we survey some the current work on skill-based robot programming, and describe how the conceptions of skills or primitives in this work relate to the low-level action models used in this book. Because action modeling is formulated as a dimension reduction problem in this book, we review current work on dimension reduction, particularly those methods based on local data models. Finally, we describe how potential uses of action models relate to current applications of reaction models, and note a few other applications which could potentially benefit through the use of action models.

Primitives and skills

The robotics literature often uses the term "skill" or "primitive" to describe a unit of functionality by which robots achieve an individual task-level objective. A "skill" is often defined as an ability to use knowledge effectively and readily, and the effectiveness of a skill for attaining a given desired result is its primary attribute [231]. The development of high-level strategies as combinations of these units is often referred to as "skill-based programming." Examples of this approach are often found in the area of dextrous manipulation, such as the work of Michelman and Allen [152], and Nagatani and Yuta [164]. Skills or primitives are also useful for encapsulating expert knowledge and for the software engineering purpose of making this expertise simple to interface into working systems by task-level programmers, as demonstrated by Morrow, et al. [160], [161] and Archibald and Petriu [9]. In the robotics literature as a whole, they are used as symbolic units of description for computer representations of task-level plans, and for mapping to human-language descriptions (e.g., "grasping," "placing," "move-to-contact") for human understanding, communication, and reasoning. In the manipulation domain, these primitives are often further defined as sensorimotor units for eliminating motion error due to modeling uncertainty in the task environment.

In contrast to this work in the manipulation domain, this book will focus on skills or primitives (the terms "low-level actions" or "component motions" will also be used) as descriptions of actions instead of as a mapping from sensed task state to actions. Thus, we will be characterizing classes of motions or subspaces of action space which correspond to hand movement when reaching, finger movement when grasping, or body movement when walking. This is more closely related to work like that of Yang, et al. [258], which uses a hidden Markov model representation to build a model of a "most-likely performance" from multiple example performances.

Dimension reduction and local learning

As discussed in Section 9.2 our approach to action modeling is to build mappings to and from reduced-dimension representations of human action data. Most work on skill learning deals with reaction learning (i.e., control) rather than action learning. One notable exception is the work of Bregler and Omohundro [34]. They present a technique for representing constraint surfaces within high-dimensional spaces. Points are sampled from the constraint surface, which is a low-dimensional subspace of the possible space of points. K-means clustering is used to generate a representative set of cluster centers for these sampled points. In the region containing each cluster, local principal component analysis (PCA) is used to linearly approximate the constraint surface and determine its local dimensionality, and the constraint-surface approximation at a given point is a blended approximation formed from the local PCA models of the nearest cluster centers. Expectation-maximization is used to refine the global model defined by the combination of all the local models. Bregler and Omohundro use this method in a speech recognition application to model a speaker's lip motions. A "snake" representation is used to model the contour of the speaker's lips. This representation consists of a chain of 40 points which track the edges of the speaker's lips within a video image, and the image coordinates of these 40 points can be represented as a single point in an 80-dimensional space. The model of the lips' motion for a given word is the surface within this huge space within which the snake model is constrained while the word is spoken. Although this method for building a reduced-dimension representation of high-dimensional data from a set of distinct local models seems fundamentally different from the methods used later in this book such as principal curves, similarities between the blended PCA models and the formulation of the nonparametric smoothers used within the principal curves algorithm actually form a strong relationship between them. Tibshirani's formulation of the principal curve [238] and Tipping and Bishop's mixtures of probabilistic principal component analysis [239] provide an interesting connection between such mixture models and the theory presented in Chapter 11.

Zhang and Knoll [261] present an interesting reaction-learning method based on a global parametric model of a high-dimensional input signal. They present a visual serving application whereby an eigenspace representation is used to reduce the dimensionality of the image used to control a robot. This reduced-dimension representation is input to a fuzzy rule-based system based on B-splines, which is trained to control the robot for performing simple manipulation tasks.

Chapter 10 of this book presents global parametric methods for action learning, including principal component analysis [111] and Kramer's nonlinear principal component analysis [124], while Chapters 11 and 12 focus on local, nonparametric methods. In the comparison of the two, it will be demonstrated that the nonparametric methods have many practical and theoretical advantages. Schaal [206], in the context of learning nonlinear transformations for control, comes to similar conclusions. He is concerned with finding learning methods which can enable autonomous systems, particularly humanoid robots [205], to develop complex motor skills in a manner similar to humans and other animals. He concludes that nonparametric models are more appropriate than parametric models when the functional form of the mapping to be learned is not known *a priori*, though the nonparametric methods can be more computationally expensive on non-parallel hardware.

Atkeson, et al. [3] survey the literature in locally weighted learning, and in the process present a good summary of the available methods for creating local models, including local regression and scatter-plot smoothers. In [4] they survey the use of these locally weighted learning techniques for control: mapping task-state information to an appropriate action. Schaal, et al. [209] compare several of these local modeling technique, using Monte Carlo simulations, and find that locally weighted partial least squares regression [68], [251] gives the best average results.

One concern mentioned in the locally-weighted learning surveys [3], [4] is the problem of high-dimensional input spaces. It is simply infeasible to collect enough training data to fully characterize all regions of high-dimensional space, and it is difficult to blend different local models together because all points begin to look equidistant in these spaces [212]. Nevertheless, Vijayakumar and Schaal [243] present a method called Locally Adaptive Subspace Regression (LASS) for learning mappings in these high-dimensional spaces. Their approach is based on empirical observations that "despite being globally high dimensional and sparse, data distributions from physical movement systems are locally low-dimensional and dense." LASS learns functional mappings in high-dimensional spaces by using locally-weighted principal-component analysis to reduce the dimension of the local model before attempting to fit the model. Its input space is a combination of local PCA patches which looks similar in some ways to the constraint surfaces of Bregler and Omohundro [34].

An important problem for building reduced-dimensional representations of data sets is choosing the dimensionality of the model. Minka [154] gives one approach to doing this for models based upon principal component analysis (PCA). His work uses the interpretation, from Tipping and Bishop [239], of PCA as a maximum-likelihood density estimation.

Uses for action models

Although reaction learning is a popular strategy for abstracting primitives from human demonstration data for use in dextrous manipulation, for example with neural networks [118], [12] and hidden Markov models [99], action learning also can build models useful for robot execution [258]. When combined with a system for recognizing a human operator's actions in a virtual environment such as that described by Ogata and Takahashi [175], such models can be used for task-level teleoperation of a remote robot (e.g., [97]). Some motions such as the brush stroke of a painter or the pen stroke of a calligrapher may be very subtle and appropriate for learning by demonstration, but may however be primarily feed-forward in nature (possibly with applied force as one dimension of the action state) and best modeled using action learning rather than reaction learning. For these motions, we probably want to learn not only the typical stroke of the painter or calligrapher, but also the ways which the stroke may vary yet remain typical of the artist. This could allow for appropriate variation in the execution of the skill to make the result look less "robotic."

Action models are also useful for applications in which skills are transferred from a human expert to a student. Nechyba and Xu [167] demonstrate a system whereby an expert's reaction skill, modeled by a cascade neural network, is used to guide a student's learning of a difficult inverted-pendulum stabilization task. They also present a stochastic similarity measure [170] for quantifying the similarity between reaction-task performances, which is useful for evaluating skill transfer experiments. The methods developed in this book are particularly applicable to action skill transfer due to their ability to *analyze* the motions of human experts, and their ability to *describe* an expert's nominal performance and likely modes of variation for comparison with student performances. For human motions such as swinging a golf club or throwing a baseball, a student's action can be compared to an expert's motion trajectory in space to form prescriptive suggestions, so an action model built around a best-fit trajectory will have important advantages over a black-box input/output model such as a neural network.

Computer animation is one of the most promising applications for action models. Instead of simply recording the motion of a human actor for later playback with an animated character, action learning can be used to characterize the full range of motions which are typical of that person's performances. These models could be used to create animations of the corresponding motions (e.g., walking) which are stochastically varied in a "human" manner to give them a more natural look. For research in physics-based animation of human models, action models could be used for evaluating different control methods. For example, Matarić, et al. [149] evaluate various methods for controlling the movements of a humanoid torso simulation in performance of the macarena. Their evaluations are based on qualitative and quantitative measures of "naturalness of motion," where their quantitative analysis is based primarily on a measure of end-effector jerk. A distance metric from an appropriate action model learned from human performances could be used to build an improved criterion for the naturalness of these simulated motions.

10

Global Parametric Methods for Dimension Reduction

10.1 Introduction

In this chapter we look at some methods for dimension-reduction which are not specific to characterizing human performance data sets, but which may however be useful for this purpose. These methods do not look at the human performance data in terms of multiple example trajectories in time, nor do they look directly at the local relationships between example points from similar parts of a given performance or the state space. They rather look at all the points from all the training examples as a single set of vectors in raw data space $\mathbf{X}_{(m \times n)} = [\mathbf{x}_0|\mathbf{x}_1|\ldots\mathbf{x}_{n-1}]$ and simply try to map them to a lower-dimensional feature space and back again while preserving as much useful information as possible. Since these methods do not look at the local structure of task performances, we call them *global* methods for dimension reduction.

In Section 10.2 we will review the general topic of parametric methods for global modeling, and in the rest of the chapter we will discuss global parametric modeling of human performance data. We will present three global methods for this purpose: principal component analysis (PCA), non-linear principal component analysis (NLPCA), and a variation on NLPCA called sequential non-linear principal component analysis (SNLPCA). Given training-data $\mathbf{X}$, these methods develop mappings $\mathbf{g} : \mathbb{R}^m \to \mathbb{R}^p$ to feature space and $\mathbf{h} : \mathbb{R}^p \to \mathbb{R}^m$ back to raw-data space. PCA generates linear mappings. The forms of the mappings resulting from both NLPCA and SNLPCA are non-linear, but they are found in slightly different ways. To demonstrate and compare these methods, we use an example human performance data set described in Section 10.3.

10.2 Parametric methods for global modeling

Reducing the dimensionality of a data set can be thought of as the process of finding correlations between its different variables or dimensions. Strong correlations between dimensions of the data may be used as constraint equations to explicitly reduce the dimensionality of their representation.

The tool boxes of statisticians are full of methods for uncovering these relationships between two or more variables in a given set of data. These methods generally assume that there is some underlying structural relationship between the variables, and also some random error or variation in one or more of the variables. If we have a data set of n observations of two variables x and y, we might first make a scatter plot of the points (x, y) to get a rough idea of how one relates to the other. Figure 10.1(a) on page 172 shows an example scatter plot. Such plots are more difficult to use for higher-dimensional data sets, but we can still learn a great deal about high-dimensional data by plotting two- and sometimes three-dimensional projections.

The most familiar methods for modeling the relationship between the different parts of a data set are parametric methods. These result in parametric models—equations relating one set of variables to another, with a set of parameters which can be adjusted to fit a given dataset. Adjusting one of the parameters of these models tends to change the model over the entire domain, and the effect is generally measured against a single scalar value indicating the goodness-of-fit over the entire set of training data. In this section we review several well-known global parametric methods, then in the rest of the chapter we will describe how similar methods may be used for modeling human performance data.

10.2.1 Polynomial regression

The class of polynomial functions is a commonly-used set of parametric models. They are most useful when we can separate the dimensions of the data we are modeling into a single *explanation* variable x and a set of *response* variables $\mathbf{y}$. We will assume a single response variable y in this discussion. All the error is assumed to appear in the response variable, while the explanation variable is assumed to be error-free.

We generally assume that the data comes from a process with an underlying systematic relation between the variables, and a separate source of random error. We model the systematic relation using a p-th order polynomial, and the random error with variable ϵ:

$$y = \sum_{j=0}^{p-1} c_j x^j + \epsilon. \tag{10.1}$$

If we assume that error ϵ comes from a Gaussian distribution with an expected value of zero, we can estimate the coefficients $\mathbf{c}^*$ of the polynomial by minimizing the sum of squared error

$$S_p = \sum_{i=0}^{n-1} \left(y_i - \sum_{j=0}^{p-1} c_j (x_i)^j \right)^2 . \tag{10.2}$$

The estimated coefficients of the polynomial $\mathbf{c}$ can be found by solving the linear system

$$\mathbf{M}_{(p \times p)} \mathbf{c}_{(p)} = \mathbf{m}_{(p)}, \tag{10.3}$$

where $\mathbf{M}$ is a symmetric matrix with terms $M_{ij} = \sum_{k=0}^{n-1} (x_k)^{i+j+1}$, and $\mathbf{m}$ is the vector such that $m_i = \sum_{k=0}^{n-1} y_k (x_k)^i$ (although this is not the most efficient solution).

The polynomial most often used for fitting data is a straight line. Figure 10.1(b) shows the result of performing linear regression (i.e., polynomial regression with $p = 2$) on the dataset in Figure 10.1(a). Because we assume that all the error is in the response variable, the error vectors, which connect each data point to its projection on the model line, are vertical. Figure 10.1(c) shows the result of a polynomial curve fit with $p = 3$, a parabolic curve fit. Again, the error vectors are vertical.

Since we have an explicit parameterization (or explanation) variable, polynomial regression of a dataset with a multi-dimensional space of response variables is equivalent to performing regression separately on each individual response variable.

10.2.2 First principal component

Sometimes we want to find the best linear model for a dataset without first choosing an explanation variable, or we might want to do a linear regression assuming that the variance of the error in the explanation variable is the same as that for the other variables. The first principal component is the proper model for these cases. If we assume for multi-dimensional dataset $\mathbf{Y} = \{\mathbf{y}_i\}$ the model

$$\mathbf{y}_i = \mathbf{u}_0 + \mathbf{a}\lambda_i + \epsilon_i, \tag{10.4}$$

and that covarience(ϵ_i) $= \sigma^2 \mathbf{I}$, then the least-squares estimate of slope vector $\mathbf{a}$ is the first principal component. Figure 10.1(d) shows the first principal component for the example dataset. Note that the error vectors are orthogonal to the model line. This is due to the fact that error is being minimized in both dimensions simultaneously. The error vectors are parallel to the second principal component.

The parameterization variable λ for the principal component is not part of the original problem formulation—it is newly introduced by the modeling equation (10.4). It can be considered a "discovered" (or "latent") explanation

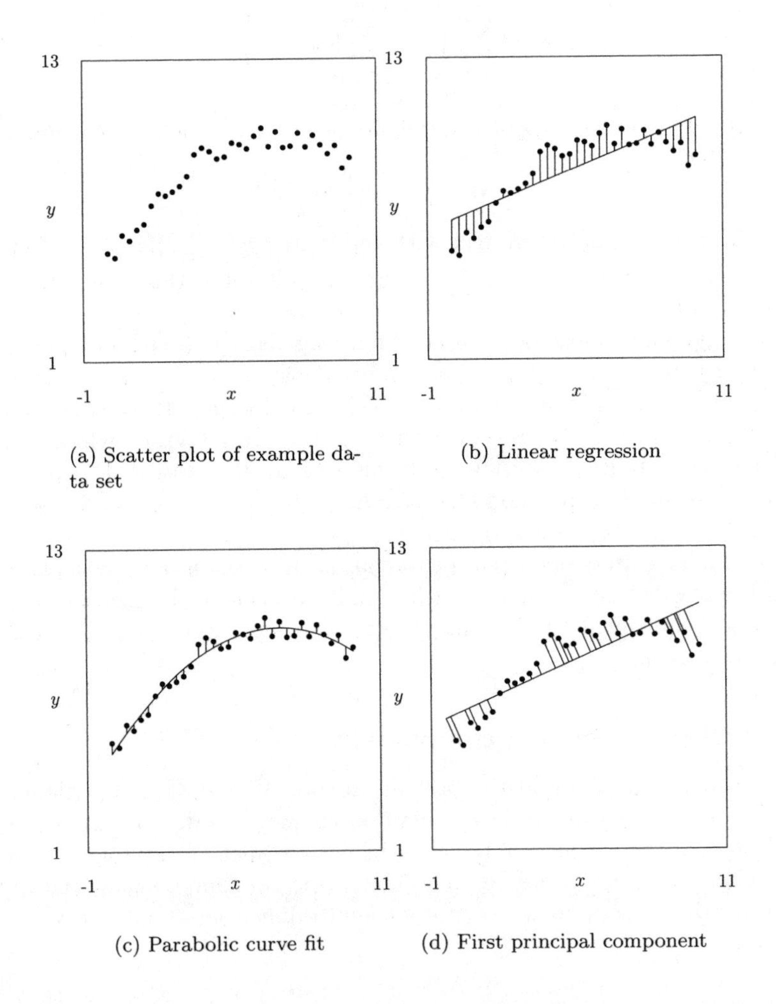

(a) Scatter plot of example data set

(b) Linear regression

(c) Parabolic curve fit

(d) First principal component

FIGURE 10.1: Various parametric models of an example dataset. Each model minimizes the sum of squared errors, where the errors are length of the vector between data point and its projection onto the model.

variable. If we rotate Figure 10.1(d) slightly clockwise to make the first principal component line horizontal, then the error bars will be vertical like in the regression plots. The regression against explanation variable λ in the rotated plot results in the horizontal-line model, a rather boring result because the principal component already explained-away the correlation.

Principal component analysis (PCA), also known as the Karhunen-Loève transform [111], is a well-understood and commonly used method which can be performed on data sets of arbitrary dimension. Because it does not need an *a priori* explanation variable to generate its models, because it is computationally inexpensive, and because it generates a linear model which is easy to understand and interpret, it is a useful first modeling technique to apply to human performance data. We will describe its use for this application in Section 10.4.

10.3 An experimental data set

In Chapter 9 we described how the process of human to robot skill transfer typically involves recording several individual human task performances by sampling performance state over time using signals from various input devices. Some example input devices include a Cyberglove or other instrumented glove, a Polhemus or other 6-DOF tracker, a joystick, a mouse, a haptic interface, or other measuring devices such as visual trackers, instrumented cockpits, etcetera. Although there may be a large number of dimensions in the resulting representations of recorded performances, the actual intrinsic dimensionality of the underlying human skill is often much lower. In this section, we present an example set of recorded human task performances whose raw-data space has a large number of dimensions, but which should be represented more appropriately using only a few independent parameters. This data is collected from the grasping phase of a ball-catching task which was demonstrated in a simple virtual environment.

A human subject performed a number of catches of a virtual ball using the system outlined in Figure 10.2. The input device was a Virtual Technologies Cyberglove which measures 18 joint angles in the fingers and wrist of its wearer, with a Polhemus sensor which returns the overall position and orientation of the wearer's hand. The feedback to the user was a rendering of the hand and ball from several perspectives, on the graphical display monitor of an SGI workstation (as shown in Figure 10.3). The recorded data was interpolated and resampled evenly at 10Hz, and stored in a performance database for later retrieval and analysis.

We manually segmented the performances into approach and grasp phases, and subjectively graded each recorded grasp on a scale from 0-9. We selected

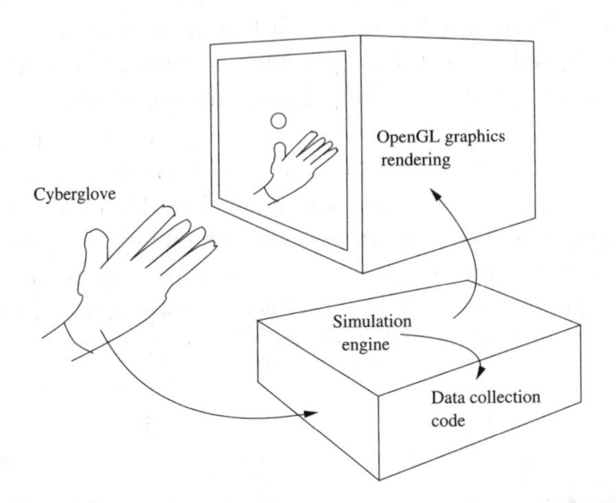

FIGURE 10.2: "Virtual reality" skill demonstration system.

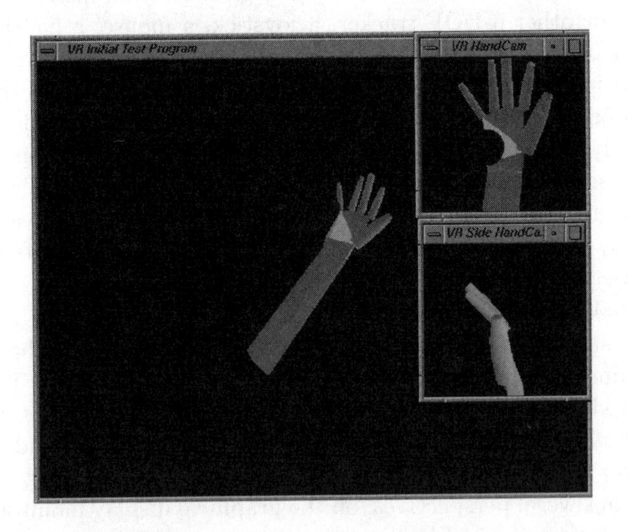

FIGURE 10.3: Graphical feedback.

the vectors representing the joint angles in the fingers and wrist during the grasp phase of all catches for which the grasp satisfied a minimum grade. This generated a data set of 461 vectors, each representing a hand configuration of 18 joint-angles during grasping motions. These vectors were randomly allocated into a training set of 155 points, a cross-validation set of 153 points, and a test set of 153 points.

10.4 Principal component analysis for modeling human performance data

As mentioned in Section 10.2.2, principal component analysis (PCA) [111] is a well-understood and useful method for modeling data sets. When applied to a set of multidimensional vectors, it finds a linear mapping between them and each lower-dimensional space such that when the vectors are mapped to a lower-dimensional space and then mapped back to the original space, the sum-of-squared error of the reconstructed vectors is minimized. In the compressed representation of a vector, we can consider each dimension a separate *feature*, and the value of that dimension of the feature-vector a *feature-score*.

To perform principal component analysis of a set of n m-dimensional zero-normed vectors $\mathbf{X}_{(m \times n)} = [\mathbf{x}_0|\mathbf{x}_2|\ldots|\mathbf{x}_{n-1}]$, we find the eigenvalues λ_i and eigenvectors $\mathbf{v}_i$ of the symmetric matrix $\mathbf{X}\mathbf{X}^T$. To generate a p-dimensional feature-vector representation of a m-dimensional vector* $\mathbf{y}$ (where $p < m$), we form a $m \times p$ matrix $\mathbf{V}_p = [\mathbf{v}_0|\ldots|\mathbf{v}_{(p-1)}]$ where $\mathbf{v}_0 \ldots \mathbf{v}_{(p-1)}$ are the eigenvectors corresponding to the p largest eigenvalues. Then the *feature-vector* $\mathbf{y}^*$ is

$$\mathbf{y}^* = \mathbf{g}(\mathbf{y}) = \mathbf{V}_p^T \mathbf{y}, \tag{10.5}$$

and the reconstructed approximation $\tilde{\mathbf{y}}$ of the original vector $\mathbf{y}$ from feature-vector $\mathbf{y}^*$ is

$$\tilde{\mathbf{y}} = \mathbf{h}(\mathbf{y}^*) = \mathbf{V}_p \mathbf{y}^* = \mathbf{V}_p \mathbf{V}_p^T \mathbf{y}. \tag{10.6}$$

The value of the i-th element of the feature vector $\mathbf{y}^*$ is the magnitude of the projection of $\mathbf{y}$ upon the corresponding eigenvector $\mathbf{v}_i$. Because eigenvalue λ_i is the variance of training data in the direction given by the eigenvector $\mathbf{v}_i$ [30], the eigenvalue indicates relative significance of feature score y_i^* for representing vectors from a distribution similar to the training data.

Performing this analysis on the data set from Section 10.3 gives us a set of eigenvectors which attempt to explain the data set in terms of linear dependencies between its dimensions. The ability of the first five eigenvalues to

*In this chapter, we will use $\mathbf{x}$ to refer to a vector which is in the training-set, and $\mathbf{y}$ to refer to a vector which is not in the training-set.

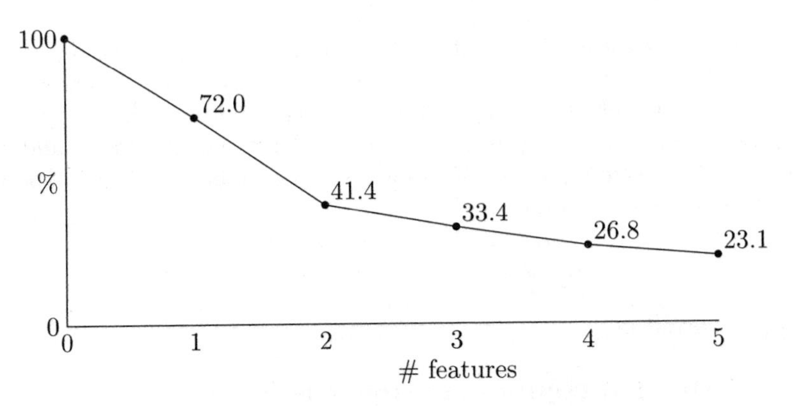

FIGURE 10.4: Error unexplained by p features of PCA.

represent the data set is summarized in Figure 10.4. The values shown are the relative magnitudes of the residuals,

$$\%\text{-Err}_p = \frac{\|\mathbf{Y} - \mathbf{V}_p \mathbf{V}_p^T \mathbf{Y}\|}{\|\mathbf{Y}\|}, \tag{10.7}$$

where $\mathbf{Y}$ is the set of testing vectors from the experiment (independent from the set $\mathbf{X}$ used to generate the eigenvectors), and $\|\cdot\|$ is the Frobenius (L_2) norm. Figure 10.4 clearly shows that the first two linear features are nearly equal in importance for explaining the configuration in the test data, while the remaining features are much less significant.

Figure 10.5 provides a graphical illustration of the effects of these first and second linear principal components, which we might call "eigenhands."† These hand configurations are generated by varying the components y_0^* and y_1^* of the feature vector, and then mapping these features to $\tilde{\mathbf{y}}$ using an interface like the one shown in Figure 10.10 on page 183. We are able to use the slider bars in this interface to look at the separate effects of the two features y_0^* and y_1^* because PCA learns a linear mapping, and thus the effects of the features are independent and additive:

$$\tilde{\mathbf{y}} = \mathbf{v}_0 y_0^* + \mathbf{v}_1 y_1^*. \tag{10.8}$$

While this linearity makes the resulting model easy to analyze and interpret, it also limits the generality of the model.

We can see that the main effect of the first principal component is to bring the fingers closer together and to bend the fingers at the knuckles (mcp), while the effect of the second principal component is to curl the fingers at the

†The thumb positions in these diagrams are slightly erroneous due to a sensor which was not working during the experiments.

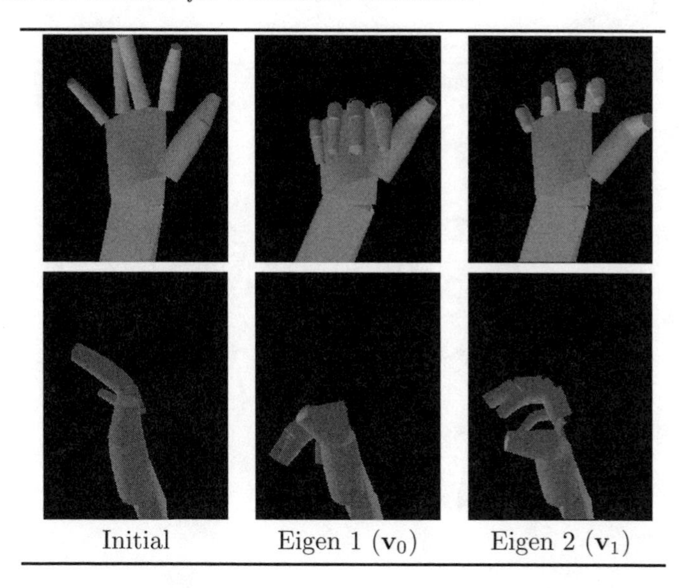

FIGURE 10.5: PCA analysis of grasp gesture. Initial position, and first and second principal "eigenhands," front and side views.

second (pip) and third (dip) joints. Moreover, we see that although the grasp positions of the hand generated from the first and second eigenvectors look plausible, the best attempt to create an adequate initial (open) position for the grasp using only the first principal component looks less plausible. Closer inspection of this configuration reveals that the pinky and index fingers overlap in space, and that it is thus physically impossible. The problem is the linear nature of the mapping. The first principal component reduces the average sum-of-squares error for all joint-angles during the grasps by fitting a straight line in the space of training configurations, but if the general trend of the performances in state space is not linear, then this principal component will fit the trend of the performances poorly at some stages. Nevertheless, we see from Figure 10.4 that PCA allows much of the 18-dimensional data set to be explained by only a few linear feature values.

10.5 NLPCA

Nonlinear principal component analysis (NLPCA) also attempts to find mappings between a multidimensional data set and a lower-dimensional feature-space while minimizing reconstruction error, but allows the mappings to be

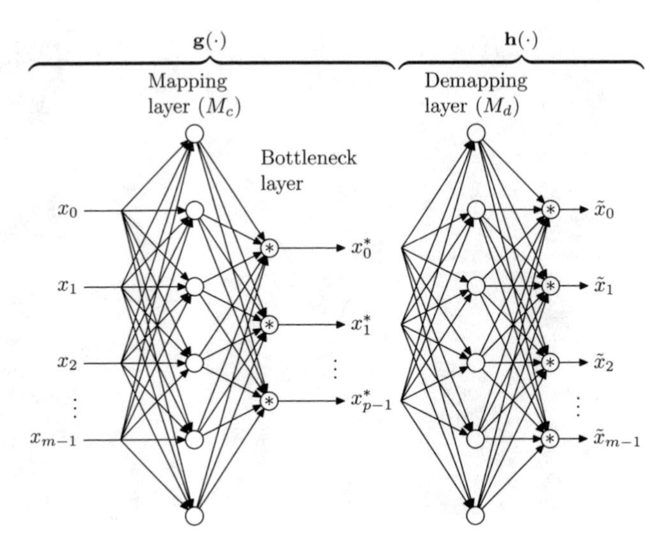

FIGURE 10.6: Neural network architecture for NLPCA. ○ indicates a sigmoidal unit, and ⊛ indicates a unit which may be either linear or sigmoidal.

nonlinear. In contrast to linear mappings (10.5) and (10.6), the nonlinear mappings are of the general form

$$\mathbf{y}^* = \mathbf{g}(\mathbf{y}) \tag{10.9}$$

$$\tilde{\mathbf{y}} = \mathbf{h}(\mathbf{y}^*) = \mathbf{h}(\mathbf{g}(\mathbf{y})) = \mathbf{n}(\mathbf{y}). \tag{10.10}$$

If the lower-intrinsic dimensionality of a data set arises from a nonlinear relationship between the different dimensions of the data set, a nonlinear principal component analysis is capable of better representing the original data set with a reduced-dimension representation than would a linear principal component analysis. Several methods proposed for performing NLPCA include the use of autoassociative neural networks, as described by Kramer [124]; principal curves analysis, as described by Hastie and Stuetzle [90], and Dong and McAvoy [56]; adaptive principal surfaces, as described by LeBlanc and Tibshirani [128]; and optimizing neural network inputs, as presented by Tan and Mavrovouniotis [234]. In this section we will focus on Kramer's method for NLPCA, and in Section 10.6 we look at a modification of that method called SNLPCA. These are global parametric methods because for a given neural-network architecture, there is a corresponding parametric equation whose parameters are adjusted to optimize a global measure of goodness-of-fit, and there is no specific relationship between parameters of the equation and local regions of the mapping space.

Kramer's method for NLPCA involves training a neural network with three hidden layers, such as the one shown in Figure 10.6. These neural networks

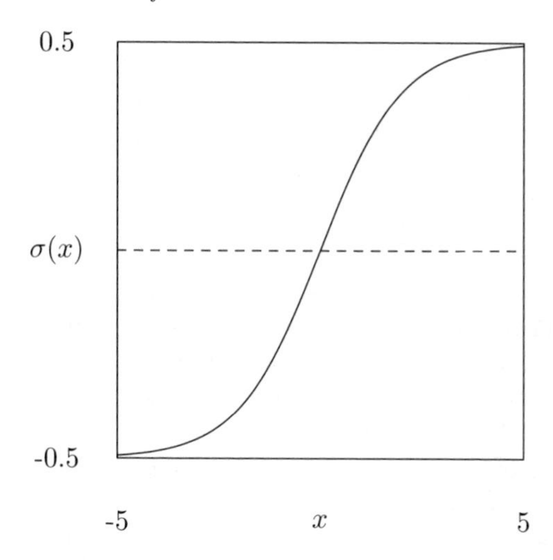

FIGURE 10.7: Sigmoid function.

are autoassociative, meaning they are trained to map a set of input vectors $\mathbf{X}$ to an identical set of output vectors. If the second hidden layer, or "bottleneck" layer, has a lower dimension than the input and output layers, then training the network creates a lower-dimensional representation $\mathbf{X}^*$ of the vectors presented to the networked inputs in the form of the activations of the units in the bottleneck layer. The mapping from the input vectors to these activations in the bottleneck layer is the "compression" transform $\mathbf{g}$, and the mapping from the bottleneck activations to the activations of the output units is the "decompression" transform $\mathbf{h}$. Using sigmoidal units of the form

$$\sigma(x) = (1 + e^{-x})^{-1} - \tfrac{1}{2} \tag{10.11}$$

in the first and third hidden layers (the sigmoid function is shown in Figure 10.7) allows the mapping function $\mathbf{g}$ and de-mapping function $\mathbf{h}$ to take the forms

$$\mathbf{g} : g_k(\mathbf{x}) = \tilde{\sigma}\left(\sum_{j=0}^{M_c} w_{kj}^{(2)} \sigma\left(\sum_{i=0}^{m} w_{ij}^{(1)} x_i\right)\right) \quad k \in 0 \dots p-1 \tag{10.12}$$

$$\mathbf{h} : h_k(\mathbf{x}^*) = \tilde{\sigma}\left(\sum_{j=0}^{M_d} w_{kj}^{(4)} \sigma\left(\sum_{i=0}^{p} w_{ij}^{(3)} x_i^*\right)\right) \quad k \in 0 \dots m-1, \tag{10.13}$$

where $\tilde{\sigma}(\cdot)$ may either be the sigmoidal function (10.11) or the identity function $\tilde{\sigma}(x) = x$ depending on whether sigmoidal or linear units are used in the bottleneck and output layers. Given enough mapping units, these functional forms may approximate any bounded, continuous multidimensional nonlinear function $v = f(u)$ with arbitrary precision [50]. Just as PCA defines a linear mapping to and from a reduced-dimension representation which minimizes the sum of squared reconstruction error $\|\mathbf{X} - \mathbf{V}^T\mathbf{V}\mathbf{X}\|^2$ for a given set of vectors, training the weights $\mathbf{W}$ of the autoassociative neural network to minimize the sum of squared error

$$e_p^2(\mathbf{X}) = \|\mathbf{X} - \mathbf{n}(\mathbf{W}, \mathbf{X})\|^2 \tag{10.14}$$

of mapping vectors $\mathbf{x}_i$ to themselves through the bottleneck layer effectively performs a nonlinear principal component analysis of the vectors.

The principal advantage of NLPCA over PCA is its ability to represent and learn more general transformations, which is necessary in cases when one wishes to eliminate correlations between dimensions in a set of data which cannot be adequately approximated by a linear dependency. However, NLPCA also has important disadvantages compared to PCA.

The trade-off for the extra-representational power of the nonlinear mapping functions $\mathbf{g}$ and $\mathbf{h}$ is that they cannot be as easily interpreted as the eigenvector-based mappings returned by PCA. In addition, NLPCA tends to require several orders more computation time than linear PCA, and because training the neural network is a high-dimensional nonlinear optimization problem over the weights of the neural network, we can guarantee only a locally optimal solution, unlike the globally optimal solution returned by PCA. In the version of NLPCA presented to this point the relative importance of each output dimension of the compression mapping cannot be determined by the training process, and there is no guarantee that any one of the output dimensions corresponds to a primary nonlinear factor of the training data. However, if an explicitly prioritized factorization is desired, Kramer's sequential NLPCA algorithm (SNLPCA), discussed in Section 10.6, may be used.

We used Kramer's NLPCA method to analyze the data set from Section 10.3. The neural networks were trained using the L-BFGS-B implementation of Byrd, et al., [37]. We used a network architecture with linear units for the bottleneck and output layers, and without direct interconnections between the input and bottleneck layers, nor between the bottleneck and output layers. Choosing linear output units for the output layer allowed the network to be trained on data which was not rescaled to fit within the output range of the sigmoidal function (although the data was zero-normed). This gave better results than rescaling the data and using sigmoidal units.

The parameter M, the number of mapping units in the first and third hidden layers of each network, was chosen by a heuristic search over the range $p \ldots \min(\frac{n}{4}, p_M - 1)$ where n is the number of training vectors, and p_M is the smallest possible value for M for which the number of weights in the network N_w will exceed the available number of values in the training matrix, (mn). M must be at least as great as p if there are no interconnections across the mapping layers; otherwise, the mapping layers would be the real bottlenecks. The selection criteria for the value of M was not cross-validation error (although it is used for early-stopping of the weight-optimization algorithm),

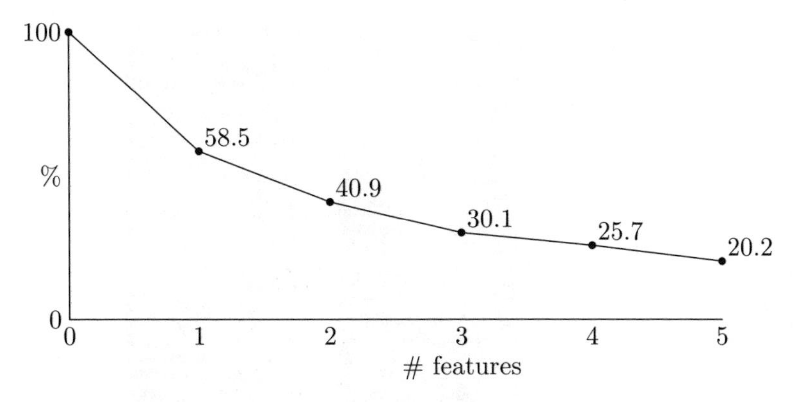

FIGURE 10.8: Error unexplained by p features of NLPCA.

but rather the information theoretic criterion described in [124]:

$$\mathrm{AIC} = \ln(e/(2m)) + 2N_w/N_d, \tag{10.15}$$

where

$$N_w = m + p + 2M(m + p + 1) \tag{10.16}$$

is the number of weights in the network, $N_d = (nm)$ is the number of training vectors times the dimension of the training vectors, and $e/(2m)$ is the average sum of squares error. This criterion penalizes network complexity, and thus tends to reduce over-fitting of the training data. At least two units are used in the mapping layers, even when $p = 1$, because a network architecture with only a single unit in each of the three hidden layers is degenerate and unable to learn significantly nonlinear principal components.

In Figure 10.8 we show the results of this NLPCA analysis. The effect of NLPCA on residual error for each number of principal nonlinear factors is calculated as

$$\%\text{-Err}_p = \frac{\|\mathbf{Y} - \mathbf{n}_p(\mathbf{Y})\|}{\|\mathbf{Y}\|}. \tag{10.17}$$

The most interesting aspect of the analysis is that although NLPCA explains significantly more of the data set than PCA when $p = 1$, the two methods perform similarly when $p > 1$. This is due to the fact that the grasping motions analyzed are relatively simple and open-loop in nature, due to a lack of haptic feedback in the virtual environment in which they were demonstrated. We can thus think of each performance as following a nominal grasping trajectory in configuration-space from an open hand to a closed hand, with some stochastic variation. The first nonlinear principal component follows the nominal grasping-trajectory, and since this trajectory is curved, it explains the data in the testing data set better than the first linear principal component.

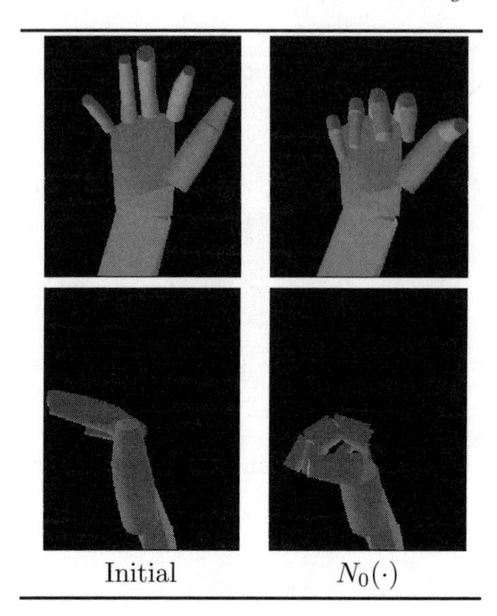

FIGURE 10.9: NLPCA analysis of grasp gesture. Initial position, and effect of first nonlinear principal component, front and side views.

However, since the basic non-stochastic structure of the grasp is adequately explained by the first nonlinear principal component, it is plausible that when a higher-dimensional feature-space is used NLPCA simply optimizes the mutual orthogonality of the resulting features to most efficiently explain the more stochastic nature of residual performance data, and the result is thus similar to PCA.

Figure 10.9 illustrates the first nonlinear principal component from our analysis. These configurations are formed by varying a scalar value y_0^* and mapping it to a hand configuration using function $\mathbf{h}_0 : \mathbb{R}^1 \rightarrow \mathbb{R}^m$. This can be accomplished using an interface with a slider-bar controlling y_0^* as in Figure 10.10. The grasping configuration of the hand looks similar to that generated by the first linear principal component in Figure 10.5, but the initial position generated from the nonlinear principal component is more plausible. Moreover, the grasping motion resulting from smoothly varying the feature value looks more natural than the motion generated by the principal linear feature.

It should be noted that in this example we started by building a representation for individual hand configurations, but this resulted in a mapping which we could use to animate a nominal grasping performance by smoothly and monotonically varying the nonlinear principal component y_0^* from an initial to a final value. This was possible because the grasping motion is a smooth directed path in configuration space, and because the forms of the

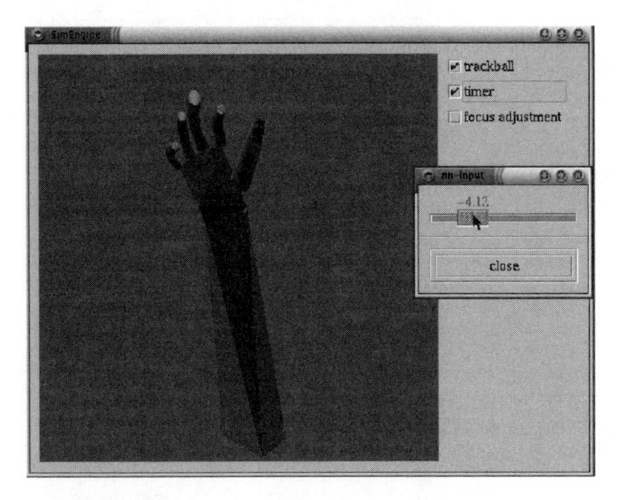

FIGURE 10.10: Interface for controlling hand configuration using y_0^*.

mapping functions (10.12) and (10.13) are well suited to learning such smooth mappings. However, we are making a substantial leap here. We are treating a nonlinear regression fit as if it were a best-fit trajectory estimate of the grasping motion—a directed path through configuration space which is typical of the example training performances. Although we have created a one-dimensional parameterization of the grasping motion, this was done using only global information, and our learning techniques learned nothing directly about the local relationships between the points in the dataset and how the hand typically moves between them in configuration space. In Section 10.8, we will discuss the difference further and motivate the methods presented in later chapters which make use of localized motion information in the training data.

10.6 SNLPCA

Kramer's sequential NLPCA algorithm (SNLPCA) [124], is a modification to the NLPCA method which produces a nonlinear factorization, and where the training process prioritizes each resulting feature as to its relative power in explaining the variations of the training set. SNLPCA performs a series of NLPCA operations, each training a neural network with a bottleneck layer consisting of a single unit. Training such a neural network on the raw performance data set explains as much of the set's variation as can be represented by a single variable, and thus trains the compression part of the network to

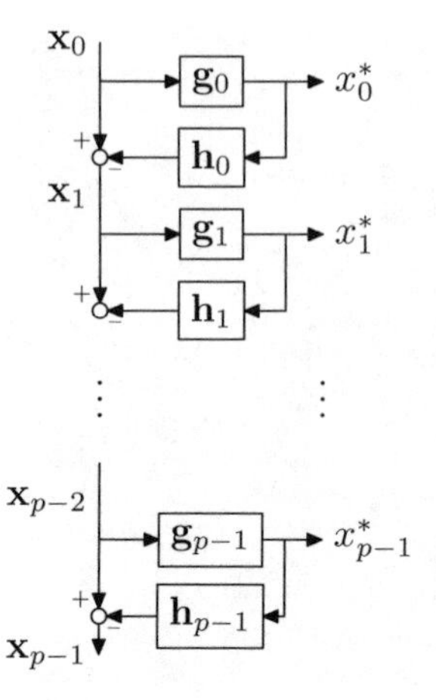

FIGURE 10.11: SNLPCA computation.

perform the primary nonlinear factorization of the data set. The next nonlinear factorization should explain the variation in the data set which is not accounted for by the first factorization, and thus a second network is trained on the residuals formed by subtracting each vector of the original data set by its estimate as calculated by the first network. This process, summarized in Figure 10.11, continues until the desired number of iterations has been reached or until enough of the original data set's variation has been explained.

The algorithm is:

1. $\mathbf{X}_0 \leftarrow \mathbf{X}, p \leftarrow 0$
2. Loop for p in $0 \ldots (F-1)$ or until $\|\mathbf{X}_p\| < \epsilon$
   (a) Train $\mathbf{W}_{p+1}$ to minimize $\|\mathbf{X}_p - \mathbf{n}_{p+1}(\mathbf{X}_p)\|^2$, where $\mathbf{n}_{p+1}(\cdot) \equiv \mathbf{n}(\mathbf{W}_{p+1}, \cdot)$
   (b) $\mathbf{X}_{p+1} \leftarrow (\mathbf{X}_p - \mathbf{n}_{p+1}(\mathbf{X}_p))$
   (c) Split $\mathbf{n}_{p+1} : \mathbb{R}^m \rightarrow \mathbb{R}^m$ into $\mathbf{g}_{p+1} : \mathbb{R}^m \rightarrow \mathbb{R}^1$ and $\mathbf{h}_{p+1} : \mathbb{R}^1 \rightarrow \mathbb{R}^m$ such that $\mathbf{h}_{p+1}(\mathbf{g}_{p+1}(\cdot)) \equiv \mathbf{n}_{p+1}(\cdot)$
   (d) $\mathbf{X}^*_{(p,\cdot)} \leftarrow \mathbf{g}_{p+1}(\mathbf{X}_p)$ (i.e., sets row p of $\mathbf{X}^*$).

This algorithm generates a reduced-dimension representation $\mathbf{X}^*$ of $\mathbf{X}$, but does not automatically generate the mapping and de-mapping functions $\mathbf{g}$ and $\mathbf{h}$. These may be formed in two ways. The first is to cascade the networks $\mathbf{g}_p$, which were trained in the SNLPCA algorithm, to form a large single network for computing $\mathbf{g}$ in a manner corresponding to the computation shown in Figure 10.11. In a similar fashion, networks $\mathbf{h}_p$ can be cascaded to form a network for computing $\mathbf{h}$. This method requires no additional training and will perform exactly the mappings calculated by the SNLPCA algorithm. The

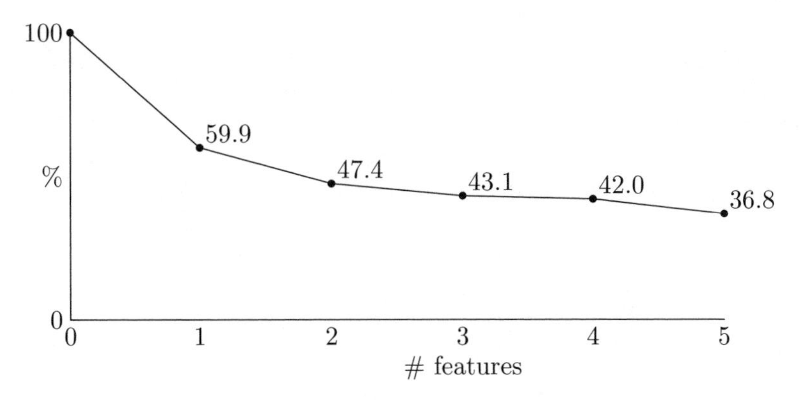

FIGURE 10.12: Error unexplained by p features of SNLPCA.

second method is to train separate neural networks to perform the mapping from $\mathbf{X}$ to $\mathbf{X}^*$ and $\mathbf{X}^*$ to $\mathbf{X}$. This method results in smaller, more computationally efficient networks, but may not necessarily converge to acceptable approximations of the desired mappings.

Figure 10.12 shows the results of SNLPCA on the data set from Section 10.3. For this data set, we see that its performance is roughly equivalent to NLPCA for $p = 1$, and worse than both NLPCA and PCA when $p > 1$ (the comparison is summarized in Table 10.1). This is due to the fact that the first iteration has eliminated almost all of the underlying structure of the grasping skill, thus leaving the residual vectors $\mathbf{X}_1$ dominated by stochastic variations in the individual human performances. Since the nonlinear factors are trained sequentially, it is difficult for the SNLPCA algorithm not to over-fit the remaining noisy residuals at each iteration, rather than learn to represent this noise by building some equivalent of an orthonormal basis of linear features in the manner of PCA. In addition, as the number of features used increases, the compression networks have increasing difficulty learning the generated compression mappings.

An additional explanation for the degradation in performance after the first iteration of the algorithm compared to NLPCA is that the residuals in $\mathbf{X}_p$ (where $p \geq 1$) are correlated with the features generated from previous iterations. This makes some intuitive sense: it seems likely that the particular ways that the sampled grasp configurations vary from the model depends largely on to which part of the grasping motion they correspond. Hand configurations at the beginning of the grasp differ from one another in different ways than configurations near the end of the grasping motion. If we consider the feature-score from the first iteration to correspond roughly with a temporal ordering of some prototypical grasp, then the variations modeled by the second iteration of the algorithm are highly dependent upon the feature-score learned by the first iteration. Since the information represented by the first

Table 10.1: Experimental results. Percent of test data set **Y** unexplained by p factors for each method, calculated by (10.7) for PCA and by (10.17) for NLPCA and SNLPCA. M is the size of the mapping layers in NLPCA, and M_c and M_d are the sizes of the mapping layers in the compression and decompression networks trained by SNLPCA.

p	PCA	NLPCA		SNLPCA		
	%-Err	%-Err	M	%-Err	M_c	M_d
1	72.0	58.5	12	59.9	14	3
2	41.4	40.9	3	47.4	14	6
3	33.4	30.1	8	43.1	12	5
4	26.8	25.7	5	42.0	15	9
5	23.1	20.2	10	36.8	19	8

feature score is not directly available to the learning process in later iterations, the modeling ability of these later iterations is handicapped. For this reason, I suspect that the results from SNLPCA may be improved by adding the output value of the preceding iterations to the dimensions of the residual data used for training in later iterations. This hypothesis is not investigated in this book, however, as we will turn our attention to local methods and away from those based upon neural networks.

10.7 Comparison

Table 10.1 compares the results from PCA, NLPCA, and SNLPCA. Based on its ability to represent the nominal trajectory of the grasping skill in configuration space, we conclude that the one-dimensional NLPCA mapping (equivalent to the one-dimensional SNLPCA mapping) is the best representation of the grasping skill of the models we generated. For faithful reconstruction of a given hand configuration from a reduced-dimension representation, PCA analysis is the simplest and most effective method. Higher-dimensional NLPCA and SNLPCA models might potentially be more appropriate for more complex skills.

Although NLPCA and SNLPCA both have the advantage that they can generate nonlinear models, these models have a black-box nature. It is difficult to understand exactly what they do and how, and this difficulty increases dramatically with the dimensionality of the input space and feature space. When NLPCA is used to model grasping motion using two feature components, and then we control the two activation values of the bottleneck layer using two slider-bars like the one shown in Figure 10.10, it is difficult to get a sense for what the two values do independently of one another. Moreover, given the functional form of the decompression network and how the entire bottleneck network is trained, there is little likelihood of any clear independent relationship.

10.8 Characterizing NLPCA mappings

A paper by Malthouse [148] discusses Kramer's NLPCA method and indicates that it has several important limitations, including an inability to model curves and surfaces that intersect themselves, and an inability to parameterize curves with discontinuities. These limitations are due to the fact that the mapping and de-mapping projections (10.12) and (10.13) are continuous functions. For our example data set, and for many typical human skills, continuous mappings to and from the feature representation are not particularly restrictive because the skills can be smoothly parameterized. Global models in general will have a difficult time learning about datasets which are highly discontinuous. Later in this book, we will show how some curves which intersect themselves in position space can be successfully modeled by other methods in phase space.

A criticism in Malthouse's paper which is more relevant to the goals of this book is that Kramer's NLPCA method tends to result in suboptimal projections, and that methods based on principal curves [56], [90], [128] tend to result in better parameterizations. This problem is demonstrated in Figure 10.13. In 10.13(a), we first randomly sample a set of points from a circular distribution, and then add uniform random noise to the x and y dimensions of the sample points. A NLPCA network, shown in 10.13(b) is used to learn to map these points to themselves through a single bottleneck node, using 10 neurons in each mapping layer. Figure 10.14(a) shows how the training points map to the learned model, and Figure 10.13(c) shows how this network projects points in the immediate vicinity of the training points onto the learned model. The circles are the input points, and the line from the center of each circle shows to where the network maps that input. The output of the decompression part of the network $\mathbf{h}_0$ over the full range of activations on the bottleneck neuron, from the minimum value generated by the training data to the maximum value, is drawn as a dashed curve. We see that although some of the mappings are plausible, particularly those in the lower-right hand side of the figure (e.g., around the $(5, -5)$ coordinate), the projections of other points are far less sensible. The closer we look toward the upper-left corner of the figure, the more distorted the mappings appear. Instead of mapping to the closest point on the dashed model-line, many inputs

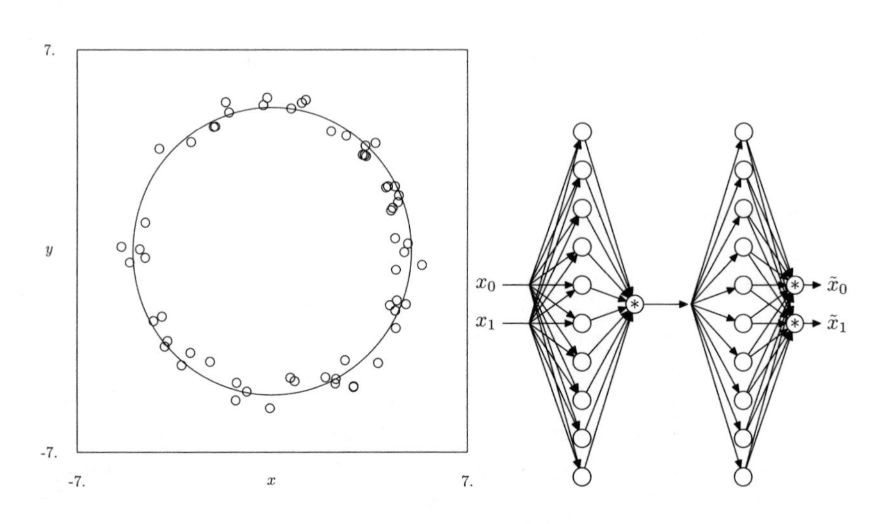

(a) Training data: Points sampled from circle with added noise.

(b) NLPCA network used for learning circle mapping.

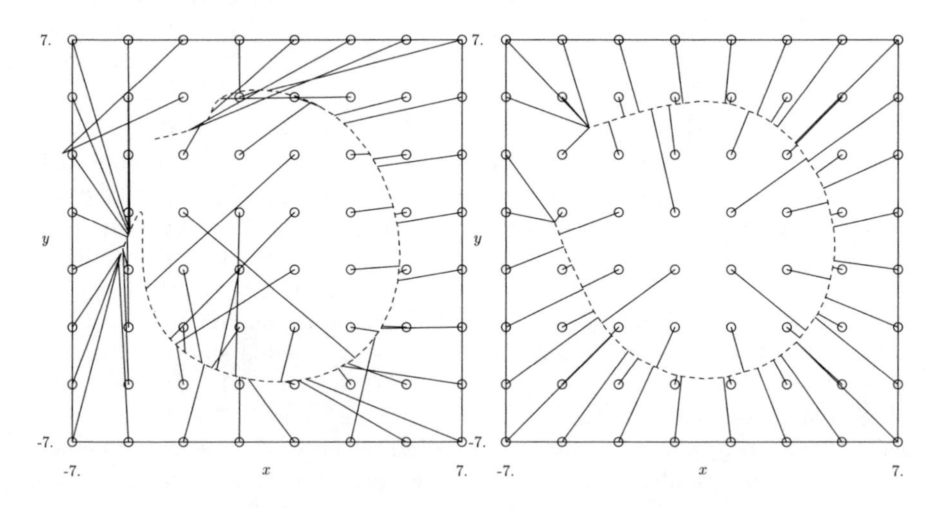

(c) Mapping generated by NLPCA.

(d) Mapping generated by principal curve.

FIGURE 10.13: Mappings to circular figure of surrounding space, generated by NLPCA and principal curves.

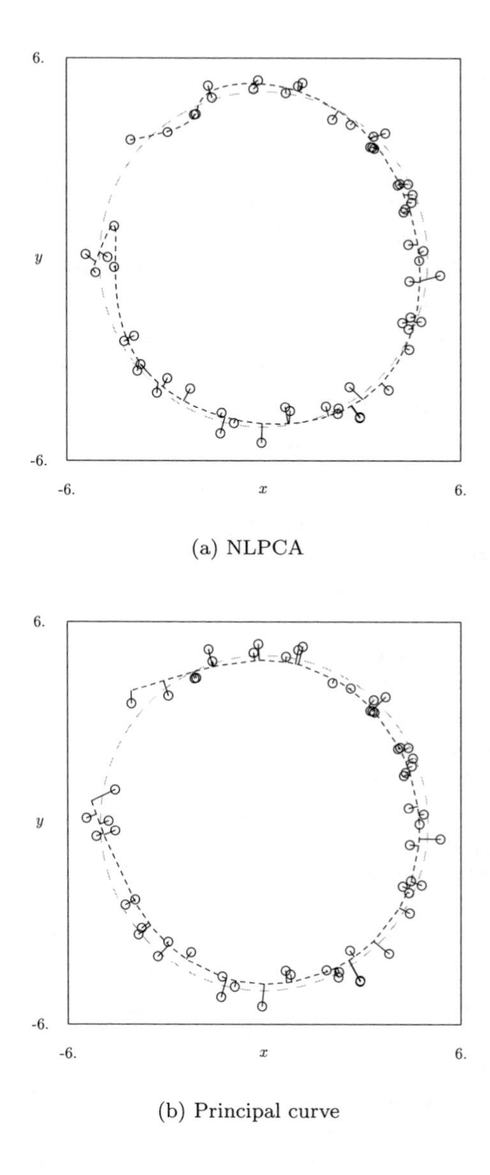

(a) NLPCA

(b) Principal curve

FIGURE 10.14: Mapping of training points to models learned by neural networks and principal curves.

project to points on the far side of the model. The input at $(3, -7)$, which is fairly close to one part of the model curve and near to inputs' points whose projections are very plausible, even maps past the far side of the model curve to the top of the plot.

These strange projection results should not be surprising given the nature of back-propagation neural networks. The power of these functional forms is that they are sufficiently nonlinear and have enough degrees of freedom that they can fit complex mappings with ease, even when trained using straightforward hill-climbing algorithms. Starting with a random mapping in this case, the network adjusts its projection of the output points until they nearly match the input points. Along the way, the learning function can easily warp the projection for any region as necessary to reduce the error measure. This ease of warping the output space, and the fact that what happens outside the training set has no effect on training error, gives us reason to believe that the resulting projection for points outside the training set may look somewhat bizarre.

There are several reasons why we are looking at plots of projections in a two-dimensional space. The first of course is that these plots are much easier to understand than higher-dimensional projections, and it is easier to make the connection between fitting the training points and modeling trajectories which might pass through them. Since building a good one-dimensional model for trajectory fitting is one of the more important uses of dimension reduction, we want our methods to perform well in this case. The complexity of the neural network mapping increases with its dimensionality, and we would expect the mappings in higher dimensions to become more rather than less strange as we look at less simple cases, so these results for two dimensions should give us pause. We are thus motivated to look for modeling techniques where the resulting projections are more understandable, and which are explicitly based on principals appropriate for fitting trajectory information.

What kind of projection function would be better than that shown in Figure 10.13(c)? The least-squares projection method of the PCA model is an appealing answer. If we associate a metric function with the raw-data space, and consider the model to be some lower-dimensional manifold in this space, then the projection of each point in the raw-data space onto the model is the nearest point in the model. Figure 10.13(d) shows a model built from the training points of 10.13(a) where the definition of projection is based on a least-squares function in a similar manner to PCA, but using a nonlinear model. The method used to generate this model is the principal curves algorithm of Hastie and Stuetzle [90]. Figure 10.14(b) shows how the training points map to the principal curve learned from them. Unlike the methods introduced in this chapter, the principal curves algorithm examines the relationship between points which are close in proximity, and thus it is not a global method.

It should be noted here, that at least one global parametric model can generate a mapping which looks more like that of the principal curve model of

Figure 10.13(d). The input-training neural network of Tan and Mavrovouniotis [234] generates a much better mapping than does Kramers's NLPCA method. For modeling human performance data, their method should thus outperform NLPCA. Their method still has the problem that it is based on a neural network mapping, however, and is thus more difficult to analyze than the methods based on local models which we present later in this book.

In the following chapters we will demonstrate how methods such as principal curves, which make use of local information, can be used to build models of human performance data. We will also exploit this local information to address another problem with the global methods: the fact that while we really would like to build parameterizations which can express typical performance trajectories, global methods can only fit individual performance data points, and thus information about how typical human performances progress from one sampled point to another is completely lost from the training data. Figure 10.13 demonstrates that while neural-network based NLPCA does a good job of fitting the training data, we should be hesitant to use the resulting projections of points between the training points, and thus we should be wary of using trajectories in feature-space to model realistic trajectories and motions in the raw-configuration space. This problem can be addressed by methods for fitting trajectories which use local information in the human performance data set.

11

Local Methods for Dimension Reduction

11.1 Introduction

This chapter introduces the use of local, non-parametric methods for dimension reduction of human performance data. The previous chapter demonstrated the use of several global methods for this purpose, methods which look at all the data points from all the training examples as a single set of vectors $\mathbf{X}_{(m \times n)} = [\mathbf{x}_0|\mathbf{x}_1|\ldots\mathbf{x}_{n-1}]$ and simply try to map them to a lower-dimensional space and back again so as to preserve maximum information. Global parametric models have a number of difficulties when it comes to modeling complex data sets such as those from human performances, however. When the models are simple enough for easy analysis, they can be too simple to adequately model the data. PCA is analytically beautiful, but the first principal component adequately describes configurations along a best-fit performance trajectory only when that trajectory is linear in configuration space. On the other hand, models like NLPCA and SNLPCA, which are flexible enough to fit a wide variety of data sets, can be very difficult to analyze and use. Methods for dimension reduction such as these, which are based upon neural networks, also tend to result in inadequate projections to and from the feature space.

Focusing on non-parametric methods will help us to construct suitable projections to and from feature space via nonlinear models, and to generate trajectories typical of human performance from multiple examples. These methods are data-driven rather than based *a priori* upon a global parametric form, and generally fit data locally rather then globally. The output value at a given domain point is typically found by a simple average or weighted regression of the values of nearby sample points. Because these local models are simple, it is easy to understand their outputs. The entire model is still flexible enough to adequately represent an almost arbitrary set of functions or mappings, however, since it is constructed from a large number of these local models.

In this chapter, we will focus on trajectory-fitting. As we discuss in the next section, a best-fit trajectory is the most basic kind of reduced-dimension action model. Two general-purpose non-parametric methods will be reviewed for this purpose: scatter plot smoothing and principal curves. We will discuss how these work, and how their use of local information helps build good trajectory models from human performance data. The next two chapters will present adaptations of these methods we have developed for the specific purpose of modeling human performance data.

11.2 Local, non-parametric methods for trajectory fitting

Non-parametric methods can allow us to explicitly design the structure of the local model for the specific purpose of modeling human performance. The simplest logical model for human action data is the "best-fit" or "most-likely" performance trajectory.

In Section 9.2, we formulated dimension reduction for a training set of action data as a feature-extraction problem. Given a set of performance data $\mathbf{X}$ in space S, we would like to convert these data to a lower-dimensional representation $\mathbf{X}^*$ in space S^* where $\mathbf{X}^* = \mathbf{g}(\mathbf{X})$. These lower-dimensional data map to a corresponding representation in the original space $\tilde{\mathbf{X}} = \mathbf{h}(\mathbf{X}^*) = \mathbf{h}(\mathbf{g}(\mathbf{X}))$. One important feature of such a mapping is that smooth paths in S should map to smooth paths in S^*, and smooth paths in S^* should map to smooth paths in S. This allows directional-derivatives in S^* to be meaningful, and it simplifies the problem of representing a trajectory in S^* which maps a plausible trajectory in S. This lets us create an animation of action-performance in a simulation environment by moving a few slider-bars representing feature-values in S^*, and it simplifies writing successful controllers that use input or output vectors from space S^*.

We have seen in Chapter 10 that it may be plausible in some cases to model an extrinsically high-dimensional dataset using just a single parameter. If we use a scalar parameter $s \in S^*$ to model a dataset from multiple examples of some action, then we want the image $\mathbf{h}(s)$ to contain representative points for all parts of a "most-likely" or "best-fit" version of the action. If a smooth non-intersecting best-fit trajectory path in S maps to a smooth path in S^* (an interval $\subset \mathbb{R}^1$), then s is a parameterization variable which can be transformed via a monotonic function to a time-parameterization of the best-fit trajectory. Parameter s represents a *temporal-ordering* of points along this best-fit trajectory. If a one-dimensional feature parameter s does not specify a temporal ordering of points along a continuous trajectory in S, then it will be much more difficult to use it to represent motion, and derivatives with respect to that parameter will not always have a meaningful interpretation. In a simulation interface such as the one shown in Figure 10.10 on page 183, moving a slider bar to vary s from one value to another at the appropriate speed should create a plausible animation representing some portion of the

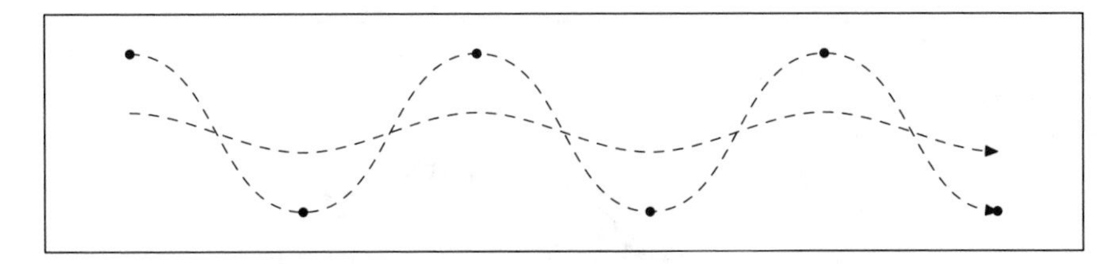

FIGURE 11.1: Explicit trade-off between fitting error and smoothness of the model.

action, and moving the slider from an s-value which is typical of a starting configuration to an s-value which is typical of an ending configuration should create a plausible animation of the entire action.

A best-fit trajectory is very useful model of a human action. Examples include the motion for a typical power grasp, Mark McGwire's typical home run swing, a typical walking or swimming motion, or a typical motion of the lips when saying a given word. Such a trajectory can be used for creating an animation for a video game or movie, for comparing a novice's actions to an expert's (e.g., "how does my golf-swing compare to Tiger Woods's"), or for gesture recognition.

We saw in the last chapter that while it is possible for global methods such as NLPCA to generate something akin to a best-fit trajectory, there is no explicit constraint requiring them to generate a satisfactory model. Global methods fit individual training points, but may not necessarily interpolate between them in a manner plausible for an actual performance. Non-parametric methods, on the other hand, can make explicit trade-offs on the local level such as balancing quality of fit versus the smoothness of the model, as depicted in Figure 11.1. Spline smoothers, discussed in the next section, are based on precisely this trade-off. Moreover, the projections of points onto the local models can be done in a principled manner by projecting to the nearest model point, as demonstrated by the principal curves in Figure 10.13(d) on page 188 and Figure 11.5 on page 203.

11.3 Scatter plot smoothing

Scatter plot smoothers allow us to balance local smoothness of the estimated model against modeling error, or to smoothly combine localized fitting into a global model, without assuming a parametric form for the model. Figure 11.2 shows the result of a smoother based on a robust locally-weighted regression [44], [45]. Note that the error vectors are vertical. As in linear regression

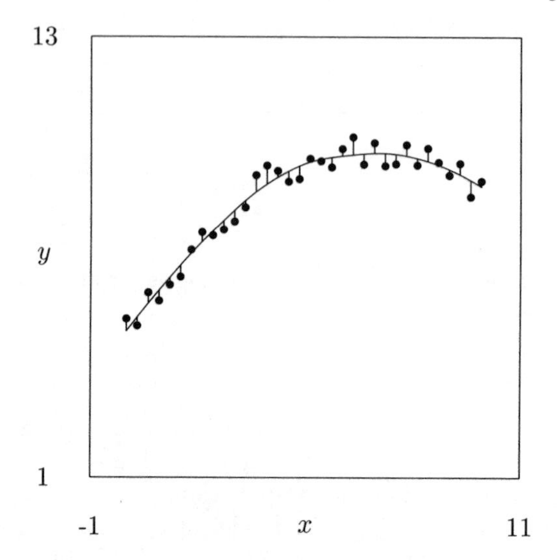

FIGURE 11.2: Modeling a dataset with a scatter plot smoother.

and polynomial fitting (Section 10.2.1), scatter plot smoothers minimize error in the response variable only.

There are several kinds of scatter plot smoothers. Kernel smoothers fit each point in the dataset using a locally weighted average, while locally-weighted regression fits local models to the data [3].

Spline smoothers [53], [198] generate a model by minimizing a cost function weighing modeling error against the integral of the d-th derivative of the model curve. The cost function is

$$S = p\sum_{i=0}^{n-1}\left(\frac{y_i - f(x_i)}{\delta y_i}\right)^2 + (1-p)\int_{x_0}^{x_{n-1}}(f^{(d)}(x))^2 dx, \qquad (11.1)$$

where p is the smoothing parameter weighing approximation error against smoothness of the curve, and the δy_i weigh individual data points. Weights δy_i are often local estimates of standard deviation. Minimizing S for a given set of values p and δy_i results in a model curve f which is a polynomial spline of order $k \equiv 2d$ with simple knots at $x_0, \ldots, x_{n-1}$ ($x_i < x_{i+1}$), and natural end conditions:

$$f^{(j)}(x_0) = f^{(j)}(x_{n-1}) \qquad \text{for } j \in \{d-1, \ldots, k-3\}. \qquad (11.2)$$

Optimization is often performed using an acceleration penalty ($d = 2$), which results in a cubic spline solution with the form

$$P_i(x) = a_i + b_i(x - x_i) + c_i(x - x_i)^2 + d_i(x - x_i)^3 \qquad (11.3)$$

between each knot.

For a given value of the smoothing parameter p, we can solve for parameters $\mathbf{c} = \{c_i\}$ using the tridiagonal system [53]

$$(6(1-p)\mathbf{Q}^T\mathbf{D}^2\mathbf{Q} + p\mathbf{R})\mathbf{c} = 3p\mathbf{Q}^T\mathbf{y}, \tag{11.4}$$

where $\mathbf{D}$ is the diagonal matrix of weights $\lceil \delta y_0, \ldots, \delta y_{n-1} \rfloor$, $\mathbf{R}$ by (12.18) on page 221, and $\mathbf{Q}^T$ by (12.19). Band matrices $\mathbf{R}$ and $\mathbf{Q}^T$ are described in Chapter 12 as we derive a spline smoother which also penalizes error in velocity information. Once $\mathbf{c}$ has been computed in this way, we can easily determine the other terms of the smoothing spine parameterization:

$$\begin{aligned} \mathbf{a} &= \mathbf{y} - 2\left(\frac{1-p}{p}\right)\mathbf{D}^2\mathbf{Qc} \\ d_i &= \frac{c_{i+1} - c_i}{3\Delta x_i} \\ b_i &= \frac{a_{i+1} - a_i}{\Delta x_i} - c_i\Delta x_i - d_i(\Delta x_i)^2, \end{aligned} \tag{11.5}$$

where $\Delta x_i = (x_{i+1} - x_i)$.

From (11.1), it might at first appear that a spline smoother is a global parametric model rather than a local model. The model is indeed defined as the minimum of a global error function, and the resulting curve may be parameterized as a spline function. However, the number of parameters defining the spline is actually greater than the number of values in the training data, and the error term of the cost function is balanced against a smoothness measure based upon derivatives, which are by definition local. In fact, Silverman [220] shows that a spline smoother is equivalent to a kernel smoother with a variable-sized kernel.

When using spline smoothers, it is important to choose a suitable value for the smoothing parameter p. Cross validation [229] is a good method to use for selecting this value for a given data set. While leave-one-out cross-validation for a data set using the simple spline solution presented in this section is computationally expensive, Craven and Wahba [48] present a closed-form method for computing the optimal solution for a given data set, and Hutchinson and de Hoog [102] show how to compute this closed-form solution in linear time.

11.4 Action recognition using smoothing splines

The error term of spline-smoother cost function (11.1) is based upon the assumption of a Gaussian error distribution at each parameterization x_i. We can use this assumption to formulate a probability that a given performance belongs to the class of actions corresponding to a given model, and to compare this probability across several models for the purpose of recognition.

Smoothing each dimension of a data set $\mathbf{Y}$ against parameterization vector $\mathbf{x}$ results in a model $(\mathbf{x}, \mathbf{f}, \{\mathbf{D}_i\})$, where $\mathbf{f} : \mathbb{R}^1 \rightarrow \mathbb{R}^m$ is the multi-dimensional spline model of the best-fit trajectory, and the diagonal elements of each matrix $\mathbf{D}_i$ are the estimated standard-deviation for each dimension of the training data at parameterization x_i. Given this model, and the assumption of a Gaussian distribution at each parameterization value, we can estimate the probability of a given example point $(x_\mathrm{e}, \mathbf{y}_\mathrm{e})$ given the model as [30]

$$p(\,(x_\mathrm{e}, \mathbf{y}_\mathrm{e}) \mid (\mathbf{f}, \{\mathbf{D}_i\})\,) = \frac{1}{(2\pi)^{m/2}|\hat{\mathbf{D}}^2|^{1/2}} \exp\left(-\frac{1}{2}(\mathbf{y}_\mathrm{e} - \mathbf{f}(x_\mathrm{e}))^T \hat{\mathbf{D}}^2 (\mathbf{y}_\mathrm{e} - \mathbf{f}(x_\mathrm{e}))\right). \quad (11.6)$$

For this evaluation, $\mathbf{f}(x_\mathrm{e})$ may either be computed exactly from the spline equation, or approximated by linear interpolation between the nearest points $\mathbf{y}_k, \mathbf{y}_{k+1}$ on the model such that $x_k < x_\mathrm{e} < x_{k+1}$, and $\hat{\mathbf{D}}$ can be estimated by linear interpolation between $\mathbf{D}_k, \mathbf{D}_{k+1}$.

While the cost function for the spline smoother relates the points in its model by the roughness penalty term, this term balances model smoothness against an error cost which treats each sample as independent of the others. When we look at data from an example performance $\mathbf{Y}_\mathrm{e}$, where n_e samples of the performance have been taken over time, we thus treat each point as independent, uncorrelated with the other samples. This gives us

$$p(\,(\mathbf{x}_\mathrm{e}, \mathbf{Y}_\mathrm{e})|\ (\mathbf{f}, \{\mathbf{D}_i\})\,) = \prod_{i=0}^{n_\mathrm{e}-1} p(\,(x_{\mathrm{e}i}, \mathbf{y}_{\mathrm{e}i}) \mid (\mathbf{f}, \hat{\mathbf{D}}(x_{\mathrm{e}i}))\,)$$

$$= \prod_i \frac{1}{(2\pi)^{m/2}|(\hat{\mathbf{D}}(x_{\mathrm{e}i}))^2|^{1/2}} \exp\left(-\frac{1}{2}(\mathbf{y}_{\mathrm{e}i} - \mathbf{f}(x_{\mathrm{e}i}))^T (\hat{\mathbf{D}}(x_{\mathrm{e}i}))^2 (\mathbf{y}_i - \mathbf{f}(x_{\mathrm{e}i}))\right). \quad (11.7)$$

For an example with many sample points, this probability will be very small, so it is useful to consider its logarithm instead:

$$\log p(\,(\mathbf{x}_\mathrm{e}, \mathbf{Y}_\mathrm{e}) \mid (\mathbf{f}, \{\mathbf{D}_i\})\,) = \sum_i \log\left(\frac{1}{(2\pi)^{m/2}|(\hat{\mathbf{D}}(x_{\mathrm{e}i}))^2|^{1/2}}\right) + \sum_i \left(-\frac{1}{2}(\mathbf{y}_{\mathrm{e}i} - \mathbf{f}(x_{\mathrm{e}i}))^T (\hat{\mathbf{D}}(x_{\mathrm{e}i}))^2 (\mathbf{y}_{\mathrm{e}i} - \mathbf{f}(x_{\mathrm{e}i}))\right), \quad (11.8)$$

which simplifies to

$$\log p(\,(\mathbf{x}_{\mathrm{e}}, \mathbf{Y}_{\mathrm{e}}) \mid (\mathbf{f}, \{\mathbf{D}_i\})\,)$$
$$= -\frac{n_{\mathrm{e}} m}{2} \log(2\pi) - \sum_{i=0}^{n_{\mathrm{e}}-1} \sum_{j=0}^{m-1} \log(\delta\hat{y}_j(x_{\mathrm{e}i})) - \frac{1}{2} \sum_{i=0}^{n_{\mathrm{e}}-1} \sum_{j=0}^{m-1} \left(\frac{y_{ji} - f_j(x_{\mathrm{e}i})}{\delta\hat{y}_j(x_{\mathrm{e}i})} \right)^2 . \tag{11.9}$$

Note that right-most terms of (11.8) and (11.9) are expressions of the Mahalanobis distance measure [30].

Bayes' theorem can then be used to express the probability that the model $(\mathbf{f}, \{\mathbf{D}_i\})$ was the cause of the data set:

$$\log p(\,(\mathbf{f}, \{\mathbf{D}_i\}) \mid (\mathbf{x}_{\mathrm{e}}, \mathbf{Y}_{\mathrm{e}})\,) = \log p(\,(\mathbf{x}_{\mathrm{e}}, \mathbf{Y}_{\mathrm{e}}) \mid (\mathbf{f}, \{\mathbf{D}_i\})\,) + \log p(\,(\mathbf{f}, \{\mathbf{D}_i\})\,) - \log p(\,(\mathbf{x}_{\mathrm{e}}, \mathbf{Y}_{\mathrm{e}})\,). \tag{11.10}$$

The probability of a real data set is 1 (because it actually happened), so we may drop the last term of (11.10).

Since the absolute magnitude of these probabilities is so low, we typically compare the ratio of probabilities between pairs of models, which corresponds to the difference between their log probabilities. If we assume that the *a priori* probabilities for all the candidate models are equal, then for recognition purposes, we need only compare the first term of (11.10), which is the probability of the data given each model.

11.5 A gesture-recognition experiment using spline smoothing

Because human performance data sets do not usually contain an appropriate explanation variable against which to smooth, the main role of spline smoothers in this book will be as the preferred smoothers for trajectory fitting using the principal curves algorithm. In some cases, however, we are able to find a suitable explanation variable. If the speed of motion is very consistent between example performances, for example, we may be able to smooth against time.

Data from finger motions in letter-signing is one potential application for building models by smoothing against time. The data set used for testing sign-language recognition using hidden Markov models has some characteristics which make it appropriate enough for modeling with smoothing splines. This data set was recorded while a user signed 14 different letters: A, B, C, D, E, F, G, I, K, L, M, U, W, and Y. The user signed each letter thirty times,

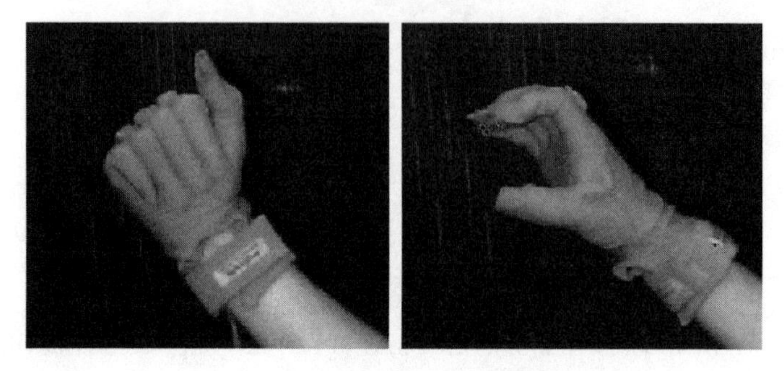

FIGURE 11.3: Final hand positions for letter-signs 'A' and 'C.'

but in a randomized order as prompted by a computer program. The program also ensured that the user was not prompted to sign the same letter twice in a row. The signer wore a Virtual Technologies "Cyberglove" which collected 18 channels of information about the configuration of the joints in the fingers, and the letters chosen for signing were a set of motions which could be unambiguously recognized using only finger motions (i.e., without data about the position, orientation, or motion of the hand). Figure 11.3 shows the final hand positions for the letter signs A and C. Because the example performances for each letter start from the final configuration of the random previous letter, the endings of these examples are much more consistent than their beginnings. As the signs are so varied near the beginning of each example, precise alignment of the examples in time is not important until the end of the gesture. Scaling the time values $\mathbf{t}$ for each gesture to fall between 0 and 1,

$$\mathbf{t}^* = \mathbf{t}/\max t_i, \tag{11.11}$$

will cause points near the end of the gestures to appear better aligned in time, and thus to correspond fairly well there.

We built a model for each letter by smoothing against t^*. For each letter, we combined the data from a set of training examples into a single matrix whose column-vectors are sorted by their corresponding values t_i^*, and smoothed each row j of this matrix against parameterization variable t^*.

The smoothing parameter p was chosen by experimentation, and the numerical stability of the smoothing solution was increased using the method for combining points with similar parameterizations documented (in a form extended to include velocity information) in Section 12.5. Initially, the variance matrices $\mathbf{D}_i^2 = \lceil \delta y_{0,i}^2, \ldots, \delta y_{m-1,i}^2 \rfloor$ were set to the identity matrix ($\delta y_{ji} = 1$), and then we used equation (12.37) on page 225 to compute equivalent values of δy_{ji} for points which need to be merged due to similar parameterizations. equation (12.38) was used to merge the corresponding values y_{ji}.

After smoothing each dimension j of the data, the local variance for each dimension as a function of time was estimated using the method suggested

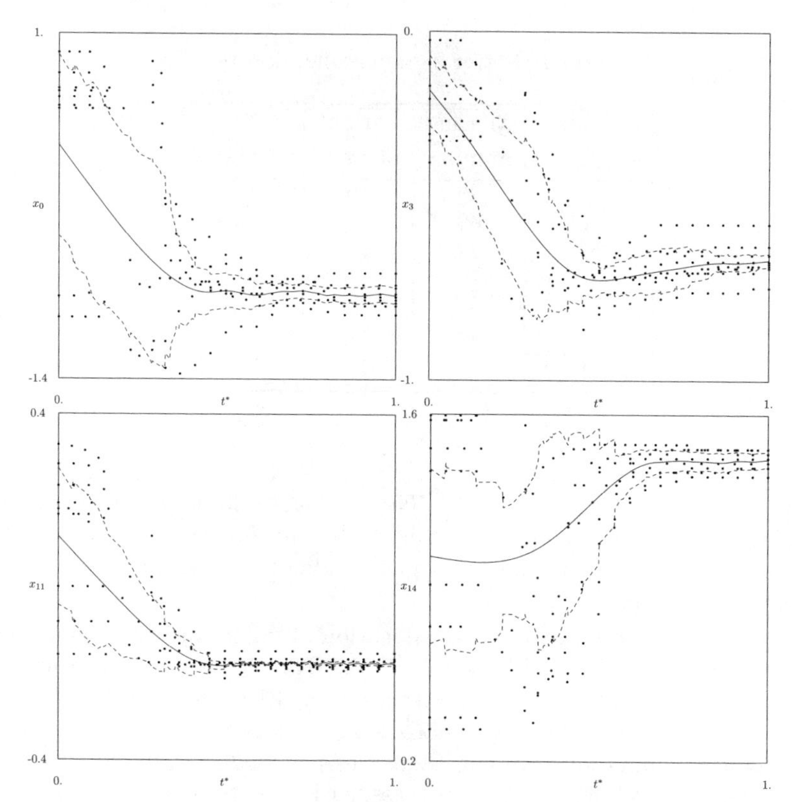

FIGURE 11.4: Spline smoother fit of four representative dimensions from the eighteen measured by a Cyberglove during signing of the letter A. Dots are sample points, the continuous line is the smoothed fit, and the dashed lines show the estimated region within one standard deviation at each time value. Time values of each example performance are rescaled to fall between 0 and 1.

by Silverman in [221] and documented (in an extended form) in Section 12.7. This variance estimate was combined via point-wise multiplication with the result from (12.37), and this new variance estimate was used to smooth the data again.

The result is a best-fit trajectory $\mathbf{f}$ for each letter-sign, and a set of variance estimates $(\delta y_{ji})^2$ for each dimension j of the motion at each rescaled time value t_i^*. We store the model for a given letter sign as $(\mathbf{t}^*, \mathbf{Y}, \{\mathbf{D}_i^2\})$, where $\mathbf{t}^*$ is the vector of rescaled time values, $\mathbf{Y}$ is composed of the column vectors $\mathbf{y}_i = \mathbf{f}(t_i^*)$, and the diagonal elements of $\mathbf{D}_i^2$ contain the corresponding variance estimates $(\delta y_{ji})^2$. Figure 11.4 plots four dimensions from the model of the sign for A. As expected, the example points from the start of each plot are widely scattered, and points near the end of the plot are clustered more tightly. This is true

Table 11.1: Letter sign classification results.

Training examples	Misclassifications number	percent	$\min(\frac{P(M_{\text{cor}})}{P(M_{\text{inc}})})$
3	46	16	-
4	69	23	-
5	24	8.2	-
6	7	2.4	-
7	8	2.7	-
8	6	2.0	-
9	0	0.0	29.9

for all 18 dimensions of the data. The estimated standard deviations are thus large at the beginning and small at the end for each dimension. The roughness of the standard deviation plots is due to the square window used for their estimation in (12.37).

These models can be used for classification of unknown performance data using the method presented in Section 11.4. This classification method was tested on 294 examples outside the training set: 21 examples for each of the 14 letter-signs. Table 11.1 * summarizes the results from using a minimum of 3 and a maximum of 9 training examples per model. After training on 6 examples per model, we see that the classifier has better than 97% reliability. After training on 9 examples per model there are no misclassifications, and the minimum ratio of the probability of the correct model to the next most likely is nearly 30 to 1.

11.6 Principal curves

Smoothing, like regression, requires an explanation variable against which to model the response variables. The principal curve, introduced by Hastie and Stuetzle [90], is a kind of smoother which can build its own parameterization against which to smooth the points in the data set. A principal curve is thus a nonlinear analogue of the first principal component discussed in Section 10.2.2. If we assume that $\mathbf{y}_i = \mathbf{f}(\lambda_i)+\epsilon_i$ where $\mathbf{f}$ is a smooth curve, and that covarience

*For the models trained using 9 training examples there were no errors in 294 classifications, and the smallest ratio of the probability of the correct model $P(M_{\text{cor}})$ to the next most likely model $P(M_{\text{inc}})$ over all the classification trials is nearly 30 to 1.

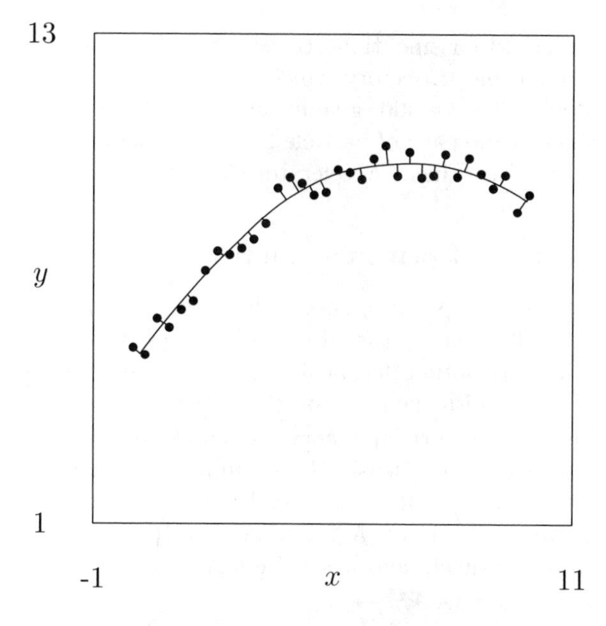

FIGURE 11.5: Principal curve.

$(\boldsymbol{\epsilon}_i) = \sigma^2\mathbf{I}$, then optimizing $\mathbf{f}$ to minimize the modeling error transforms it into an estimated principal curve of the data set.

Figure 11.5 shows a principal curve model of the example dataset from Figure 10.1(a). Figure 10.14(b) on page 189 also shows a principal curve, this time resulting from a noisy dataset sampled from a circular two-dimensional distribution. In these plots, we see that the error vectors are orthogonal to the model curve. This is because the principal curves algorithm runs a smoother on each dimension of the data (the *conditional expectation* step, which will be described later), and then each point is modeled by the nearest point on the resulting curve (the *projection* step).

Because every dimension of the data is smoothed, we need an additional parameterization against which to smooth them. The parameterization variable λ arises here in a manner similar to the way it does in the principal component model of Section 10.2.2. For the principal component model, λ was a discovered explanation variable against which variables in the orthogonal direction could be modeled. This turned a problem of finding a linear model with two response variables into a (trivial) regression of one response variable against one explanation variable. In a similar manner, the projection step of the principal curves algorithm gives us a parameterization variable λ against which to use a scatter plot smoother for each dimension of the data.

Because the principal curves algorithm can use a smoother which balances local quality of fit against the smoothness of the model curve, it is intuitively appealing as a method for smoothing trajectories. Figures 11.5 and 10.14(b) each suggest a plausible trajectory which fits a set of configuration points. In fact, Hastie and Stuetzle [90] give an example of where principal curves were used to smooth the path of particle beams through the Stanford Linear Collider at the Stanford Linear Accelerator Center.

11.6.1 Definition of principal curves

Principal curves are those smooth curves that are *self consistent* for a distribution or dataset. This means that if we pick any point on the curve, collect all the data in the distribution that project onto that point, and average them, this average coincides with the point on the curve.

To be more precise, the principal curve is defined as follows [90]. Let X be a random vector in $\mathbb{R}^m$ with density h and finite second moments. Assume, without loss of generality, that $E(X) = 0$. Let $\mathbf{f}$ be a smooth C^∞ unit speed curve in $\mathbb{R}^m$, parameterized over $\Lambda \subseteq \mathbb{R}^1$, a closed (possibly infinite) interval, that does not intersect itself, and has finite length inside any ball in $\mathbb{R}^m$.

The *Projection index* $\lambda_{\mathbf{f}}$: $\mathbb{R}^m \to \mathbb{R}^1$ is

$$\lambda_{\mathbf{f}}(\mathbf{x}) = \sup_\lambda \left[\lambda : \|\mathbf{x} - \mathbf{f}(\lambda)\| = \inf_\mu \|\mathbf{x} - \mathbf{f}(\mu)\|\right]. \tag{11.12}$$

The curve $\mathbf{f}$ is called *self-consistent* or a *principal curve* of h if

$$f(\lambda) = E(X \mid \lambda_{\mathbf{f}}(X) = \lambda) \tag{11.13}$$

for a.e. λ.

In general, we do not know for what kinds of distributions principal curves exist, nor the properties of the curves for these distributions. For some cases, however, we can give answers to these questions. For ellipsoidal distributions, the first principal component is a principal curve. For spherically-symmetric distributions, any straight line through the center of the distribution is a principal curve.

11.6.2 Distance property

Principal curves are critical points of the distance from observations. Let $\mathfrak{G}$ be a class of curves parameterized over Λ. For $\mathbf{g} \in \mathfrak{G}$ define $\mathbf{f}_t \equiv \mathbf{f} + t\mathbf{g}$. Then, curve $\mathbf{f}$ is called a critical point of the distance function D for variations in the class $\mathfrak{G}$ if:

$$\left.\frac{dD^2(h, \mathbf{f}_t)}{dt}\right|_{t=0} = 0 \quad \forall g \in \mathfrak{G}. \tag{11.14}$$

A nice property of the principal curve is that this distance property is similar to the minimization of the cost function in a spline smoother.

11.6.3 Principal curves algorithm for distributions

Principal curves are actually defined over continuous distributions rather than discrete data sets. Thus before we discuss how to approximate the principal curve from a given set of data, we present the algorithm for distributions. We can roughly state the algorithm for finding a principal curve of a given distribution as:

1. Starting with any smooth curve (usually the largest principal component), check whether the curve is self-consistent by projecting and averaging.

2. If it is not, repeat the procedure using the new curve obtained by averaging as a starting guess.

3. Iterate until the estimate (hopefully) converges.

More precisely, the algorithm is

1. INITIALIZATION:
   Set $\mathbf{f}^{(0)}(\lambda) = \bar{\mathbf{x}} + \mathbf{a}\lambda$ where $\mathbf{a}$ is the first linear principal component of h. Set $\lambda^{(0)}(\mathbf{x}) = \lambda_{\mathbf{f}^{(0)}}(\mathbf{x})$.

2. Repeat, over iteration counter j:

   (a) CONDITIONAL EXPECTATION:
   Set $\mathbf{f}^{(j)}(\cdot) = E(X \mid \lambda_{\mathbf{f}^{(j-1)}}(X) = \cdot)$.

   (b) PROJECTION:
   Define $\lambda^{(j)}(\mathbf{x}) = \lambda_{\mathbf{f}^{(j)}}(\mathbf{x})\ \forall \mathbf{x} \in h$; transform $\lambda^{(j)}$ so that $\mathbf{f}^{(j)}$ is unit speed.

   (c) ERROR EVALUATION:
   Calculate $D^2(h, \mathbf{f}^{(j)}(\mathbf{x})) = E_{\lambda^{(j)}} E[\|X - \mathbf{f}(\lambda^{(j)}(X))\|^2 \mid \lambda^{(j)}(X)]$

   Until the change in $D^2(h, \mathbf{f}^{(j)}(\mathbf{x}))$ is below some threshold.

Unfortunately, there is no guarantee that the curves produced by the conditional expectation step are differentiable (especially at the ends). Therefore, convergence cannot be proved.

However, the algorithm converges well in practice. Some justifications for why it generally works include that by definition, principal curves are fixed-points of the algorithm. Also, assuming that each iteration is well defined and differentiable, the distance does converge. Finally, if we use straight lines for conditional expectation, the algorithm converges to the first principal component.

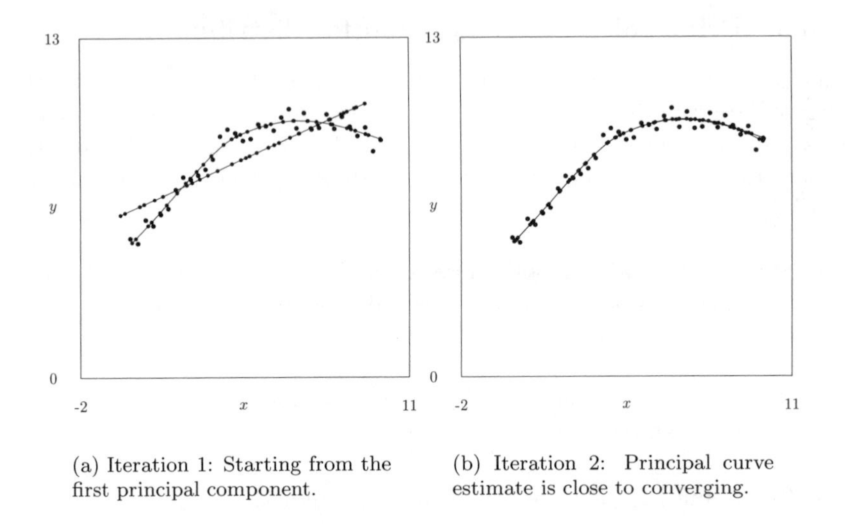

(a) Iteration 1: Starting from the first principal component.

(b) Iteration 2: Principal curve estimate is close to converging.

FIGURE 11.6: Result of first two iterations of the principal curves algorithm, starting from a projection onto the first principal component. The data points, the larger dots, are from a *sine* curve with some added noise. The smaller dots are the projections of the data points onto the principal curve estimates.

11.6.4 Principal curves algorithm for data sets: projection step

Although principal curves are defined in terms of distributions, in practice we are generally concerned with finding principal curves of datasets containing a finite number of points sampled from a distribution. The algorithm for finding a principal curve through a dataset is roughly analogous to that for finding the curve through a distribution, with a projection step and a conditional expectation step. The conditional expectation step, however, necessarily involves smoothing to estimate the effects of the distribution from which the data was sampled.

Figure 11.6 shows the result of the first two iterations of the principal curves algorithm on the example data set from Figure 10.1(a). We see in Figure 11.6(b) that the effect of the second iteration is small, indicating that it is converging to a self-consistent curve. The initial estimate, the straight line in Figure 11.6(a), is the first principal component.

In the projection step, we parameterize each data point in terms of a distance from some starting point along the estimated principal curve from the last step. The assumption here is that the probability that a given data point corresponds to a given point on a model-trajectory is a monotonically decreasing function of the distance between these points. Thus, for fixed $\mathbf{f}^{(j)}(\cdot)$ we

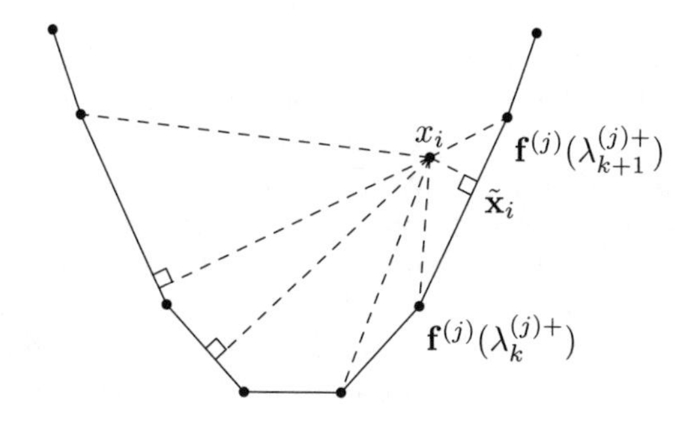

FIGURE 11.7: Projection step

find for each $\mathbf{x}_i$ in the sample the value

$$\lambda_i = \lambda_{\mathbf{f}^{(j)}}(\mathbf{x}_i). \tag{11.15}$$

The most obvious way to do this is to find the λ-parameterization of the nearest point $\tilde{\mathbf{x}}_i$ of the set of $(n-1)$ points nearest to $\mathbf{x}_i$ on the line segments comprising the current estimate of the principal curve. Let $\boldsymbol{\lambda}^{(j)+}$ be the vector of $\lambda_i^{(j)}$ values sorted into order of increasing magnitude. The current estimate of the principal curve is then the set of line segments with endpoints $\mathbf{f}^{(j)}(\lambda_k^{(j)+})$ and $\mathbf{f}^{(j)}(\lambda_{k+1}^{(j)+})$. If λ_{ik}^{+} is the parameterization of the closest point to $\mathbf{x}_i$ on the line segment between $\mathbf{f}^{(j)}(\lambda_k^{(j)+})$ and $\mathbf{f}^{(j)}(\lambda_{k+1}^{(j)+})$, and d_{ik} is the distance $\|\mathbf{x}_i - \mathbf{f}^{(j)}(\lambda_{ik}^{(j)+})\|$, then $\tilde{\mathbf{x}}_i$ is the point $\mathbf{f}^{(j)}(\lambda_{ik}^{(j)+})$ for which index $k \in (0, \ldots, n-2)$ corresponds to smallest value of d_{ik} (see Figure **??**). The λ-parameterization for $\mathbf{x}_i$ can then be computed as $\|\mathbf{f}^{(j)}(\lambda_k^{(j)+}) - \tilde{\mathbf{x}}_i\| + \lambda_k^{(j)+}$. Note that it is an approximation to consider the principal curve estimate a sequence of line segments. The actual form of the principal curve estimate between consecutive points depends on the kind of smoother used in the conditional expectation step. For a spline smoother which penalizes acceleration, the functional form of the curve-estimate is actually a cubic spline. The line segment approximation tends to work well enough in practice, however.

Unfortunately, finding each $\tilde{\mathbf{x}}_i$ in this way involves computing the nearest point to each of $(n-1)$ line segments, so the entire projection step is an $O(n^2)$ operation. If we are willing to assume that the line segment containing $\tilde{\mathbf{x}}_i$ has at least one endpoint in the set of l points $\mathbf{f}^{(j)}(\lambda_k^{(j)+})$ that are closest to $\mathbf{x}_i$, we can dramatically reduce the expense of the search. We can use a method such as kd-trees to find the l nearest neighbors in $\mathbf{f}^{(j)}(\lambda_k^{(j)+})$, then assemble all segments with at least one of these points as endpoints (there will not be less than $l-2$ and not more than $2l$ such segments), and select $\tilde{\mathbf{x}}_i$ as the closest of the points on those segments closest to $\mathbf{x}_i$. We chose to use this approximation when we coded the implementation of the principal curves algorithm for data sets, generally using $l = 3$.

The efficiency of this method is determined by the cost of building the kd-tree once and then doing n searches for the l closest points in the tree. Building the kd-tree is an $O(n \log n)$ operation [70]. The cost of the l-nearest-neighbor search depends on the intrinsic dimensionality of the set of points in the kd-tree. Generally, if the intrinsic dimensionality is approximately 8 or greater, the naïve $O(n^2)$ search is more efficient, and if the intrinsic dimension is much lower than 8, then building and searching the kd-tree to do the projection step is close to a cost of $O(n \log n)$. Since the points in the kd-tree are from a representative subset of a smooth one-dimensional curve, we can expect the kd-tree method to be much faster than the naïve search.

11.6.5 Principal curves algorithm for datasets: conditional expectation step

In the conditional expectation step, we make a new estimate of the principal curve. The goal is to estimate

$$\mathbf{f}^{(j+1)}(\lambda) = E(\mathbf{X} \mid \lambda_{\mathbf{f}^{(j)}} = \lambda), \tag{11.16}$$

for the values $\lambda \in \{\lambda_0 \ldots \lambda_{n-1}\}$ from the projection step. Because we have a data set rather than a distribution, the process for each data point $\mathbf{x}_i$ is actually to find the point which is the expected value at $\lambda_i^{(j)} \equiv \lambda^{(j)}(\mathbf{x}_i)$ of the distribution most likely to have generated the data in this vicinity of λ-space. That is, given the points which are near $\mathbf{x}_i$ in terms of their λ-values, we estimate a local distribution over λ then determine what the value of that distribution is at $\lambda_i^{(j)}$. Fortunately, as described in Section 11.3, this is exactly what a scatter plot smoother does. The projection step of the principal curves algorithm thus generates a parameterization against which a smoother can be run for each dimension of the data set to generate the next estimate of the principal curve.

The fact that the conditional-expectation step of the principal curves algorithm uses a scatter plot smoother is what makes the principal curve of a data set a local model. Section 11.2 noted that local models should be able to use local information in the example data to build good trajectory models while balancing trajectory smoothness against modeling error. By choosing an appropriate smoother, we can leverage this information to use the principal curves algorithm to find most-likely or best-fit trajectories from multiple example human performances.

11.7 Expanding the one-dimensional representation

There are some significant problems with dimension reduction by trajectory fitting as it has been presented so far. The most obvious, discussed in this section, is that a best-fit trajectory is only a one-dimensional parameterization. The second, discussed in the next section, is the problem of "branching" in the distribution of example trajectories in the training data. The problem of principal curves over-fitting the data set is discussed in Section 11.9.

Trajectory fitting generates a one-dimensional action model, with a parameterization representing the temporal ordering of points in the best-fit trajectory. For sufficiently complex action skills, this is an inadequate parameterization. If the motion of the fingers while we grasp the handles of different mugs varies consistently in particular ways depending on the shape of the particular handle or just due to stochastic variation, then there are other important variables besides temporal ordering necessary for modeling the grasping motion.

The solution to this problem is to construct additional parameterizations locally around the temporal-ordering. A simple first step is to estimate the local variance of the training points which map to each given point on the model trajectory. For a principal curve $\mathbf{f}$, and a distribution of training points X, this is

$$\sigma^2(\lambda) = E(\|X - \mathbf{f}(\lambda)\|^2 \mid X = \lambda), \tag{11.17}$$

where σ is the standard deviation. Such an estimate identifies those parts of the model trajectory corresponding to portions of the example trajectories which are most consistent, and also those parts of the model corresponding to portions of the example trajectories which are highly variable.

This variance estimate is also helpful for gesture recognition applications. When comparing an unknown gesture to the model, the variance estimate tells you how to weigh the distance between each sample point and the closest point on the gesture model. The use of variance here is similar to its use for gesture recognition using smoothing splines in Section 11.5.

Though useful, variance estimates do not actually increase the dimensionality of the model—the number of feature values assigned to a given data point to show where it projects onto the model. Using localized principal component analysis along the model trajectory can increase the dimensionality of the model, however. A small enough portion of a smooth model curve looks like a straight line segment, which should approximate the first principal component of the data that is nearby with respect to parameterization λ. Principal component analysis of the residuals from the nearby data points $[\mathbf{x}_i - \mathbf{f}(\lambda(\mathbf{x}_i))]$, weighted by their distance in λ-space, should give the directions of greatest variation which are orthogonal to the local model curve. If each point along the model curve is associated with one or more local principal component vectors, then the dimension of the model increases by this number of local principal components. To project a data point $\mathbf{x}$ onto this augmented model, one first projects it onto the model trajectory to get parameterization value λ, then computes the parameter $a_{\lambda,j}$ for each local principal component vector $\mathbf{v}_{\lambda,j}$ by projecting the residual vector onto that principal component vector

$$a_{\lambda,j} = [\mathbf{x} - \mathbf{f}(\lambda)] \cdot \mathbf{v}_{\lambda,j}. \tag{11.18}$$

This approach is related to other work on local dimensionality reduction, particularly that of Bregler and Omohundro [34] who blend local PCA models from neighboring patches in a high-dimensional space. Tibshirani [238] presents an alternative definition for principal curves which is very related to such a blend of locally linear models. The work on mixtures of probabilistic principal component analyzers of Tipping and Bishop [239] is also very similar.

Augmenting the dimensionality of a one-dimensional trajectory model by using local PCA is like adding hyper-ellipsoidal "flesh" to a one-dimensional "skeleton" model. Use of one local principal component turns the one-dimensional curve model into a ribbon-shaped manifold in the raw-data space, and adding two local principal components turns the model into a snake-like shape. The associated singular values of the local principal components describe the length of each diameter of the local hyper-ellipsoid, serving an analogous function to the local variance value discussed earlier, but weighing an associated direction in "residual space."

As opposed to the multi-dimensional feature-models generated by NLPCA and SNLPCA, this type of multi-dimensional model is much easier to interpret. The first dimension of the model corresponds to the temporal ordering of points from the model (i.e., what part of the motion), and the remaining dimensions represent the projections of the point in the local directions of greatest variance from the best-fit trajectory. The model describes a great deal about the nature of the performances it was trained from, and it is easy to understand the meaning of the feature-scores for a particular raw data point projected onto the model.

The problem with this approach of growing a higher-dimensional data representation from the "skeleton" of a best-fit trajectory is the issue of whether the meaning of the second or third feature score is similar in different regions of the model. For instance, imagine a nearly cylindrical distribution, where the first feature score s_0 is roughly a length along the axis of the cylinder, and the second feature score s_1 is the projection of the residual vector onto the most significant principal component of a nearly two-dimensional circular distribution of residual vectors $[\mathbf{x} - \mathbf{h}_0(s_0)]$. Since this distribution of residual vectors is nearly cylindrical, the orientation of the first principal component of the residuals may be a highly erratic function of the first feature score. Thus, there is no consistent interpretation of the second feature score in this case. This is an issue for future research.

11.8 Branching

The second major problem with dimension reduction by trajectory-fitting is branching. What if there is no single best-fit trajectory in the set of example performances, but instead two or more distinct prototypical trajectories? Imagine that the action we are modeling is the path as we walk from one side of a room to another, and that there is a table in the center of the room. Figure 11.8(a) shows a simple illustration of my paths across the room. This data was actually entered using the interface we developed, with the obstacle drawn on-screen. Sometimes we may walk to the left of the table, and sometimes to the right, but of course never through.

When we try to model a best-fit trajectory of my room-crossing action using a principal curve, one of two things can happen. The first is that the model assigns equivalent parameterization values for points in both branches of my paths as they pass the table. In this case, demonstrated in Figure 11.8(b), the trajectory smoother will draw the model trajectory through the table as it takes the expected value of the points with similar parameterizations. One method for diagnosing this problem is to examine the distribution of residual vectors from training points in each region of the trajectory. If the distribution over some region of the λ-space has two or more distinct clusters rather than a roughly Gaussian shape, this is an indication that the principal curve model has averaged two branches.

Another possibility is that the branching of paths will completely confuse the parameterization step of the model-building procedure. This will result in a failure to build the model—the principal curves algorithm will not converge.

To address the problem of branching, we need to treat each branch as an action to be modeled separately, as demonstrated in Figure 11.8(c). This solution, however, may result in two further difficulties. The first is that we need sufficient example points in each branch to build a quality model, but are subdividing a fixed pool of training data. The second difficulty is the number of models we may need to deal with. If we are modeling letter-signing, for instance, each signing gesture is a path in hand-configuration space from the final configuration of one letter-sign to another. Thus, instead of learning 26 models we (theoretically) need to learn 26^2 models to show each possible transition, which is an infeasible undertaking. This is the problem we avoided in Section 11.5 by eliminating the parameterization step of the principal curves algorithm and instead smoothing against time.

The multiple starting points were handled by the high estimated variance values at the beginning of the gestures. This approach is more useful for building models for recognition rather than for performance.

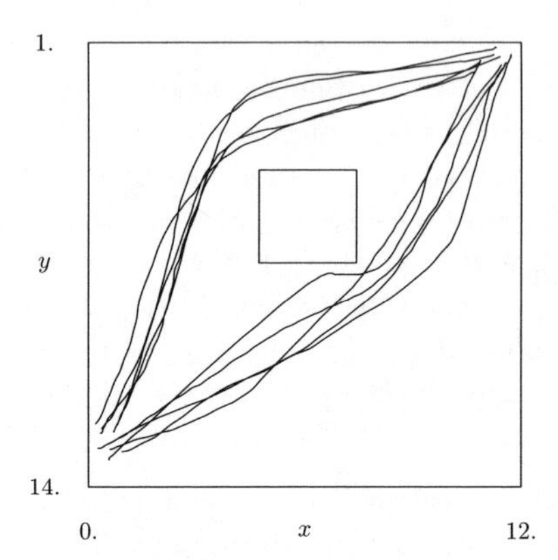

(a) Paths crossing a room, going around a table.

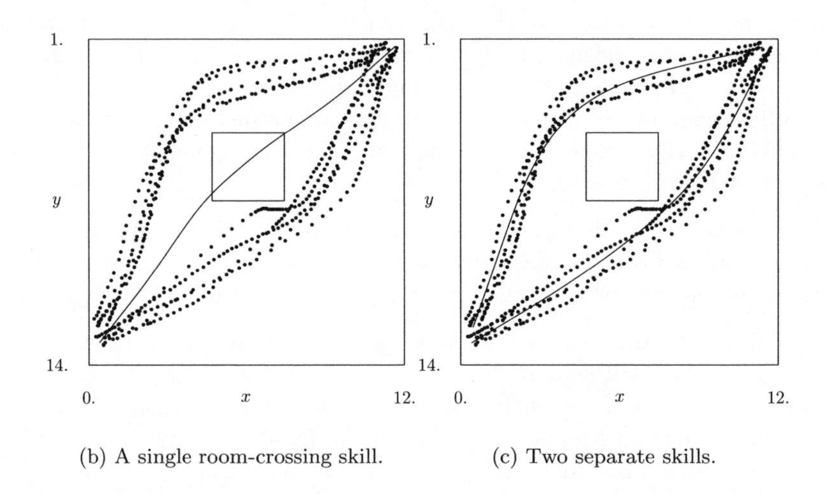

(b) A single room-crossing skill.

(c) Two separate skills.

FIGURE 11.8: Room crossing as an illustration of the branching problem for local modeling. A table in the center of the room is an off-limits part of the state space for the room-crossing skill. Using a single model for the skill gives a result by which the path is averaged into the state-space obstacle. Using separate models for each path results in acceptable models.

11.9 Over-fitting

We fit the grasping data from the previous chapter with a principal curve, using a spline smoother for conditional expectation. The difficulty in comparing the result of this model with the one-dimensional models from PCA and NLPCA, summarized in Table 10.1, is that principal curves can easily over-fit the data. Using a fairly crude method for cross-validation of the smoothing spline (compared to generalized cross-validation), we were easily able to reduce the error of the testing data to 43.1% while still reducing cross-validation error. Hastie and Stuetzle [90] note that when finding principal curves, using cross-validation to weigh the smoothness penalty versus approximation error tends to over-fit the data.

The problem with using cross validation to determine the smoothing parameter when using the principal curves algorithm is that over-fitting the data tends to lengthen the principal curve as it wiggles through space to come near to each data point. The fact that there is a greater length of curve means that it is more likely that there is a point on that curve which happens to be close to a given cross-validation point. Although Hastie and Stuetzle recommend selecting the smoothing weight manually or using early stopping of the principal curves algorithm, it is difficult to do this for fitting data in a very high-dimensional space, which is by nature difficult to visualize.

It is important to note that while an over-fit principal curve can approximate the individual points in a human performance data set very accurately, such a curve will not look like any of the example performances. In particular, the local direction of the curve at its approximation of a given example point will likely be very different than the direction of the trajectory in configuration space from which the point was sampled. The greater the over-fitting in configuration space, the greater the error we would see if we plotted the velocity data from the training data against the local derivative of the principal curve. Fortunately, the methods presented in the next two chapters will allow us to fit both the position and velocity components of the training data simultaneously, and this will greatly reduce the problem of over-fitting.

12

A Spline Smoother in Phase Space for Trajectory Fitting

12.1 Trajectory smoothing with velocity information

Many important problems in robotics and related fields can be addressed by fitting smooth trajectories to datasets containing several example trajectories. Chapter 11 describes some methods based upon local, non-parametric methods which can be used to model such trajectories. These models can be used for robot programming by demonstration, gesture recognition, animation, and comparison of a novice's motions to that of an expert.

Although using non-parametric models allows us to improve the quality of best-fit trajectory estimates by accounting for effects such as the trade-off between local fitting-error and the smoothness of the model curve, in the chapters to this point we have been using a collection of static positions to learn to model trajectories of dynamic systems. We have not used any explicit information about what happens *between* these points. By considering position information only, we have also been limited to using only part of the state-space information of most dynamical systems. The state space for physical systems typically includes both position and velocity variables.

Using velocity information in addition to configuration information enables us to build better trajectory models. When multiple example performances have been combined into a single data set without explicit information indicating which points correspond to which example performance, for instance, we can sometimes exploit the close structural relationship between local position and velocity information to reconstruct the individual performances. Figure 12.1 depicts how local velocity information can aid in finding a best-fit trajectory from two example trajectories. A smoother operating in phase space rather than configuration space, used alone or with the principal curves algorithm, can thus potentially build better trajectory models.

Conventional smoothers such as kernel smoothers, locally weighted regression [44], and spline smoothers [53], [198] are not suitable for building smoothed curves in phase space. These smoothers are designed for performing nonlinear regression—fitting response variables to explanation variables. They

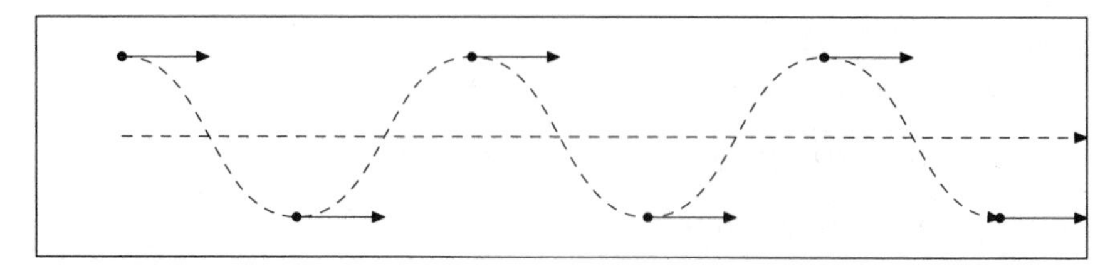

FIGURE 12.1: Velocity information can help a smoother determine that these are two parallel straight paths which should be averaged rather than interpolated between.

are not designed to operate in multi-dimensional spaces where dimensions are tightly coupled on a local scale, as are position and velocity variables within phase space vectors. Elementary calculus gives us this relationship for smooth curves: $\frac{1}{\epsilon}(\mathbf{x}_{t+\epsilon} - \mathbf{x}_t) \approx \dot{\mathbf{x}}_t$.

Because they cannot deal directly with such coupling, we could try two different methods for modeling phase space data with conventional smoothers: (a) smooth only in position space, then estimate the velocity values from the smoothed model (this is easiest with spline smoothers), or (b) smooth the full phase space data as if all its dimensions were independent. Method (a) has the disadvantage that all potentially useful state information in the velocity data is lost. Method (b) has the problem that although we are using the velocity information, we are not considering the close structural relationship between position and local velocity. Thus the result of method (b) will be an inconsistent trajectory through phase space where the smoothed paths in position space and velocity space are not compatible. That is, if we integrate the path in the velocity subspace, there is no reason why we should exactly get the corresponding path in the position subspace, and if we differentiate the position-path, there is no reason why that should exactly match the velocity-path.

For this reason, we have derived a spline smoother which can smooth data in phase space by correctly modeling the relationship between velocity and position variables. This derivation is presented in this chapter. As discussed in Section 11.5, where we used a conventional spline smoother to model letter-signing motions, smoothers are only directly useful for modeling human performance data when we have a suitable explanation variable to smooth against. The principal curves algorithm can often generate such a suitable parameterization if one cannot be specified *a priori*. In fact, our smoother can be used as part of the principal curves algorithm to find principal curves in phase space for trajectory modeling.

The formulation is based upon a smoothing spline because the polynomial form of the spline sections expresses the relationship between position and velocity in a straightforward manner. A spline function is also an explicit representation of a smooth trajectory, which is our desired output. It is a convenient form for use in interpolation, and thus well suited for tasks such as animation, or for measuring the distance between the smoothed trajectory and points sampled from other trajectories.

12.2 Problem formulation

Assume we are attempting to find a "best-fit" or "characteristic" trajectory f^*, over a given domain, of a given process $\frac{dy}{dx} = p_{\boldsymbol{\eta}}(x, y)$. The state equation of the process is unknown, and it is controlled by a set of unknown parameters $\boldsymbol{\eta}$ which vary stochastically between trials, so that we cannot expect two trials of the process over the given domain to result in the same trajectory. Suppose the process's trajectory to be smooth over each trial, and traversing through roughly the same region of the phase space within each region of the domain, but with some stochastic variation in position y and velocity $v \equiv \frac{dy}{dx}$ (e.g., Figure 12.2(a)).

During a number of trials we sample y and v over a representative set of the domain x, and call the samples $(\mathbf{x}_{ts}, \mathbf{y}_{ts}, \mathbf{v}_{ts})$ where t indicates the trial index and s indicates the index of the sample within a given trial. Instead of explicitly preserving the information about which samples correspond to which trial, we sort the data points from the multiple trials into a single list (x_i, y_i, v_i) such that $x_0 < x_1 < \cdots < x_{n-1}$. The data now resembles a set of samples in a vector field (e.g., Figure 12.2(b)). We want to find the path through the field which best balances smoothness against the quality of fit. We achieve this by minimizing the cost function

$$S = p_1 \sum_{i=0}^{n-1} \left(\frac{y_i - f(x_i)}{\delta y_i} \right)^2 + p_2 \sum_{i=0}^{n-1} \left(\frac{v_i - f'(x_i)}{\delta v_i} \right)^2 + (1 - p_1 - p_2) \int_{x_0}^{x_{n-1}} (f^{(d)}(x))^2 dx, \quad (12.1)$$

which weighs accuracy in position and velocity against smoothness of the curve, where smoothness is defined as the integral of the square of the d-th derivative over the trajectory. Thus minimizing S for $p_1 = 1, p_2 = 0$ interpolates the data (i.e., strong over-fitting), while any curve f for which $f^{(d)}(x) = 0$ over $x_0 \leq x < x_{n-1}$ (e.g., a straight line for $d = 2$) will minimize S when $p_1 = p_2 = 0$. If we assume that $(y_i - f(x_i))$ and $(v_i - f'(x_i))$ are drawn from roughly Gaussian distributions at each point x_i, then δy_i and δv_i should be the corresponding estimated standard deviations at x_i. Equation (12.1) is a simple extension of the cost function for a conventional spline smoother (11.1).

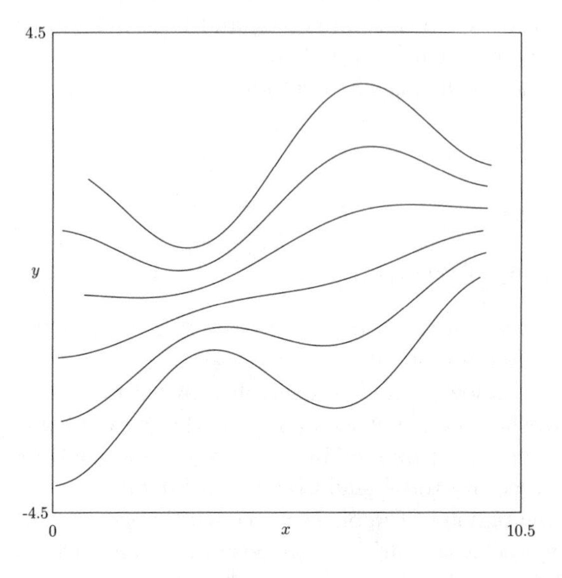

(a) Example trajectories

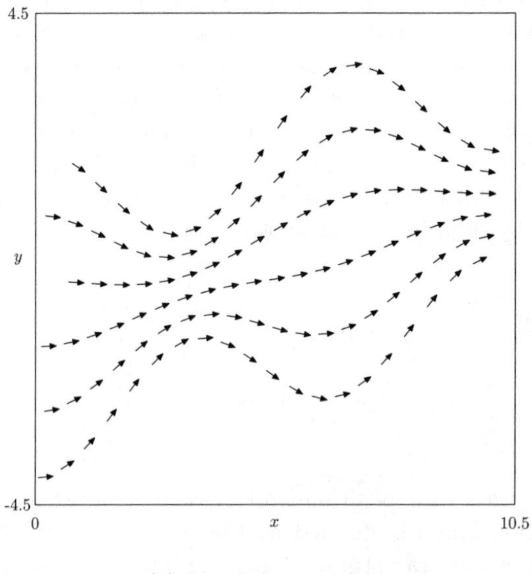

(b) Sampled vector field

FIGURE 12.2: Artificial trajectories generated over domain x using equation $y = 2\beta - \cos\left(\frac{\pi}{10}x\right) + \beta\cos\left(\frac{3\pi}{10}x\right)$ for $\beta \in \{1.0, 0.6, 0.2, -0.2, -0.6, -1.0\}$.

The function $f = f^*$ which minimizes S is a spline of order $k \equiv 2d$ with simple knots at $x_0, \ldots, x_{n-1}$ $(x_i < x_{i+1})$, and natural end conditions:

$$f^{(j)}(x_0) = f^{(j)}(x_{n-1}) = 0 \qquad \text{for } j \in \{d, \ldots, k-2\}. \tag{12.2}$$

Although there may be some arguments for minimizing jerk ($d = 3$) for certain human performance datasets, we will focus on penalizing acceleration ($d = 2$). Spline smoothing is similar to smoothing with a variable kernel, and it has been demonstrated [220] (in the conventional smoother case) that the equivalent kernel for $d = 3$ is very similar to that for $d = 2$, so we don't expect to see a dramatic difference in results between the two. The result of using $d = 2$ is that the optimal curve f^* is a cubic spline with free end conditions:

$$f''(x_0) = f''(x_{n-1}) = 0. \tag{12.3}$$

The following derivation of this smoothing spline starts in a similar manner to that presented by De Boor [53] for the conventional case which does not consider velocity information [53],[198].

12.3 Solution

Given a set of n data points $(x_i, g(x_i))$ for which $x_0 < x_1 < \cdots < x_{n-1}$, the cubic spline that interpolates these points is

$$f(x) = P_i(x) \qquad \text{for } x_i \leq x < x_{i+1}, \tag{12.4}$$

where each $P_i(x)$ is a fourth-order polynomial function, and $f(x)$ is continuous in position, velocity, and acceleration. This gives rise to the following smoothness constraints:

$$P_{i-1}(x_i) = P_i(x_i) = g(x_i) \tag{12.5}$$

$$P'_{i-1}(x_i) = P'_i(x_i) \tag{12.6}$$

$$P''_{i-1}(x_i) = P''_i(x_i). \tag{12.7}$$

Polynomials P_i may be expressed in Newton form

$$P_i(x) = a_i + b_i(x - x_i) + c_i(x - x_i)^2 + d_i(x - x_i)^3, \tag{12.8}$$

where the coefficient terms may be expressed in terms of divided differences with the knot sequence (x_i, x_i, x_i, x_{i+1}) *:

$$P_i(x) = P_i(x_i) + (x - x_i)[x_i, x_i]P_i + (x - x_i)^2[x_i, x_i, x_i]P_i + (x - x_i)^3[x_i, x_i, x_i, x_{i+1}]P_i. \quad (12.9)$$

The divided differences in equation (12.9) are determined by this divided differences table.

	$[\ ]P_i$	$[\ ,\ ]P_i$	$[\ ,\ ,\ ]P_i$	$[\ ,\ ,\ ,\ ]P_i$
x_i	$g(x_i)$			
		b_i		
x_i	$g(x_i)$		c_i	
		b_i		$\frac{([x_i,x_{i+1}]g-b_i-c_i\triangle x_i)}{(\triangle x_i)^2}$
x_i	$g(x_i)$		$\frac{([x_i,x_{i+1}]g-b_i)}{\triangle x_i}$	
		$[x_i, x_{i+1}]g$		
x_{i+1}	$g(x_{i+1})$			

This gives us an expression for d_i in terms of b_i and c_i

$$d_i = ([x_i, x_{i+1}]g - b_i - c_i\triangle x_i)/(\triangle x_i)^2 \quad (12.10)$$

which we can solve for b_i:

$$b_i = [x_i, x_{i+1}]g - c_i\triangle x_i - d_i(\triangle x_i)^2. \quad (12.11)$$

Applying smoothness constraint (12.7) to (12.8) gives us another expression for d_i:

$$\begin{aligned} c_i + 3d_i\triangle x_i &= c_{i+1} \\ d_i &= \frac{1}{3\triangle x_i}(c_{i+1} - c_i). \end{aligned} \quad (12.12)$$

Using (12.11) and (12.12), we can express b_i as

$$\begin{aligned} b_i &= [x_i, x_{i+1}]g - \tfrac{2}{3}\triangle x_i c_i - \tfrac{1}{3}\triangle x_i c_{i+1} \\ &= \frac{\triangle a_i}{\triangle x_i} - \tfrac{2}{3}\triangle x_i c_i - \tfrac{1}{3}\triangle x_i c_{i+1} \end{aligned} \quad \text{for } i \in \{0, \ldots, n-2\}. \quad (12.13)$$

*Following [53], we define the *k-th divided difference of a function g at the points* $x_i, \ldots, x_{i+k}$ (written $[x_i, \ldots, x_{i+k}]g$) to be the leading coefficient (i.e., the coefficient of x^k) of the polynomial of order $k+1$ which agrees with g at the points $x_i, \ldots, x_{i+k}$. We say that *function p agrees with function g at points* $\tau \in \{\tau_0, \ldots, \tau_n\}$ if, for each point τ which occurs m times in the sequence $\{\tau_0, \ldots, \tau_n\}$, p and g agree m-fold at τ, i.e.,

$$p^{(i)} = g^{(i)} \qquad \text{for } i \in \{0, \ldots, m-1\}.$$

Applying smoothness constraint (12.6) to (12.8) gives us the equation

$$b_{i-1} + 2\triangle x_{i-1}c_{i-1} + 3(\triangle x_{i-1})^2 d_{i-1} = b_i. \tag{12.14}$$

Using (12.13) and (12.12), we can write this constraint in terms of c_i for $i \in \{1, \ldots, n-2\}$:

$$\begin{aligned}([x_{i-1}, x_i]g - \tfrac{2}{3}\triangle x_{i-1}c_{i-1} - \tfrac{1}{3}\triangle x_{i-1}c_i)& \\ + 2\triangle x_{i-1}c_{i-1} + 3(\triangle x_{i-1})^2 &\left(\frac{1}{3\triangle x_{i-1}}(c_i - c_{i-1})\right) \\ = [x_i, x_{i+1}]g - \tfrac{2}{3}\triangle x_i c_i - \tfrac{1}{3}\triangle x_i c_{i+1},&\end{aligned}$$

which simplifies to

$$\triangle x_{i-1}c_{i-1} + 2(\triangle x_{i-1} + \triangle x_i)c_{i-1} + \triangle x_i c_{i+1} = 3\left(\frac{\triangle a_i}{\triangle x_i} - \frac{\triangle a_{i-1}}{\triangle x_{i-1}}\right). \tag{12.15}$$

Since (12.3) tells us that

$$c_0 = c_{n-1} = 0, \tag{12.16}$$

we can write the relationship between $\mathbf{a} \equiv (a_i)_0^{n-1}$ and $\mathbf{c} \equiv (c_i)_1^{n-2}$ in matrix form

$$\mathbf{Rc} = 3\mathbf{Q}^T\mathbf{a} \tag{12.17}$$

where $\mathbf{R}_{(n-2\times n-2)}$ is the symmetric tridiagonal matrix having general row

$$[\triangle x_{i-1}, \qquad 2(\triangle x_{i-1} + \triangle x_i), \qquad \triangle x_i]$$

$$\mathbf{R} = \begin{bmatrix} 2(\triangle x_0 + \triangle x_1) & \triangle x_1 & \cdots & \mathbf{0} \\ \triangle x_1 & 2(\triangle x_1 + \triangle x_2) & \triangle x_2 & \vdots \\ & \triangle x_2 & \ddots & \\ \vdots & \ddots & \ddots & \triangle x_{n-3} \\ \mathbf{0} & \cdots & \triangle x_{n-3} & 2(\triangle x_{n-3} + \triangle x_{n-2}) \end{bmatrix} \tag{12.18}$$

and $\mathbf{Q}^T_{(n-2\times n)}$ the tridiagonal matrix with general row

$$[1/\triangle x_{i-1}, \qquad -1/\triangle x_{i-1} - 1/\triangle x_i, \quad 1/\triangle x_i]$$

$$\mathbf{Q}^T = \begin{bmatrix} \frac{1}{\triangle x_0} & -\frac{1}{\triangle x_0} - \frac{1}{\triangle x_1} & \frac{1}{\triangle x_1} & \cdots & & \mathbf{0} \\ & & & & & \vdots \\ \vdots & \ddots & \ddots & & \ddots & \\ \mathbf{0} & \cdots & \frac{1}{\triangle x_{n-3}} & -\frac{1}{\triangle x_{n-3}} - \frac{1}{\triangle x_{n-2}} & & \frac{1}{\triangle x_{n-2}} \end{bmatrix}. \tag{12.19}$$

We can determine b_{n-1} using the derivative of (12.8), with (12.13) and (12.12) and the fact that $c_{n-1} = 0$:

$$\begin{aligned} b_{n-1} &= b_{n-2} + 2\triangle x_{n-2} c_{n-2} + 3(\triangle x_{n-2})^2 d_{n-2} \\ &= ([x_{n-2}, x_{n-1}]g - \tfrac{2}{3}\triangle x_{n-2} c_{n-2}) + 2\triangle x_{n-2} c_{n-2} \\ &+ 3(\triangle x_{n-2})^2 \frac{1}{3\triangle x_{n-2}}(-c_{n-2}) \\ &= \frac{-1}{\triangle x_{n-2}} a_{n-2} + \frac{1}{\triangle x_{n-2}} a_{n-1} + \tfrac{1}{3}\triangle x_{n-2} c_{n-2}. \end{aligned} \tag{12.20}$$

Combining (12.20) and (12.13), and writing in matrix form, we have

$$\mathbf{b} = \mathbf{W}\mathbf{a} - \mathbf{Z}\mathbf{c} \tag{12.21}$$

where

$$\mathbf{W}_{n\times n} = \begin{bmatrix} \frac{-1}{\triangle x_0} & \frac{1}{\triangle x_0} & & \cdots & \mathbf{0} \\ & \frac{-1}{\triangle x_1} & \frac{1}{\triangle x_1} & & \vdots \\ \vdots & & \ddots & \ddots & \\ & & & \frac{-1}{\triangle x_{n-2}} & \frac{1}{\triangle x_{n-2}} \\ \mathbf{0} & \cdots & & \frac{-1}{\triangle x_{n-2}} & \frac{1}{\triangle x_{n-2}} \end{bmatrix} \tag{12.22}$$

and

$$\mathbf{Z}_{n\times n-2} = \frac{1}{3}\begin{bmatrix} \triangle x_0 & & \cdots & \mathbf{0} \\ 2\triangle x_1 & \triangle x_1 & & \vdots \\ & \ddots & \ddots & \\ \vdots & & 2\triangle x_{n-3} & \triangle x_{n-3} \\ & & & 2\triangle x_{n-2} \\ \mathbf{0} & \cdots & & -\triangle x_{n-2} \end{bmatrix}. \tag{12.23}$$

Using (12.17), we can write this in terms of $\mathbf{a}$ only

$$\mathbf{b} = (\mathbf{W} - 3\mathbf{Z}\mathbf{R}^{-1}\mathbf{Q}^T)\mathbf{a} = \mathbf{F}\mathbf{a} \tag{12.24}$$

where $\mathbf{F} \equiv (\mathbf{W} - 3\mathbf{Z}\mathbf{R}^{-1}\mathbf{Q}^T)$.

Now, we simplify the form of the integral term in (12.1). Over each interval (x_i, x_{i+1}), the smoothness term in (12.1) is the integral of expression $(2c_i + 6d_i(x - x_i))^2$, which is the square of the area under the straight line segment with endpoints $(2x_i, 2c_i)$ and $(2x_{i+1}, 2c_{i+1})$. Since for any straight line l

$$\int_0^h l^2(x)dx = (h/3)(l^2(0) + l(0)l(h) + l^2(h), \tag{12.25}$$

we can write the integral term as

$$(1 - p_1 - p_2)\int_{x_0}^{x_{n-1}} (f''(x))^2 dx = \tfrac{4}{3}(1 - p_1 - p_2)\sum_{i=0}^{n-2} \triangle x_i (c_i^2 + c_i c_{i+1} + c_{i+1}^2). \tag{12.26}$$

Using (12.24) and (12.26) we can express cost function (12.1) in vector form, in terms of $\mathbf{a}$ and $\mathbf{c}$

$$S = p_1(\mathbf{y}-\mathbf{a})^T\mathbf{D}_p^{-2}(\mathbf{y}-\mathbf{a}) + p_2(\mathbf{v}-\mathbf{Fa})^T\mathbf{D}_v^{-2}(\mathbf{v}-\mathbf{Fa}) + \tfrac{2}{3}(1-p_1-p_2)\mathbf{c}^T\mathbf{Rc}, \quad (12.27)$$

where $\mathbf{D}_p$ is the diagonal matrix $\lceil \delta y_0, \ldots, \delta y_{n-1} \rfloor$ and $\mathbf{D}_v$ is $\lceil \delta v_0, \ldots, \delta v_{n-1} \rfloor$. By (12.17) we can rewrite (12.27) as a function of $\mathbf{a}$ only:

$$\begin{aligned} S(\mathbf{a}) =& p_1(\mathbf{y}-\mathbf{a})^T\mathbf{D}_p^{-2}(\mathbf{y}-\mathbf{a}) + p_2(\mathbf{v}-\mathbf{Fa})^T\mathbf{D}_v^{-2}(\mathbf{v}-\mathbf{Fa}) \\ &+ 6(1-p_1-p_2)(\mathbf{R}^{-1}\mathbf{Q}^T\mathbf{a})^T\mathbf{R}(\mathbf{R}^{-1}\mathbf{Q}^T\mathbf{a}). \end{aligned} \quad (12.28)$$

Because $\mathbf{D}_p^{-2}$, $\mathbf{D}_v^{-2}$, and $(\mathbf{R}^{-1}\mathbf{Q}^T\mathbf{a})^T\mathbf{R}(\mathbf{R}^{-1}\mathbf{Q}^T\mathbf{a})$ are positive definite, the cost function is minimized when $\mathbf{a}$ satisfies

$$\begin{aligned} &12(1-p_1-p_2)(\mathbf{R}^{-1}\mathbf{Q}^T)^T\mathbf{R}(\mathbf{R}^{-1}\mathbf{Q}^T\mathbf{a}) \\ &\quad - 2p_1\mathbf{D}_p^{-2}(\mathbf{y}-\mathbf{a}) - 2p_2\mathbf{F}^T\mathbf{D}_v^{-2}(\mathbf{v}-\mathbf{Fa}) = 0. \end{aligned} \quad (12.29)$$

Solving for $\mathbf{a}$,

$$\begin{aligned} &\left[p_1\mathbf{D}_p^{-2} + p_2\mathbf{F}^T\mathbf{D}_v^{-2}\mathbf{F} + 6(1-p_1-p_2)(\mathbf{R}^{-1}\mathbf{Q}^T)^T\mathbf{R}(\mathbf{R}^{-1}\mathbf{Q}^T)\right]\mathbf{a} \\ &\quad = [p_1\mathbf{D}_p^{-2}\mathbf{y} + p_2\mathbf{F}^T\mathbf{D}_v^{-2}\mathbf{v}], \end{aligned} \quad (12.30)$$

and simplifying using $(\mathbf{R}^{-1})^T = \mathbf{R}^{-1}$ (because $\mathbf{R}$ is symmetric), we can solve the following linear system for $\mathbf{a}$:

$$\begin{aligned} \left[p_1\mathbf{D}_p^{-2} + p_2\mathbf{F}^T\mathbf{D}_v^{-2}\mathbf{F} + 6(1-p_1-p_2)\mathbf{Q}\mathbf{R}^{-1}\mathbf{Q}^T\right]\mathbf{a} &= [p_1\mathbf{D}_p^{-2}\mathbf{y} + p_2\mathbf{F}^T\mathbf{D}_v^{-2}\mathbf{v}] \\ \mathbf{Ma} &= \mathbf{z}. \end{aligned} \quad (12.31)$$

Once we know $\mathbf{a}$, we can use (12.17) to solve for $\mathbf{c}$, and by (12.24) we can compute $\mathbf{b} = \mathbf{Fa}$. Finally, we can use (12.12) to determine $\mathbf{d} \equiv (d_i)_{i=0}^{n-2}$. Knowing $\mathbf{a}$, $\mathbf{b}$, $\mathbf{c}$, and $\mathbf{d}$, we now can use (12.4) and (12.8) to compute the trajectory $f^*(x)$ over the domain $x_0 \leq x \leq x_{n-1}$. Figure 12.3 shows the result of this computation for the dataset plotted in Figure 12.2 (b), using the method of Section 12.7 for estimating variances $\mathbf{D}_p$ and $\mathbf{D}_v$.

12.4 Notes on computation and complexity

One major problem with the computation in (12.31) is the matrix $\mathbf{R}^{-1}$, which is always used in conjunction with $\mathbf{Q}^T$. Letting $\mathbf{H} = \mathbf{R}^{-1}\mathbf{Q}^T$, we can solve the following linear system for $\mathbf{H}$

$$\mathbf{RH} = \mathbf{Q}^T, \quad (12.32)$$

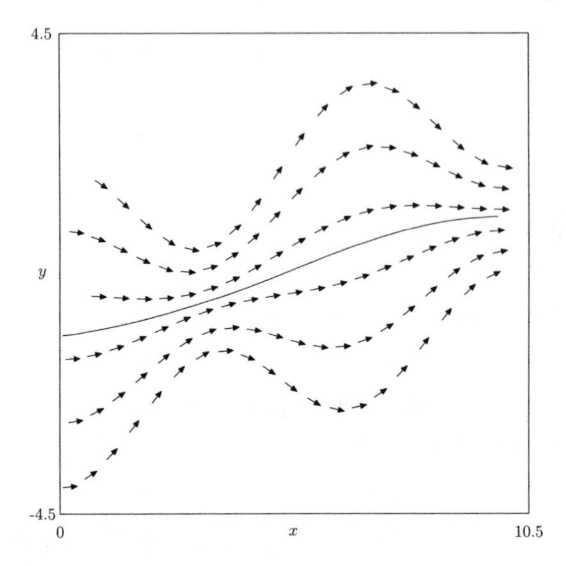

FIGURE 12.3: Computed interpolation using $p_1 = p_2 = 0.1$, with variance estimation.

such that we can use $\mathbf{H}$ to eliminate explicit computation of $\mathbf{R}^{-1}$ in (12.31) and (12.24). Because $\mathbf{R}$ is a tridiagonal symmetric matrix composed of $\triangle x_i$ elements (see (12.18)), it turns out that although $\mathbf{H}$ is not a band matrix, it is highly band-dominated.

If we can adequately approximate $\mathbf{H}$ by a band matrix $\tilde{\mathbf{H}}$, then this bandedness will propagate through all terms composing $\mathbf{M}$ in (12.31) to give us a band matrix approximation $\tilde{\mathbf{M}}$. In this case, inspection of equation (12.31) shows us that matrix $\tilde{\mathbf{M}}$ not only can be computed in linear time, but also is a positive definite symmetric band matrix. Cholesky factorization of $\tilde{\mathbf{M}}$ can be performed in linear time and space [81], and then approximations to $(\mathbf{a}, \mathbf{b}, \mathbf{c}, \mathbf{d})$ can be also found in linear time and space.

If we do not approximate $\mathbf{H}$ with a band matrix, then $\mathbf{M}$ will be a non-banded positive-definite symmetric matrix, and Cholesky decomposition will be an $O(n^3)$ operation [81], dominating the cost of the smoothing solution. Since $\mathbf{M}$ is an $(n \times n)$ matrix, storage cost of the computation will be $O(n^2)$.

The main difference in computational expense between the smoothing algorithm in [53] and our smoother stems from the fact that without velocity information the linear system may be computed in terms of $\mathbf{c}$ rather than $\mathbf{a}$ using (11.4), which is introduced in Chapter 11 but repeated here for convenience:

$$(6(1-p)\mathbf{Q}^T\mathbf{D}^2\mathbf{Q} + p\mathbf{R})\mathbf{c} = 3p\mathbf{Q}^T\mathbf{y}.$$

This reduces the amount of computation necessary for computing the smoothed function through the data compared to the trajectory smoother presented here. Equation (11.4) is a tridiagonal linear system which can be solved in linear time and space.

12.5 Combining points with similar parameterizations

Matrices $\mathbf{Q}^T$ and $\mathbf{W}$ are built from terms of the form $1/\triangle x_i$. This is problematic when multiple points in the dataset have the same or nearly the same parameterization value x. The solution is to replace each set of points having similar parameterization values with a single point such that the function $f = f^*$ which minimizes cost function (12.1) remains unchanged. Thus, for each set of points for which

$$|x_k - x_*| < \epsilon \qquad \text{for } k \in (a, a+1, \ldots, b) \tag{12.33}$$

we compute parameters $(y_*, \delta y_*, v_*, \delta v_*)$, of a replacement point such that

$$\frac{\partial}{\partial f(x_*)}\left(\frac{y_* - f(x_*)}{\delta y_*}\right)^2 = \frac{\partial}{\partial f(x_*)}\sum_{k=a}^{b}\left(\frac{y_k - f(x_*)}{\delta y_k}\right)^2 \tag{12.34}$$

and

$$\frac{\partial}{\partial f'(x_*)}\left(\frac{v_* - f'(x_*)}{\delta v_*}\right)^2 = \frac{\partial}{\partial f'(x_*)}\sum_{k=a}^{b}\left(\frac{v_k - f'(x_*)}{\delta v_k}\right)^2. \tag{12.35}$$

Simplifying each side of (12.34) gives us

$$\frac{-2(y_* - f(x_*))}{(\delta y_*)^2} = -2\left(\sum_{k=a}^{b}\frac{y_k}{(\delta y_k)^2} - f(x_*)\sum_{k=a}^{b}\frac{1}{(\delta y_k)^2}\right). \tag{12.36}$$

Setting

$$\frac{1}{(\delta y_*)^2} = \sum_{k=a}^{b}\frac{1}{(\delta y_k)^2}, \tag{12.37}$$

$$y_* = (\delta y_*)^2\sum_{k=a}^{b}\frac{y_k}{(\delta y_k)^2} \tag{12.38}$$

ensures the equality in (12.36). A similar argument from (12.35) gives us

$$\frac{1}{(\delta v_*)^2} = \sum_{k=a}^{b}\frac{1}{(\delta v_k)^2}, \tag{12.39}$$

$$v_* = (\delta v_*)^2\sum_{k=a}^{b}\frac{v_k}{(\delta v_k)^2}. \tag{12.40}$$

Thus to smooth data $(\mathbf{x}, \mathbf{y}, \mathbf{v})$ given $(\delta\mathbf{y}, \delta\mathbf{v})$, we map these vectors via (12.37)-(12.40) to $(\mathbf{x}_*, \mathbf{y}_*, \mathbf{v}_*)$ and $(\delta\mathbf{y}_*, \delta\mathbf{v}_*)$ such that $x_{*i} + \epsilon < x_{*i+1}$, then smooth the mapped points. Many practical uses of the smoother will require mapping the smoothed results back to the order of the original data, which may require mapping individual smoothed points to more than one location in the original ordering.

In most cases $\delta v_* = \delta v_k$ for $k \in (a, \ldots, b)$, because δv is an estimated variance which is a function of x. Then, (12.40) and (12.39) reduce to

$$\frac{1}{(\delta v_*)^2} = (b - a + 1)(\delta v_a)^2, \tag{12.41}$$

$$v_* = \frac{1}{b - a + 1} \sum_{k=a}^{b} v_k, \tag{12.42}$$

and the same happens for the position values. This happens for a posteriori computations of variance, as in Section 12.7.

12.6 Multi-dimensional smoothing

When smoothing data of more than one dimension, the cost function becomes

$$S = p_1 \sum_{i=0}^{n-1} \|(\mathbf{y}_i - \mathbf{f}(x_i))\mathbf{D}_{\text{p}i}^{-1}\|^2 + p_2 \sum_{i=0}^{n-1} \|(\mathbf{v}_i - \mathbf{f}'(x_i))\mathbf{D}_{\text{v}i}^{-1}\|^2 + (1 - p_1 - p_2) \int_{x_0}^{x_{n-1}} \|\mathbf{f}^{(d)}(x)\|^2 dx \tag{12.43}$$

where $\mathbf{D}_{\text{p}i}$ and $\mathbf{D}_{\text{v}i}$ are diagonal weighting matrices. The process of minimizing this expression can be split into a set of separate minimizations of the form (12.1), one for each dimension of the data. Thus, there is no significant difference in method of solution. The method for combining points with similar parameterizations (Section 12.5) can also be performed separately for each dimension.

Note that if we use the same weights for each dimension, $\mathbf{D}_{\text{p}i} = (\delta y_i)\mathbf{I}$ and $\mathbf{D}_{\text{v}i} = (\delta v_i)\mathbf{I}$, then we need only compute matrix $\mathbf{M}$ from (12.31) once for the entire multi-dimensional smoothing problem. Moreover, we can compute its Cholesky decomposition once, and use this decomposition to smooth each dimension of the data. Thus for each separate dimension of the smoothing problem, we need only multiply a vector by a matrix (a band matrix, if we are using the band matrix approximation $\tilde{\mathbf{H}}$), and sum the pairwise product of two vectors to form $\mathbf{z}$, then use forward and backward substitution to solve for $\mathbf{a}_i$.

12.7 Estimation of variances

In the problem formulation (Section 12.2), we said that if we assume that $(y_i - f(x_i))$ and $(v_i - f'(x_i))$ are drawn from roughly Gaussian distributions at each point x_i, then δy_i and δv_i should be the corresponding standard deviations at x_i. The variances are thus $(\delta y_i)^2$ and $(\delta v_i)^2$. Because we rarely know these variances *a priori*, we need some way to estimate them from the training data. If we assume that the variances are smooth functions of domain x, then we can estimate them by locally weighing error residuals from an unweighted iteration of the smoothing procedure [221]. In this case, we first smooth data $(\mathbf{y}, \mathbf{v})$ using unit weights $\delta y_i = \delta v_i = 1$ to obtain $(\mathbf{a}_\mathrm{u}, \mathbf{b}_\mathrm{u})$, and calculate the unweighted residuals $\mathbf{r}_\mathrm{pu} = (\mathbf{y} - \mathbf{a}_\mathrm{u})$, $\mathbf{r}_\mathrm{du} = (\mathbf{v} - \mathbf{b}_\mathrm{u})$. Then, the variances are estimated with a local moving average of squared residuals

$$\begin{aligned} (\delta y_{1i})^2 &= (n_i - m_i + 1)^{-1} \sum_{k=m_i}^{n_i} r_{\mathrm{pu}k}^2 \\ (\delta v_{1i})^2 &= (n_i - m_i + 1)^{-1} \sum_{k=m_i}^{n_i} r_{\mathrm{du}k}^2, \end{aligned} \tag{12.44}$$

where

$$m_i = \max(0, i - k) \qquad \text{and} \qquad n_i = \min(n - 1, i + k). \tag{12.45}$$

Silverman [221], in the context of conventional spline smoothing, explains that this method is generally reliable because highly accurate estimates of the variances are generally not necessary. He suggests that $k = 5$ has produced good results for data sets of moderate size, and this has produced reasonable results for our spline smoother as well. Figure 12.4 shows the effect of this kind of variance estimation on the example dataset, with $k = 5$.

If desired, another iteration of variance re-estimation can also be performed by smoothing with our previous estimates $(\delta y_{1i}, \delta v_{1i})$ to obtain $(\mathbf{a}_1, \mathbf{b}_1)$ and residuals $\mathbf{r}_\mathrm{p1} = (\mathbf{y} - \mathbf{a}_1)$, $\mathbf{r}_\mathrm{d1} = (\mathbf{v} - \mathbf{b}_1)$:

$$\begin{aligned} (\delta y_{2i})^2 &= (n_i - m_i + 1)^{-1} (\delta y_{1i})^2 \sum_{k=m_i}^{n_i} r_{\mathrm{p1}k}^2 \\ (\delta v_{2i})^2 &= (n_i - m_i + 1)^{-1} (\delta v_{1i})^2 \sum_{k=m_i}^{n_i} r_{\mathrm{d1}k}^2. \end{aligned} \tag{12.46}$$

If we are performing multi-dimensional smoothing and want to use a single variance estimate for all dimensions of each point, we perform the unweighted smoothing on data $(\mathbf{Y}, \mathbf{V})$ using weights $\mathbf{D}_\mathrm{pu} = \mathbf{D}_\mathrm{vu} = \mathbf{I}$ to obtain $(\mathbf{A}_\mathrm{u}, \mathbf{B}_\mathrm{u})$,

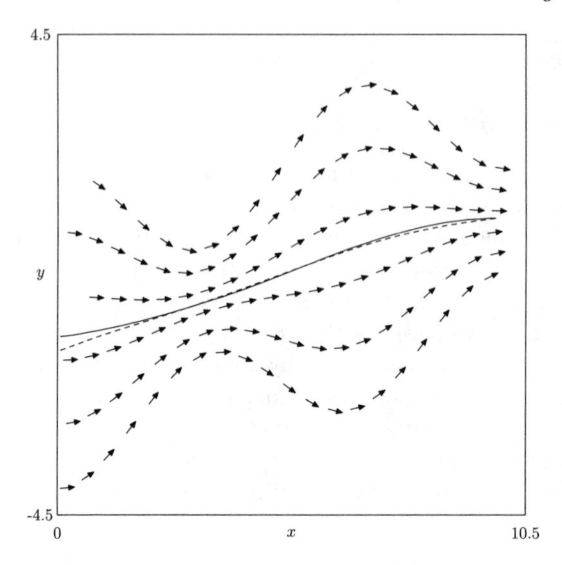

FIGURE 12.4: Effect of variance estimation with $p_1 = p_2 = 0.1$, $k = 5$. Dashed line unweighted, solid line weighted.

and calculate the unweighted residuals $\mathbf{r}_{\mathrm{pu}i} = (\mathbf{y}_i - \mathbf{a}_{\mathrm{u}i})$, $\mathbf{r}_{\mathrm{du}i} = (\mathbf{v}_i - \mathbf{b}_{\mathrm{u}i})$. The variances are then estimated as

$$\begin{aligned}
(\delta y_{1i})^2 &= (n_i - m_i + 1)^{-1} \sum_{k=m_i}^{n_i} \|\mathbf{r}_{\mathrm{pu}k}\|^2 \\
(\delta v_{1i})^2 &= (n_i - m_i + 1)^{-1} \sum_{k=m_i}^{n_i} \|\mathbf{r}_{\mathrm{du}k}\|^2 \\
\mathbf{D}_{\mathrm{p}1i} &= (\delta y_{1i})\mathbf{I} \\
\mathbf{D}_{\mathrm{v}1i} &= (\delta v_{1i})\mathbf{I},
\end{aligned} \tag{12.47}$$

A second iteration can be performed by smoothing with $(\mathbf{D}_{\mathrm{p}1}, \mathbf{D}_{\mathrm{v}1})$ to obtain $(\mathbf{A}_1, \mathbf{B}_1)$ and residuals $\mathbf{r}_{\mathrm{p}1i} = (\mathbf{y}_i - \mathbf{a}_{1i})$, $\mathbf{r}_{\mathrm{d}1i} = (\mathbf{v} - \mathbf{b}_{1i})$:

$$\begin{aligned}
(\delta y_{2i})^2 &= (n_i - m_i + 1)^{-1} (\delta y_{1i})^2 \sum_{k=m_i}^{n_i} \|\mathbf{r}_{\mathrm{p}1k}\|^2 \\
(\delta v_{2i})^2 &= (n_i - m_i + 1)^{-1} (\delta v_{1i})^2 \sum_{k=m_i}^{n_i} \|\mathbf{r}_{\mathrm{d}1k}\|^2 \\
\mathbf{D}_{\mathrm{p}2i} &= (\delta y_{2i})\mathbf{I} \\
\mathbf{D}_{\mathrm{v}2i} &= (\delta v_{2i})\mathbf{I}.
\end{aligned} \tag{12.48}$$

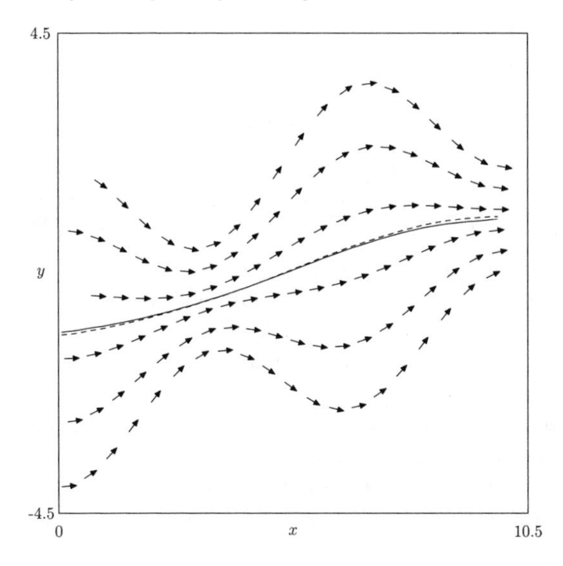

FIGURE 12.5: Effect of windowing the variance estimates. Dashed line is without windowing, solid with windowing.

12.8 Windowing variance estimates

Because the recorded trajectories may not all begin and end at the exact same locations in the domain, the data at the beginning and end of the smoothing domain will often be sparser and less representative of the underlying process than data in the middle of the domain. Therefore, we may want to adjust the weights of the points at the extremes of the domain to make them count less in the smoothing process. This can be easily accomplished by applying a windowing function, such as the Hamming function, to the first and last k points in the dataset:

$$\begin{aligned} \delta y_i &\leftarrow (0.54 + 0.46\cos(\pi(k-j)/k))\delta y_i \\ \delta v_i &\leftarrow (0.54 + 0.46\cos(\pi(k-j)/k))\delta v_i, \end{aligned} \tag{12.49}$$

where

$$j = \begin{cases} i & \text{for } i \in \{0, \ldots, k-1\} \\ n-i-1 & \text{for } i \in \{n-1-k, \ldots, n-1\}. \end{cases}$$

The effects of windowing the example dataset are shown in Figure 12.5. In this case, the effect is minimal because the variance estimation procedure has solved most of the problem on its own.

12.9 The effect of velocity information

The effects of smoothing with and without velocity information are demonstrated in Figures 12.6 and 12.7. In 12.6(a), the effects of the velocity information are most noticeable at the ends of the plot. Since the trajectories starting at the lowest y values tend to start slightly before those with higher initial y values, we see that the trajectory fitted using only position information starts low, with a positive first-derivative which does not match those of the sampled points around it. The end of the trajectory has a similar problem, but it is less pronounced since the local variance of the sample points is smaller. The fitted trajectory which weighs velocity information has a more plausible beginning and ending, as we would expect. Figure 12.6(b) shows an example where a random half of the data points are removed. The trajectory fitted using velocity information more closely tracks a single example trajectory.

12.10 Cross-validation

Recall from Section 11.3 that cross-validation [229] is typically used to automatically select the smoothing parameter p for a conventional spline s-moother, and that the work of Craven and Wahba [48] and Hutchinson and de Hoog [102] show how to compute this parameter in closed-form and in linear time as part of the smoothing process.

We have not attempted to derive a similar generalized cross-validation method for the smoother presented in this chapter, and suspect that it may not be a simple extension of the generalized cross-validation formulation for conventional spline smoothing. For deriving generalized cross-validation, conventional spline smoothers are generally formulated using the mathematics of reproducing kernel Hilbert spaces [11], [246]. We do not currently know whether the spline smoother in phase space can be formulated in this manner.

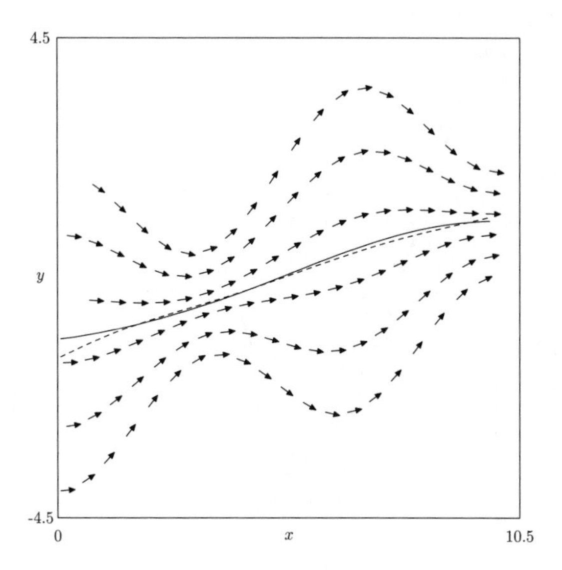

(a) Solid: $p_1 = p_2 = 0.05$, dashed: $p_1 = 0.1$, $p_2 = 0$.

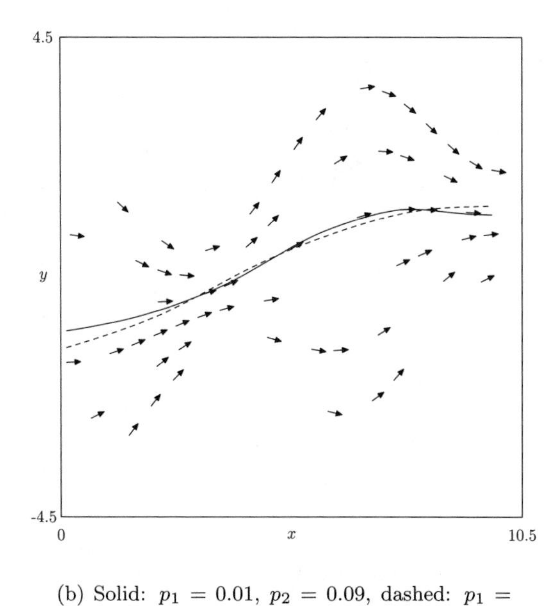

(b) Solid: $p_1 = 0.01$, $p_2 = 0.09$, dashed: $p_1 = 0.1, p_2 = 0$.

FIGURE 12.6: Effect of derivative information. Plots use local variance estimation ($k = 5$).

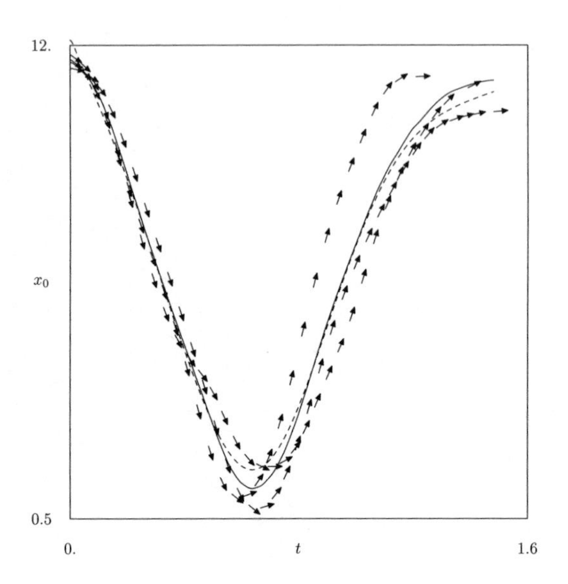

(a) Smoothing of horizontal coordinate

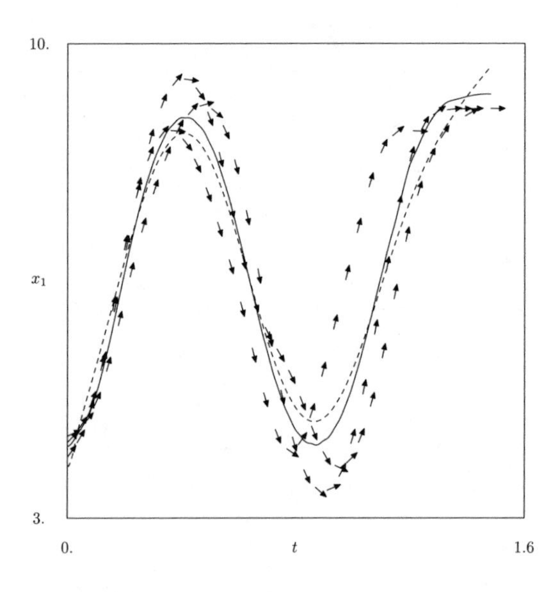

(b) Smoothing of vertical coordinate

FIGURE 12.7: Smoothing data from multiple examples of a drawing motion. Solid line: $p_1 = 0.645$, $p_2 = 0.35$. Dashed: $p_1 = 0.995$, $p_2 = 0.0$. No variance estimation.

13

Analysis of Human Walking Trajectories for Surveillance

13.1 Introduction

In this chapter, we study an application in which the actions of humans are modeled for analysis. * We apply artificial intelligence and statistical analysis techniques towards real-time observation of people, leading to the classification of their walking patterns. The needs for intelligent human motion understanding in various areas are immediate, especially in applications of video surveillance for security purposes. It has become an important global concern as many nations have experienced tragic terrorist activities in recent years. One of the fundamental issues in many security systems is the understanding of human motions. Although video cameras and monitoring systems are cheap and commonplace in countless locations, the costs involved in getting the suitable people to observe and analyze the recorded materials are still very high. Also, instead of investigating the scenes after the activities have already finished, we can actually push the technology forward by interpreting the human actions intelligently in real-time. Real-timeliness is especially important for the application of surveillance, in which crimes and abnormal activities can only be prevented and stopped if the system is equipped with the capability to understand human actions at the moment when they are performed.

Various systems have been developed to tackle the problem of detecting unusual events for visual surveillance applications [67], [200]. The analysis of human motions in outdoor scenes is one of the aspects that has many important applications.

Foresti and Roli [71] developed a system that can locate, track, and classify dangerous events. Three types of potentially dangerous events can be recognized by building a statistical model of the object trajectory in order to obtain high-level descriptions of the behaviors of the detected object in the scene. Detected trajectories are mapped to Bezier curves and a number of param-

 "Analysis of Human Walking Trajectories for Surveillance," *Int. J. of Information Acquisition*, vol. 1, no. 2, pp. 169-189, 2004.

eters of the fitted curves are considered as trajectory features. The features are then passed to a binary-neural-tree-based network for classification.

Owens and Hunter [180] applied an approach based on a self-organizing map for trajectory classification. The method classifies trajectory normality based on point information only. The global features of the trajectories are not well represented.

Fraile and Maybank [72] presented a technique for finding smooth approximation to vehicle trajectories. The measured trajectory is divided into overlapping segments. In each segment the trajectory is approximated by a smooth function and then assigned to one of four categories: ahead, left, right, or stop. The string of symbol is then fed into a hidden Markov model for classification.

The video surveillance system developed by Ivanov, et al. [108] can track the motion of cars and humans in the scene. The system maps the object tracks into a set of pre-determined discrete events. It can perform segmentation and labeling of surveillance video of a parking lot and identifies person-vehicle interactions. However, it does not analyze the detailed configuration of the trajectories. Only the startpoints and endpoints of the tracks are used for analysis.

Johnson and Hogg [110] developed a model of the probability density functions of possible instantaneous movements and trajectories within a scene. The probability density functions are represented by the distribution of prototype vectors that are placed by a neural network implementing vector quantization. But it can only determine the typicality of the observed trajectories after they have completed.

The previous researches have provided valuable insights into the development of successful human trajectory analysis systems. However, further research can still be pursued in order to achieve better results. Based on the limitation of the previous work, we would like to develop a new system with the following points taken into consideration.

1. Both the global and local information of the trajectories are important to the classification task. Relying on only one of the two cannot give us comprehensive information to determined the normality of the trajectories.

2. Since the local and global trajectory information have different characteristics, they should be handled using separate intelligent subsystems, so that the information provided by the two types of data can be better utilized.

3. Instead of waiting for the completion of the whole trajectory, the normality of the trajectory should be determined at the moment when sufficient abnormal information has been collected.

In this research, we propose an approach that separates the local and global information of the trajectories and handles them by two specialized learning mechanisms. Also, temporal and sequential characteristics of the human trajectories will be taken into account.

13.2 System overview

We have developed an intelligent system that can analyze the walking trajectories of pedestrians in outdoor environments and report abnormal or suspicious activities in real-time. The system takes into account both the local and global information of the trajectories for decision making. Two separate learning mechanisms are running in parallel to process the local and global trajectory information at the same time. Walking humans are extracted from the observed scene using adaptive background subtraction techniques and the trajectories of the humans are recorded and examined.

In order to understand the local information of the human trajectory, support vector learning is utilized to analyze the normality of trajectory points (the local trajectory point classifier). Features are retrieved from a set of normal and abnormal trajectories and are used for the training of support vector machines (SVM). The trained SVMs can classify the data points into normal or abnormal cases. A counter is used to store the number of accumulated abnormal positions detected.

The long-term behavior of the human in the scene is analyzed using hidden Markov models (the global trajectory similarity estimator). We have stored a number of trajectories that are demonstrated as normal for particular scenes in a database and the observed trajectory is compared against the trajectories in the database using a hidden-Markov-model-based stochastic similarity measure. Each comparison will yield a value and the number of similarity values generated is the same as the number of normal trajectories in the database. The highest similarity value will become the global normality index for the observed trajectory.

The outputs of the local trajectory point classifier and the global trajectory similarity estimator will be passed to a rule-based trajectory normality classifier to determine whether the walking motion of the observed human should be classified as normal or abnormal.

13.3 Background subtraction

For indoor applications, it might be reasonable to assume that the background is stationary for a long period of time. However, for our outdoor application, background variations such as illumination change, swaying of trees and object deposition, etc., can produce wrong recognition results. An adaptive background model is adopted to prevent the changes in the background from affecting the human tracking performance [88]. In order to handle the illumination changes in the environment, the background model is updated periodically. Physical changes in the background scene such as the parking and removal of cars will be absorbed by the background model if the effect of the change is present for a long period of time.

Preprocessing steps are applied on the foreground pixels to enhance the performance of tracking. The foreground pixels close to each other are aggregated using a connected component approach. Morphological dilation and erosion operations help to fill up the small holes in the foreground object and remove the interlacing anomalies. The contour of the foreground object in the scene is obtained using a border-following algorithm. The centroid (x_c, y_c) of the object boundary is found using the following equations and is recorded for feature extraction and trajectory analysis: $x_c = \frac{1}{N}\sum_{i=1}^{N} x_i$ and $y_c = \frac{1}{N}\sum_{i=1}^{N} y_i$, where (x_i, y_i) represent the points on the human boundary and there are a total of N points on the boundary.

The recorded human walking trajectory points are converted into features for analysis. The following features are used in the learning process of SVM: mean X position, mean Y position, mean X velocity, mean Y velocity, mean X acceleration, mean Y acceleration, distance traveled, mean orientation, and variance of orientation. These values are averaged over a short period of time τ.

13.4 Local trajectory point classification

The local trajectory point classifier is responsible for classifying the normality of the motion features extracted during the time of trajectory observation. It involves two steps: the learning stage (Figure 13.1) and the classification stage (Figure 13.2). We utilize support vector machines to classify the normality of the features of the trajectory points. Support vector machine [211] is a recently developed statistical learning technique which has found applications in various areas and we have successfully applied SVM in applications such as robot control skills learning [179] and on-line fault diagnosis of intelligent manufacturing systems [253]. SVM can be considered as a linear method in a high-dimensional feature space nonlinearly related to the input space. By the use of kernels, all input data are mapped nonlinearly into a high-dimensional features space. Separating hyperplanes are constructed with maximum margins which yields a nonlinear decision boundary in the input space. Using appropriate kernel functions, it is possible to compute the separating hyper-

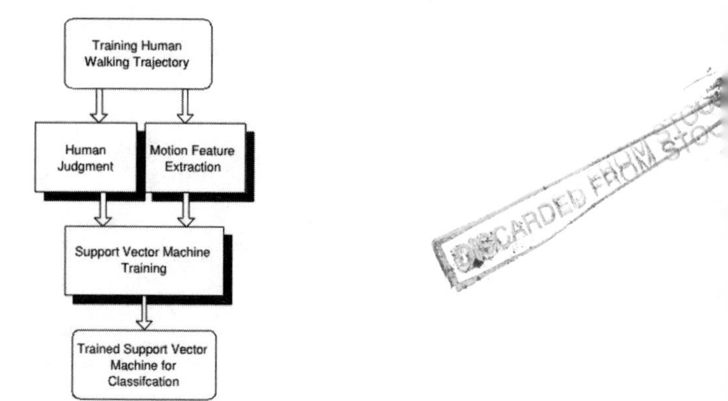

FIGURE 13.1: Learning the normality of local trajectory points using support vector machine.

planes without explicitly carrying out the map into the feature space. The problem will turn into one of quadratic programming in terms of kernels.

The classification estimate has the form

$$f(x) = sgn(\sum_{i=1}^{l} v_i \cdot k(x_i, x) + b) \tag{13.1}$$

where $f \in \{-1, 1\}$ represents the output class, v_i are coefficients computed as solutions of the quadratic programming problem, b is the bias term, l is the number of vectors in the testing set, and x_i are the set of support vectors selected from the training data. Two types of kernels $k()$ will be used in the experiment: Polynomial $k(x, y) = ((x \cdot y) + 1)^d$ and Radial Basis Function $k(x, y) = exp(-\gamma|x - y|^2)$. Various values of polynomial order (d) and Gamma (γ) will be applied.

By summarizing a short segment of the trajectory and retrieving features from it, we can obtain some local information which is useful for determining the normality of the human movement in real-time. Each input vector of the local trajectory point classifier has a dimension of 9. In the training process, the classifier is supplied with training trajectory data manually marked as normal and abnormal in a point-by-point basis. The trajectory vectors will be marked as abnormal at the point where the human in the scene has demonstrated some abnormal walking motions. In the training process, the characteristics of the abnormal motion points will be captured by the SVM. The criteria embedded in the trained SVM will be applied to the classification of new observed data in real-time.

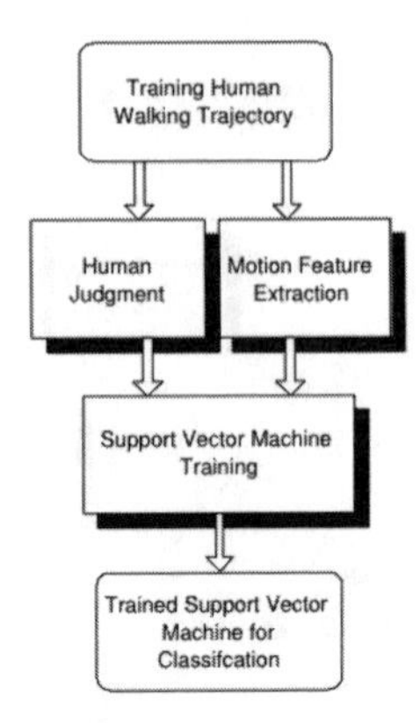

FIGURE 13.2: Local trajectory point classifier.

13.5 Global trajectory similarity estimation

Human walking trajectories are dynamic, nonlinear, and stochastic. For the step of global trajectory similarity estimation (Figure 13.3), we propose to compare the overall similarity between the observed trajectories and the demonstrated normal trajectories using a probabilistic model that takes the information of the whole trajectory into account.

We have developed a hidden Markov model (HMM) based similarity measure [170] to compare the similarity between different trajectories. HMMs are doubly stochastic trainable statistical models and no *a priori* assumption is made about the statistical distribution of the data. This measure is particularly suitable for our application because a high degree of sequential structure can be encoded within the model.

HMM [192] has a number of states and it can move between different states. At each state, an output symbol will be randomly produced. The HMM λ can be specified by three matrices. $\lambda = \{A, B, \pi\}$, where A is the state transition matrix which shows the probability of transitions between different states at any given state. B is the output probability matrix which shows the probability of producing different output symbols at any given state. π is the initial state probability distribution vector. For a given λ, it is capable of producing a series of output symbols which we call observation sequence O.

There are two operations we can do using HMM in our trajectory comparison task: (1) Given an observation sequence O, we can train a HMM λ (maximize $P(\lambda|O)$ using Baum-Welch Expectation Maximization Algorithm). (2) We can also calculate the probability that a given observation sequence O is generated from a HMM model λ (calculate $P(O|\lambda)$ using Forward-Backward Algorithm).

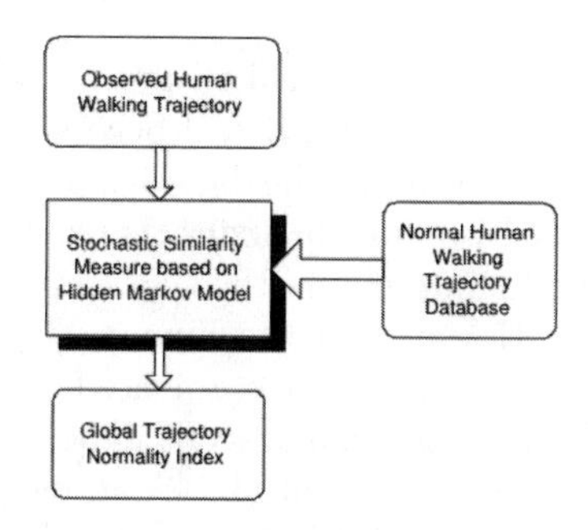

FIGURE 13.3: Global trajectory similarity estimator.

The similarity measure σ between two observation sequences O_1, O_2 are calculated using Equations (13.2) and (13.3). We let $P_{ij} = P(O_i|\lambda_j)^{1/T_i}$ denotes the probability of the observation sequence O_i given the model λ_j, normalized with respect to T_i, the length of the sequence. In practice, P_{ij} can be calculated using Equation (13.3) to prevent numerical underflow for long observation sequences.

$$\sigma(O_1, O_2) = \sqrt{\frac{P_{21}P_{12}}{P_{11}P_{22}}} \tag{13.2}$$

$$P_{ij} = 10^{logP(O_i|\lambda_j)/T_j} \tag{13.3}$$

Equations (13.4) to (13.6) show the properties of our similarity measure between two sequences O_1 and O_2.

$$\sigma(O_1, O_2) = \sigma(O_2, O_1) \tag{13.4}$$

$$0 \leq \sigma(O_1, O_2) \leq 1 \tag{13.5}$$

$$\sigma(O_1, O_2) = 1 \quad iff \quad \lambda_1 = \lambda_2 \tag{13.6}$$

We convert the real-valued human trajectory data into sequences of discrete symbols by data preprocessing. The human trajectory data is firstly normalized to be between $[-1, 1]$. It is then segmented into possibly overlapping window frames. Hamming window is applied to each frame to minimize spectral leakage caused by data windowing. Discrete Fourier Transform is used to convert the real vectors to complex vectors. Using power spectral density estimation (Fourier), a feature matrix V is created. By applying LGB vector quantization algorithm, the feature vector V is converted to L discrete symbols such that the total distortion between the symbols and the quantized vectors can be minimized. The quantized sequence of human trajectory data will be used to train a five-state Bakis HMM for similarity measure.

The entire observed trajectory is matched with each of the demonstrated normal trajectories stored in the database. If there are M normal trajectories in the database, each comparison will yield a value σ_i, where $i = \{1...M\}$. The highest value of σ_i will be selected and it will become the value of the global trajectory normality index. Normally, the data of a pedestrian analyzed by the global trajectory similarity estimator includes the whole trajectory beginning from the point where the human enters the scene up to the point where the human leaves the scene. However, if the human stays in the scene for an extended period of time, the data will be segmented and sent to the estimator for processing. The segmentation time is scene dependent. To achieve better results, the trajectories stored in the database can be formed by averaging a number of normal trajectories recorded over a period of time.

13.6 Trajectory normality classifier

The trajectory normality classifier combines the information obtained from the local trajectory point classifier and the global trajectory similarity estimator to arrive at a conclusion whether the observed trajectory should be considered as normal or not (Figure 13.4).

The trajectory is classified as abnormal and attention is required if one of the following three conditions is satisfied.

1. $C > \Omega$
2. $S < \Phi$
3. $C > \omega$ and $S < \phi$

where C is the number of abnormal trajectory points accumulated so far and S is $max\{\sigma_i\}$. Ω, ω, Φ, and ϕ are constant values dependent on the nature of the scene and application, and $\Omega > \omega$ and $\Phi > \phi$.

13.7 Experiment 1: Trajectory normality classifier

The selection of normal walking trajectories is scene dependent. For our experimental scene, we decide that there are six different normal trajectories, including walking up the road (Dn-U), walking down the road (U-Dn), leaving the door and walking down the road (Dr-Dn), leaving the door and walking up the road (Dr-U), approaching the door from the lower part of the scene along the road (Dn-Dr), and approaching the door from the upper part of the scene along the road (U-Dr). The normal trajectories used for training the support vector machines are shown in Figures 13.5 to 13.7.

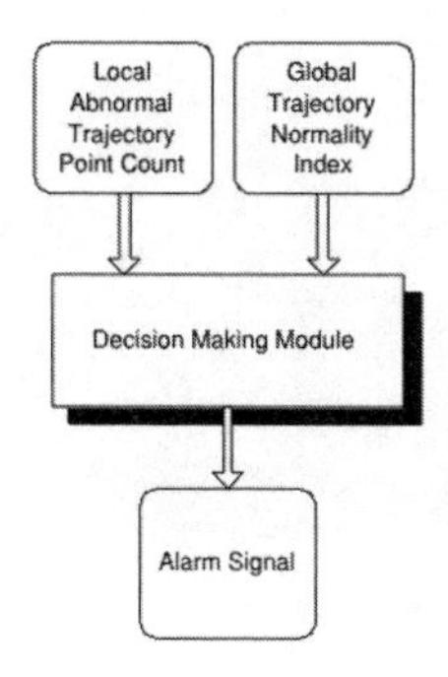

FIGURE 13.4: Trajectory normality classifier.

Table 13.1: Performance of different support vector models in classifying the test data.

	Poly ($d = 2$)	Poly ($d = 3$)	Poly ($d = 4$)	Poly ($d = 5$)	RBF ($\gamma = 1$)	RBF ($\gamma = 5$)	RBF ($\gamma = 10$)
False Alarm	15	23	43	28	31	26	67
Missed Alarm	11	9	27	19	59	32	44

Table 13.2: Overall system performance over test trajectories.

Number of Normal Test Trajectories	30
Number of Abnormal Test Trajectories	15
Number of False Alarms	4
Number of Missed Alarms	2
Success Rate	86.7%

FIGURE 13.5: Normal trajectories of walking up and down the road.

FIGURE 13.6: Normal trajectories of walking towards the door.

FIGURE 13.7: Normal trajectories of walking out from door.

We have performed experiments to test the performance of the support vector machines in classifying trajectory features, as well as the overall system performance.

The total training data set includes six normal trajectories, containing 1399

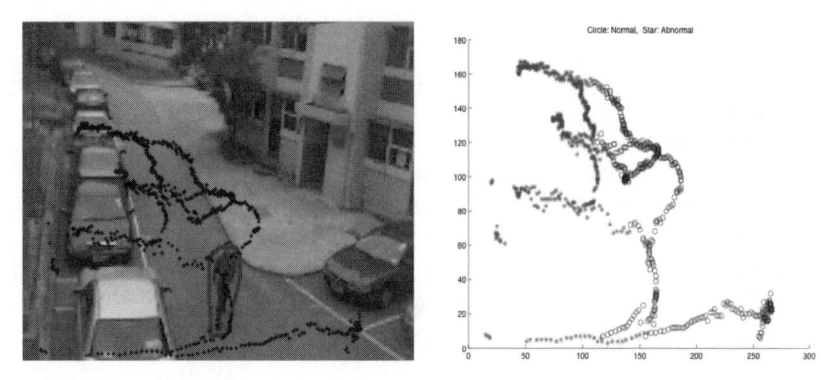

FIGURE 13.8: Detection of abnormal trajectory points using SVM (polynomial power 2 kernel).

FIGURE 13.9: Two of the test trajectories classified as abnormal.

points, and six abnormal trajectories in which 873 out of 1265 points are manually marked as abnormal. The normal and abnormal points in all of the 12 trajectories are used in the training of the SVMs. The six normal trajectories are stored in the normal trajectory database for similarity measure of the entire trajectory.

Different types of support vector machines are used for learning the classification of trajectory features, including SVMs with polynomial (Poly) kernel of orders 2 to 5, and SVMs with radial basis function kernels (RBF) of γ values 1, 5, and 10.

For testing the feature classification capability of the SVMs, the trained SVM models are used to analyze a test trajectory consisting of 1538 points. The position points of the testing trajectory are also manually marked for comparing with the model-generated results. The number of false alarms and missed alarms are shown in Table 13.2. The test trajectory and the classification result of the best model (polynomial kernel of order 2) is shown in Figure 13.8.

Using a USB camera (320 × 240 pixels) and a Pentium III 1GHz PC, the local trajectory classifier system can run at a speed of 10Hz. The manual classification of normal and abnormal training points is subjective and fine-tuning of classification parameters ϕ, Φ, ω, and Ω are required to achieve better performance. The overall performance of the system is verified using 30 normal walking trajectories and 15 abnormal trajectories which include a total of 11253 test points. The threshold values used for the trajectory normality classifier are $\Omega = 100, \omega = 50, \Phi = 0.05$, and $\phi = 0.1$. The results are shown in Table 13.2. Two trajectories classified by the system as abnormal are shown in Figure 13.9.

13.8 Further analysis on global trajectory similarity based on LCSS

In the previous sections, we have discussed an approach that separates the local and global information of walking trajectories and handles the data by two specialized learning mechanisms. Now we apply an approach based on longest common subsequences (LCSS) [244] to further analyze the global properties of the observed walking trajectories for classification. In order to properly determine the boundaries for classification, we propose to incorporate support vector regression and cascade neural networks in establishing the matching boundaries. We compare the classification results for the LCSS approach based on different threshold establishing methods. The performance of the systems will also be compared against the similarity measure based on hidden Markov model [170] that we established earlier.

The main purpose of this research is to classify the walking trajectory patterns observed in a scene and determine whether the trajectories are normal or not. For any path in a scene, people may display different normal walking styles. People may walk at various speeds and their trajectories may have different shapes. In order to achieve this classification, we aim to determine the allowable position and speed boundary limits within which the walking trajectories can be considered as normal.

13.9 Methodology used in boundary modeling

We adopt the learning by demonstration approach, in which many normal trajectory samples are demonstrated for a given path. Before the learning commences, we first record the demonstrated trajectories of different classes.

For each point in the trajectory, we determine the percentage of points passed. The lengths of all the demonstrated trajectories are normalized to be 100 and the average trajectory is found. Each point in a normalized trajectory will be assigned an index value between 1 and 100. In the real-time analysis, given the position index of the observed trajectory, the corresponding boundary limits will be determined.

The goal of the mapping is to find out the relationship between the position of points in the trajectory and the boundary limits. The inputs of the mapping include:

1. X-position of the point
2. Y-position of the point
3. Index of the point in the trajectory (1...100)

The outputs of the mapping include:

1. X-position boundary limit
2. Y-position boundary limit
3. Time boundary limit in x-direction
4. Time boundary limit in y-direction

13.9.1 Trajectory similarity based on LCSS

The similarity analysis considers one-dimensional data moving on the (x, y)-plane in discrete time. Assume that A and B are two trajectories, where $A = ((a_{x1}, a_{y1}), ..., (a_{xn}, a_{yn}))$ and $B = ((b_{x1}, b_{y1}), ..., (b_{xn}, b_{yn}))$. Let $Head(A)$ be the subsequence of A such that $Head(A) = ((a_{x1}, a_{y1}), ..., (a_{x(n-1)}, a_{y(n-1)}))$.

$$LCSS_{\delta,\epsilon}(A, B) = \tag{13.7}$$
$$\begin{cases} 0, \quad \text{if} \quad A \quad \text{or} \quad B \quad \text{is empty} \\ 1 + LCSS_{\delta,\epsilon}(Head(A), Head(B)), \\ \quad \text{if} \quad (|a_{x,n} - b_{x,m}| < \epsilon_x \quad \text{or} \quad |a_{y,n} - b_{y,m}| < \epsilon_y) \quad \text{and} \quad |n - m| \leq \delta \\ max(LCSS_{\delta,\epsilon}(Head(A), B), LCSS_{\delta,\epsilon}(A, Head(B))), \quad \text{otherwise} \end{cases}$$

where δ and ϵ are two boundary limits. δ specifies the allowable difference in unnormalized time that a given point in one trajectory can match to a corresponding point in the second trajectory, and ϵ is the matching threshold in terms of position. The value of $LCSS_{\delta,\epsilon}(A, B)$ will be increased by 1 if the corresponding points in the two trajectories are close enough. In this LCSS-based similarity measure, time stretching is taken into account.

The similarity measure σ_{LCSS} is then defined as

$$\sigma_{LCSS}(\delta, \epsilon, A, B) = \frac{LCSS_{\delta,\epsilon}(A, B)}{min(n, m)} \tag{13.8}$$

where m is the length of sequence A and n is the length of sequence B. In our application, the lengths of all the trajectories are normalized to be the same.

We define the similarity measure between two observed 2-D trajectories T_1 and T_2 in a scene as the average of the X and Y similarity components:

$$\sigma_{LCSS}(T_1, T_2) = \frac{\sigma_{LCSS-X(\delta_x, \epsilon_x, T_{1x}, T_{2x})} + \sigma_{LCSS-Y(\delta_y, \epsilon_y, T_{1y}, T_{2y})}}{2} \tag{13.9}$$

There are a number of advantages in using LCSS as a similarity measure for pedestrian trajectories. Firstly, it can handle trajectories recorded at different sampling rates, because not all of the values in the trajectories are matched. Secondly, this method is not very sensitive to outliers, such as the jerky motion produced during the visual tracking process. Thirdly, it can deal with trajectories at different lengths. Finally, this approach is computationally simple.

The LCSS method calculates the similarity between two trajectories based on the given position and time boundary limits. If more corresponding points in the two sequences fall within the boundaries, the trajectories will be considered as more similar.

The selection of time threshold δ and distance threshold ϵ are very important in the usage of this similarity measure. The efficiency and accuracy of the similarity results depend highly on the proper selection of these two boundaries. In the following sections, we investigate the performance of this similarity measure in relation to the method of determination of the boundary values.

13.10 LCSS boundary limit establishment

We will study and compare the classification power exhibited by the following boundary establishing methods:

1. Thresholds learnt from support vector regression
2. Thresholds learnt from cascade neural networks
3. Fixed thresholds
4. Variable thresholds

The results will also be compared with the similarity measure based on hidden Markov model that we established earlier.

13.10.1 LCSS thresholds learnt from support vector regression

We propose using support vector regression to learn the boundary limits for LCSS. For position boundary learning, we establish a mapping between the trajectory time index and the X-Y position boundary of the training trajectories. For the learning of time boundary limits, the relationship between the trajectory time index and the time shift from the corresponding points in the mean trajectory is modeled.

Assume we have a set of training samples, S_f, which is independent and identically distributed: $S_f = \{(x_1, y_1), (x_2, y_2), ..., (x_n, y_n)\}$, where $x \in \Re^N$ denotes the input vector and $y \in \Re$ is the output value.

The nonlinear support vector regression (SVR) estimate has the form

$$y = f(x) = \sum_{i=1}^{n} (\alpha_i - \alpha_i^*) \cdot k(x_i, x) + b \qquad (13.10)$$

where α_i, α_i^* are coefficients, and b is a bias term that are computed as solutions of the following quadratic optimization problem:

$$\min_{\alpha_i, \alpha_i^*} -\sum_{i=1}^{n} (\alpha_i - \alpha_i^*) y_i + \frac{1}{2} \sum_{i,j=1}^{n} (\alpha_i - \alpha_i^*)(\alpha_j - \alpha_j^*) k(x_i, x_j) + \qquad (13.11)$$

$$\rho \sum_{i=1}^{n} (\alpha_i + \alpha_i^*)$$

$$\text{such that} \sum_{i=1}^{n} (\alpha_i - \alpha_i^*) = 0, 0 \leq \alpha_i, \alpha_i^* < C, i = 1, 2, ..., n.$$

where n is the number of vectors in the input set, ρ is the expected level of precision, and C is a constant that provides a trade-off between the flatness of function f and the amount to which output deviation is larger than ρ. There are only a few parameters α that take non-zero values and the corresponding x_i are known as the support vectors.

13.10.2 LCSS thresholds learnt from cascade neural networks

We also apply cascade neural networks to learn the boundary limits for LCSS. The cascade neural network (CNN) differs from other neural network models in a number of aspects. In particular, the network structure and size are not fixed, but they can be dynamically adjusted during the training process (Figure 13.10). New hidden units receive connections from all input units as well as all previous hidden units, that is why the network is named as cascade neural network.

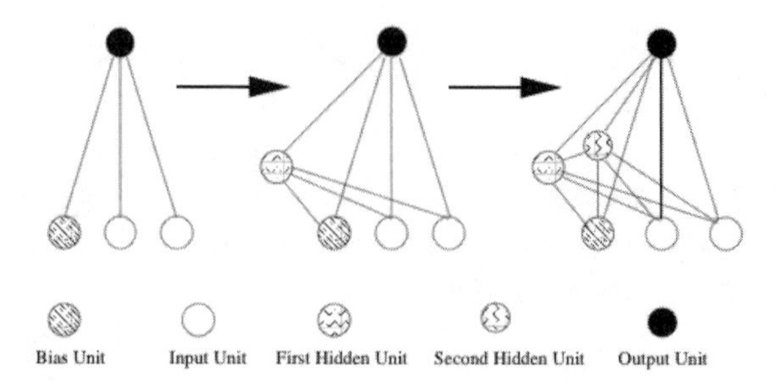

FIGURE 13.10: Training of cascade neural network.

For a given network node, the kind of activation function which gives the best performance will be installed. During the training process, the kind of activation function which reduces the RMS error the most is selected. The available types of activation functions include Sigmoid, Gaussian, Bessel, and sinusoids of various frequencies.

Node-Decoupled Extended Kalman Filtering (NDEKF) is used for network training. Since the network gradually organizes its structure to optimize the performance, no a priori model assumption about facial expression intensity is required. Network size is minimized as it grows from a simple two-layer network. In NDEKF training, only part of the network is trained at any one time. It converges within fewer number of training epochs and is able to achieve smaller RMS error compared to the quick-prop algorithm.

In our research group, cascade neural networks [171] have been successfully applied in modeling various human functions including human control strategies in simulated vehicle driving [166], human skill in controlling a single wheel robot [254], and human sensation in virtual environments [134]. This architecture has demonstrated good performance in modeling continuous human data, so it is a suitable candidate for our application.

13.10.3 Fixed LCSS thresholds

In this method, the values of the LCSS thresholds are fixed integers determined by the system operator. In our experiment, three different sets of values are selected: 5, 10, and 20. For each set, the boundary values will be the same.

13.10.4 Variable LCSS thresholds

In this approach, the LCSS thresholds are derived from the d training trajectories T_i, where $i = 1...d$. Assume that all the trajectories are normalized to have a fixed length m and the mean trajectory is T_{mean}. Each point in the trajectory has an index number $i = 1...m$. The boundary T_{th} is then found by calculating the training trajectory point furthest away from the mean training trajectory.

$$T_{mean} = \{(\bar{x}_{i1}, \bar{y}_{i1}), (\bar{x}_{i2}, \bar{y}_{i2}), ..., (\bar{x}_{in}, \bar{y}_{in})\} \quad (13.12)$$

$$T_{th} = \{(\max\{x_{i1}\} - \bar{x}_{i1}, \max\{y_{i1}\} - \bar{y}_{i1}), (\max\{x_{i2}\} - \bar{x}_{i2}, \max\{y_{i2}\} - \bar{y}_{i2}), ..., (\max\{x_{i100}\} - \bar{x}_{i100}, \max\{y_{i100}\} - \bar{y}_{i100})\} \quad (13.13)$$

In order to find the time boundary, we first need to find out how each point in the training trajectory mean is matched with each of the points in the training trajectory. For each point in the training trajectory mean, we look for the corresponding point in the training trajectories which has the maximum shift in time. The maximum index difference for each point in the training trajectory mean becomes the time boundary for that point.

13.11 Experiment 2: Boundary modeling

We use the same scene setting and normal trajectory types as in section 13.7. For each trajectory type, 20 trajectory samples are used for training and 10 trajectory samples are used for testing. In the previous sections, we explored the power of the system to classify between normal and abnormal walking trajectories. The following experiment is aimed at improving the performance of the global trajectory classifier subsystem in which the proposed LCSS approaches are used to classify the six types of trajectories in the scene.

In the rest of this section, we will first present the boundary modeling result for one class of walking trajectory as an example, then the performance of the boundary modeling methods will be discussed.

Let's consider the case where the walking pedestrians go from the top-left corner of the scene and exit at the lower-right corner (U-Dn). Figures 13.11 and 13.12 show the maximum distance between the individual training trajectory points and the mean training trajectory points in both x and y directions. The average training trajectory is shown by crosses and the boundaries are represented by the circles. Some of the training data for the time boundary in x and y directions is displayed in Figures 13.13 and 13.14, and some of the training data for the position boundaries in x and y directions is displayed in Figures 13.15 and 13.16. The training data are modeled by support vector machines and cascade neural networks. The mappings between the time index and the time and position boundaries at each time index are learnt. The model outputs for position boundaries are shown in Figures 13.17 and 13.18, and the model outputs for time boundaries are shown in Figures 13.19 and 13.20. These model outputs will be used as position and time boundaries in the LCSS similarity measure for trajectory classification.

Table 13.3: Similarity between trajectory classes using fixed LCSS boundary value 5

	U-Dn	U-Dr	Dr-U	Dn-U	Dr-Dn	Dn-Dr
U-Dn	0.58	0.19	0.11	0.07	0.17	0.05
U-Dr	0.19	0.51	0.06	0.05	0.07	0.07
Dr-U	0.11	0.06	0.61	0.11	0.07	0.03
Dn-U	0.07	0.05	0.11	0.54	0.08	0.10
Dr-Dn	0.17	0.07	0.07	0.08	0.49	0.08
Dn-Dr	0.05	0.07	0.03	0.10	0.08	0.57

Different values of overall similarity between the trajectories are obtained from different boundary establishing methods. The results are shown in Tables 13.3 to 13.12. Tables 13.3 to 13.5 show the similarity results for using fixed boundary values 5, 10, and 20, respectively. As the threshold values increase, the overall similarities between all the trajectories also increase. However, the optimal combination of boundary values is very difficult to determine arbitrarily. Table 13.6 shows the similarity results for boundary value based on the maximum distance between the individual training trajectory points and the mean training trajectory points. This table shows that trajectories having the same starting and ending positions are considered as very similar to each other, and it is difficult to distinguish between different types of classification errors. For example U-Dn versus U-Dn and U-Dn versus U-Dr are very close to each other, and Dr-U versus U-Dn and Dr-U versus U-Dr are very similar. The same problem happens to the similarity measure based on HMM. For instance, in Table 13.9, U-Dn versus Dn-U is very similar to U-Dn versus. Dr-U.

Tables 13.10 and 13.11 show the similarity results in x and y directions using boundaries generated by SVM and CNN models, respectively. The mean values between x and y directions are shown in Tables 13.8 and 13.7. For the CNN approach, there is some inconsistency between the similarity values in the x and y directions. For cases such as Dn-U versus Dr-U and Dr-Dn versus U-Dn, there is a big difference in the similarity values for x and y directions.

In comparison, the SVM boundary learning approach is superior than the other boundary establishing methods for this application and it has the following advantages: (1) This method is based on learning by demonstration, thus avoiding the need for subjective selection of values. (2) The similarity for the components in x and y directions is consistent. (3) It can also achieve a better classification for different kinds of errors, and the dissimilarity between different mismatches is high.

Table 13.4: Similarity between trajectory classes using fixed LCSS boundary value 10

	U-Dn	U-Dr	Dr-U	Dn-U	Dr-Dn	Dn-Dr
U-Dn	0.80	0.28	0.10	0.06	0.34	0.10
U-Dr	0.28	0.78	0.11	0.10	0.07	0.12
Dr-U	0.10	0.11	0.76	0.18	0.03	0.12
Dn-U	0.06	0.10	0.18	0.81	0.04	0.13
Dr-Dn	0.34	0.07	0.03	0.04	0.79	0.03
Dn-Dr	0.10	0.12	0.12	0.13	0.03	0.79

Table 13.5: Similarity between trajectory classes using fixed LCSS boundary value 20

	U-Dn	U-Dr	Dr-U	Dn-U	Dr-Dn	Dn-Dr
U-Dn	0.92	0.37	0.17	0.10	0.41	0.21
U-Dr	0.37	0.91	0.15	0.07	0.11	0.22
Dr-U	0.17	0.15	0.94	0.29	0.19	0.27
Dn-U	0.10	0.07	0.29	0.94	0.17	0.26
Dr-Dn	0.41	0.11	0.19	0.17	0.89	0.24
Dn-Dr	0.21	0.22	0.27	0.26	0.24	0.93

Table 13.6: Similarity between trajectory classes using variable LCSS boundary value

	U-Dn	U-Dr	Dr-U	Dn-U	Dr-Dn	Dn-Dr
U-Dn	0.99	0.92	0.20	0.17	0.79	0.19
U-Dr	0.92	0.98	0.20	0.15	0.23	0.37
Dr-U	0.20	0.20	0.99	0.91	0.61	0.32
Dn-U	0.17	0.15	0.91	0.99	0.25	0.41
Dr-Dn	0.79	0.23	0.61	0.25	0.99	0.39
Dn-Dr	0.19	0.37	0.32	0.41	0.39	0.99

Table 13.7: Similarity between trajectory classes using LCSS boundary learnt from CNN

	U-Dn	U-Dr	Dr-U	Dn-U	Dr-Dn	Dn-Dr
U-Dn	0.89	0.64	0.11	0.11	0.59	0.13
U-Dr	0.64	0.81	0.31	0.21	0.23	0.17
Dr-U	0.11	0.31	0.85	0.61	0.55	0.24
Dn-U	0.10	0.21	0.61	0.90	0.26	0.54
Dr-Dn	0.59	0.23	0.55	0.24	0.88	0.12
Dn-Dr	0.13	0.17	0.24	0.54	0.12	0.84

Table 13.8: Similarity between trajectory classes using LCSS boundary learnt from SVM

	U-Dn	U-Dr	Dr-U	Dn-U	Dr-Dn	Dn-Dr
U-Dn	0.88	0.53	0.16	0.13	0.41	0.18
U-Dr	0.53	0.87	0.22	0.11	0.16	0.39
Dr-U	0.16	0.22	0.81	0.53	0.32	0.11
Dn-U	0.13	0.11	0.53	0.83	0.24	0.49
Dr-Dn	0.41	0.16	0.32	0.24	0.92	0.17
Dn-Dr	0.18	0.39	0.11	0.49	0.17	0.89

Table 13.9: Similarity between trajectory classes using HMM

	U-Dn	U-Dr	Dr-U	Dn-U	Dr-Dn	Dn-Dr
U-Dn	0.89	0.31	0.28	0.27	0.37	0.21
U-Dr	0.31	0.95	0.26	0.34	0.30	0.31
Dr-U	0.28	0.26	0.89	0.43	0.28	0.39
Dn-U	0.27	0.34	0.43	0.89	0.30	0.40
Dr-Dn	0.37	0.30	0.28	0.30	0.94	0.29
Dn-Dr	0.21	0.31	0.39	0.40	0.29	0.93

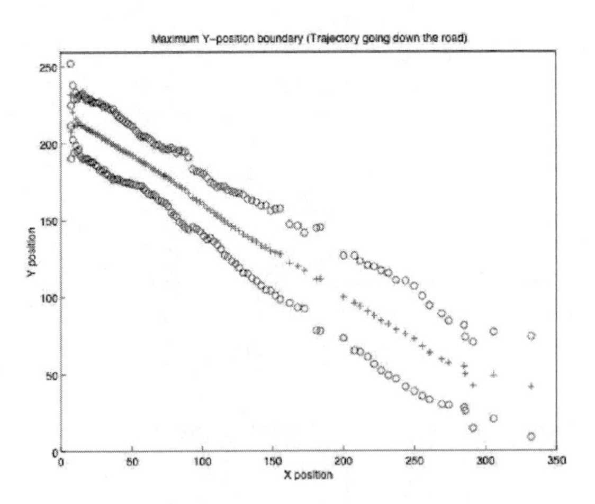

FIGURE 13.11: Variable boundary along the x-direction: Maximum distance between the individual training trajectory points and the mean training trajectory points.

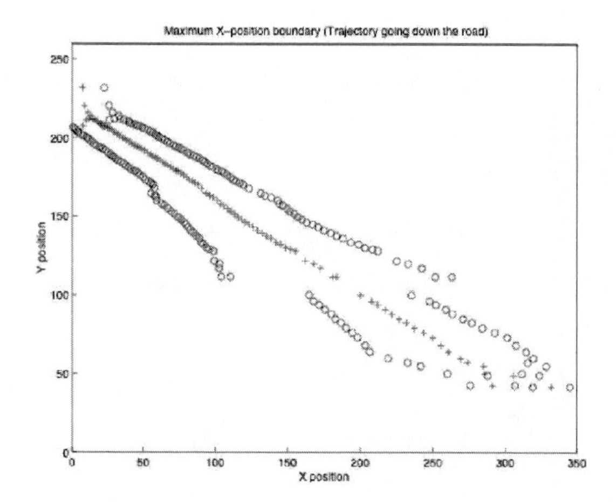

FIGURE 13.12: Variable boundary along the x-direction: Maximum distance between the individual training trajectory points and the mean training trajectory points.

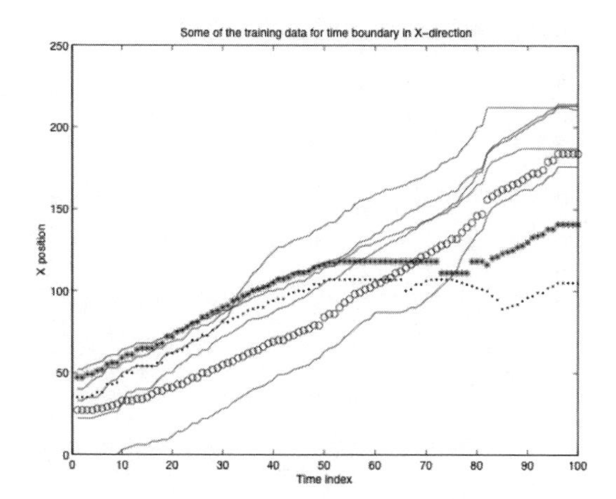

FIGURE 13.13: Some of the training data for time boundary along the x-direction.

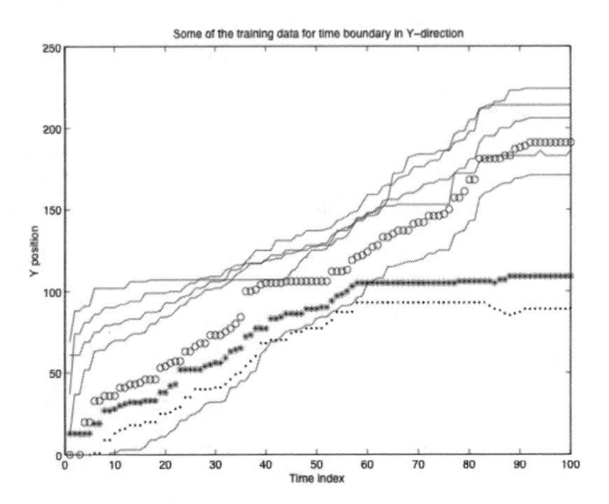

FIGURE 13.14: Some of the training data for time boundary along the y-direction.

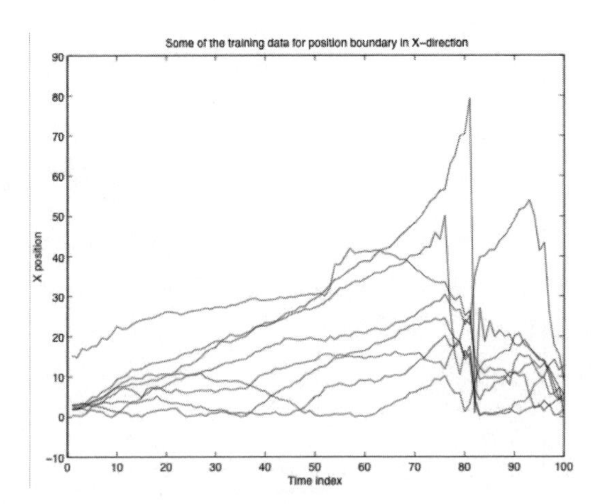

FIGURE 13.15: Some of the training data for position boundary along the x-direction.

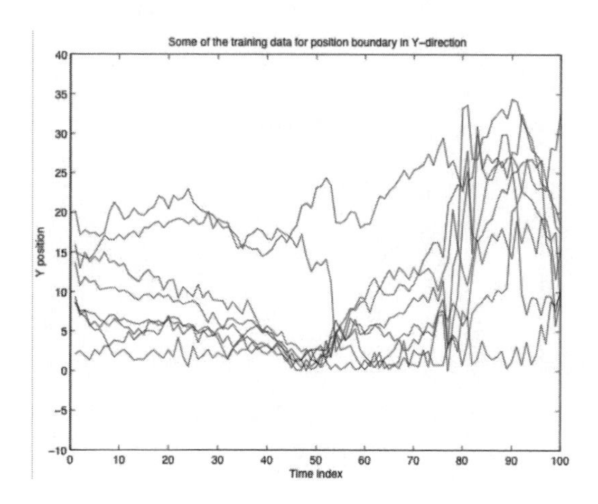

FIGURE 13.16: Some of the training data for position boundary along the y-direction.

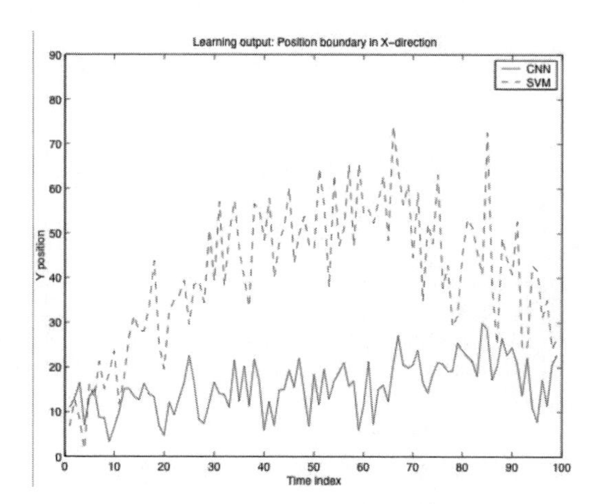

FIGURE 13.17: Model output of CNN and SVM: Position boundaries along the x-direction.

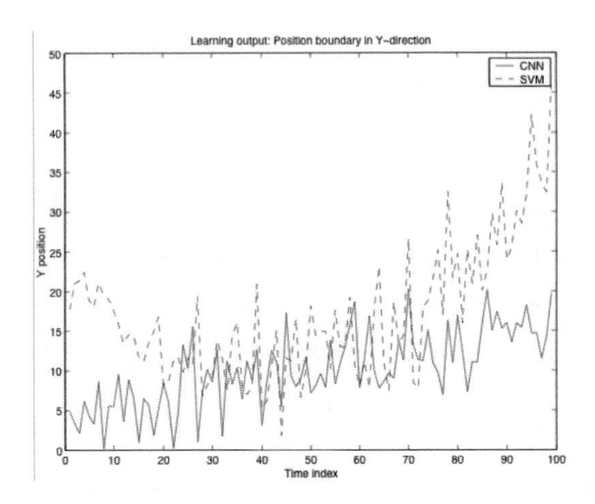

FIGURE 13.18: Model output of CNN and SVM: Position boundaries along the y-direction.

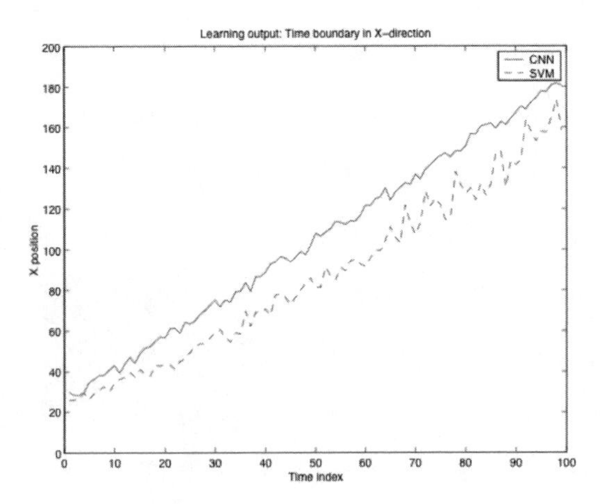

FIGURE 13.19: Model output of CNN and SVM: Time boundaries along the x-direction.

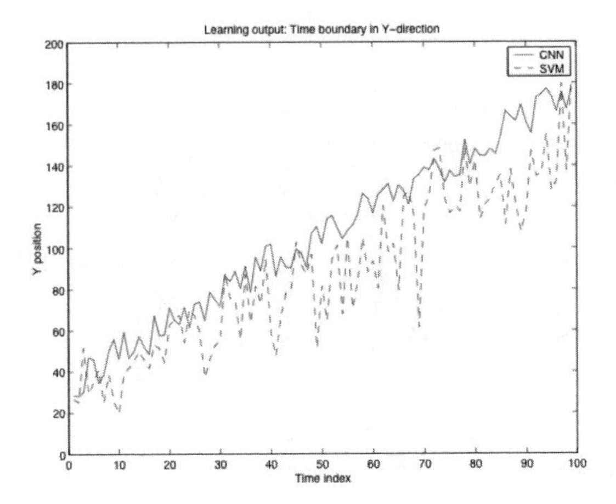

FIGURE 13.20: Model output of CNN and SVM: Time boundaries along the y-direction.

Table 13.10: Similarity between trajectory classes using LCSS boundary leant from SVM (x and y components).

	U-Dn		U-Dr		Dr-U		Dn-U		Dr-Dn		Dn-Dr	
	x	y	x	y	x	y	x	y	x	y	x	y
U-Dn	0.94	0.82	0.52	0.54	0.18	0.14	0.15	0.11	0.42	0.39	0.19	0.17
U-Dr	0.52	0.54	0.90	0.84	0.20	0.24	0.07	0.15	0.13	0.19	0.37	0.39
Dr-U	0.18	0.14	0.20	0.24	0.87	0.75	0.58	0.48	0.34	0.29	0.17	0.06
Dn-U	0.15	0.11	0.07	0.15	0.58	0.48	0.81	0.85	0.27	0.20	0.47	0.52
Dr-Dn	0.42	0.39	0.13	0.19	0.34	0.29	0.27	0.20	0.95	0.89	0.15	0.19
Dn-Dr	0.19	0.17	0.37	0.39	0.17	0.06	0.47	0.52	0.15	0.19	0.88	0.90

Table 13.11: Similarity between trajectory classes using LCSS boundary leant from CNN (x and y components)

	U-Dn		U-Dr		Dr-U		Dn-U		Dr-Dn		Dn-Dr	
	x	y	x	y	x	y	x	y	x	y	x	y
U-Dn	0.91	0.87	0.79	0.48	0.18	0.25	0.11	0.17	0.79	0.39	0.15	0.11
U-Dr	0.79	0.48	0.83	0.79	0.34	0.28	0.11	0.31	0.22	0.24	0.08	0.26
Dr-U	0.18	0.25	0.34	0.28	0.90	0.80	0.82	0.40	0.67	0.43	0.11	0.37
Dn-U	0.11	0.17	0.11	0.31	0.82	0.41	0.95	0.85	0.18	0.29	0.42	0.66
Dr-Dn	0.79	0.39	0.22	0.24	0.67	0.43	0.18	0.29	0.90	0.86	0.17	0.07
Dn-Dr	0.15	0.11	0.08	0.26	0.11	0.37	0.42	0.66	0.17	0.07	0.81	0.87

Table 13.12: Similarity between trajectory classes using HMM (x and y components)

	U-Dn		U-Dr		Dr-U		Dn-U		Dr-Dn		Dn-Dr	
	x	y	x	y	x	y	x	y	x	y	x	y
U-Dn	0.88	0.90	0.31	0.30	0.33	0.28	0.32	0.23	0.41	0.32	0.23	0.19
U-Dr	0.31	0.30	0.95	0.95	0.32	0.21	0.33	0.34	0.40	0.20	0.37	0.24
Dr-U	0.33	0.23	0.32	0.21	0.91	0.87	0.42	0.04	0.27	0.28	0.37	0.41
Dn-U	0.32	0.23	0.33	0.34	0.42	0.04	0.89	0.89	0.32	0.29	0.39	0.42
Dr-Dn	0.41	0.32	0.40	0.20	0.27	0.28	0.32	0.29	0.96	0.92	0.33	0.25
Dn-Dr	0.23	0.19	0.37	0.24	0.37	0.41	0.39	0.42	0.33	0.26	0.93	0.93

13.12 Discussion

One problem with this tracking system is that it is not yet able to handle shadows. If the shadow is large or changes frequently in size, meaningful results may not be obtained. Also, it is assumed that the whole body silhouette is detected. If the person is occluded by other objects such as a car, error may occur in the estimation of the body centroid.

Apart from hidden Markov model, other popular methods that can be used to model the similarity between time-series include Euclidean distance [136] and dynamic time warping.

Euclidean distance is not a very accurate technique because its performance is very sensitive to the small variation in time axis and the measurement degrades rapidly in the presence of noise. Therefore, flexible similarities cannot be easily captured. Dynamic time warping has been applied in a number of application areas which involve data comparison, including DNA sequencing [21], handwritten text classification [196], and fingerprint classification [121]. The weakness of DTW lies in the fact that it requires the matching of all the points in the trajectories, which also include noise and outliers. Therefore, the performance of DTW also deteriorates in the presence of noise.

The advantage of LCSS is that it allows some of the elements in the trajectories to be unmatched, which can avoid the problem of outliers as in the cases of Euclidean distance and DTW. Also, the algorithm involved in LCSS is computationally efficient, especially compared to the computation required to determine the L_p norm in DTW and the iterative algorithm of HMM. However, the longest common subsequence approach also has its own limitation. It has difficulties in differentiating the sequences that have the same length of longest common subsequences but different gap sizes in between. In our implementation, the "gaps" in the trajectories occur when a person walks away from the expected normal path for a short while and then returns to the normal path. This short-term abnormality can be picked up by the SVM-based trajectory point normality classifier. Therefore the limitation of LCSS does not affect the performance of our overall performance to a significant extent.

For global human walking trajectory classification, the performance of the HMM-based similarity measure is not as good as the best LCSS-based approach. Compared to LCSS, HMM has a number of disadvantages. First of all, in the similarity measure, two HMM models need to be constructed for each comparison. Since iterative operations are unavoidable in the HMM modeling process, the measure is very computationally expensive. Also, real-valued data is converted to sequences of symbols in the preprocessing step. The calculations add to the computation load of the system.

For future work, in order to improve the robustness of this prototype surveillance system, we will try to investigate the possibility of combining other subsystems that can characterize other human features, such as face recognition and gait recognition. The realistic interaction between humans and cars in the scene will also be studied.

13.13 Conclusion

This work is largely based on a hypothesis that non-alive biometrics may be copied digitally; therefore, they are not unique and not secure. Our research question is how to capture the aliveness features involved in human walking trajectories and bound the variation digitally. We have developed an intelligent surveillance system that can automatically detect abnormal pedestrian walking trajectories in real-time using learning models by demonstration. By using support vector classification we can identify the trajectory points at which the observed pedestrian is performing abnormal walking motions. By utilizing a stochastic similarity measure based on hidden Markov model, the normality of the shape of the entire trajectory can be determined. The outputs of both learning mechanisms are combined by a rule-based module to arrive at a more reasonable and robust conclusion. Furthermore, we utilize the longest common subsequence approach to compare the similarity between different types of walking trajectories in order to achieve better classification. We have compared a number of methods to determine the position and time boundary limits required for the similarity measure. Support vector regression is found to exhibit the best performance. The system has potential for applications in situations where human walking trajectory monitoring is needed.

14

Modeling of Facial and Full-Body Actions

In the previous chapter we modeled human walking trajectories for surveillance. In this chapter, we present two more examples of human action modeling, namely the modeling of human facial expression intensity and body limb actions.

Changing facial expressions is a natural and powerful way of conveying personal intention, expressing emotion, and regulating interpersonal communication. Automatic estimation of human facial expression intensity is an important step in enhancing the capability of human-robot interface. In this research, we have developed a system that can automatically estimate the intensity of facial expressions in real-time. Based on isometric feature mapping, the intensity of expression is first extracted from training facial transition sequences. Then, intelligent models, including cascade neural networks and support vector machines, are applied to model the relationship between the trajectories of facial feature points and expression intensity level. Based on these techniques, we have implemented a vision system that can estimate the expression intensities of happiness, anger, and sadness in real-time.

In the second example of this chapter, a tracking system that is capable of locating the head and hand positions of moving humans has been developed. We classify the motion trajectories of humans in the scene by using support vector classification. Since the data size of human motion trajectories is large, we apply principal component analysis (PCA) and independent component analysis (ICA) for data reduction. We have successfully applied the developed technique on two different applications: motion recognition of table tennis players, and detection of human fighting motions.

14.1 Facial expression intensity modeling

Facial expressions serve many different purposes in our day-to-day interpersonal communication. Through these, we can express communicative signals of intent, express emotional inner states, activate emotions, and regulate communication. This information is very valuable to human-robot interaction.

Based on the research in human social interaction, we understand that facial expressions play a very important role in the regulation of communication and emotion. People interpret the facial expressions of others as important cues that help to regulate their own behavior. We monitor the facial expressions of others when we attempt to regulate their emotions. Thus, when we want to assess people's attitude, their facial expressions provide us with very useful information.

There are a number of reasons why we need to estimate facial expression intensity in human-robot interaction. Firstly, expression intensity helps intelligent robots to understand people's emotions. Perceived emotional intensity varies linearly with the manipulated physical intensity of the expression. In the absence of situational information, people tend to view emotional facial expression as relatively direct read-outs of an underlying emotional state [93].

Secondly, intensity estimation aids automatic emotion classification. Emotion category decoding accuracy varies linearly to a large extent with the manipulated physical intensity of the expression for expressions such as anger, disgust, and sadness. For instance, a qualitatively distinct emotional state by the expresser cannot be correctly perceived by other people or robots unless the intensity of the muscular displacement involved exceeds a certain level [93].

Thirdly, there are differences in the ways that different individuals express the same facial expression. The facial expression transition for different people tends to differ in terms of speed, magnitude and duration. There is a need to develop personal computational models for facial expression that can adapt to personal style and deal with user-specific expressions.

Fourthly, the perception of facial expression intensity is dependent on many different factors such as sex, cultural background, and communication context, etc. People tend to rate anger and disgust expressed by male actors as more angry or more disgusting, whereas we rate happiness expressed by female actors as more happy. Psychologists also suggest that individuals perceive in a face an emotional expression that is congruent with their present emotional state [173]. Cultures also differ in the absolute level of intensity they attribute to faces. One study reveals that the Americans tend to perceive greater intensity in facial display than the Japanese. The difference may be caused by the existence of different cultural decoding rules [150]. It will, therefore, be beneficial if we can develop a method to automatically construct the intensity spectrum of facial expression transition based on some computational criteria regardless of the variance in personal perception.

Finally, facial expression changes tend to be very fast. Rapid micro expressions are typically missed by observers. A real-time facial expression analysis system can allow intelligent robots to detect subtle changes in expression on human faces.

By intensity of facial expression we mean the relative degree of displacement, away from a neutral or relaxed facial expression, of the pattern of muscle movements involved in emotional expressions of a given sort [93]. In order to enhance the capability of perceptual interface for human-robot interaction, we develop a method to quantitatively measure the intensity of facial expression for different emotions. The interpretation of the expression intensity will depend on the type of application. We hope to provide intelligent robots with the ability to adapt to peoples' natural behaviors instead of forcing the users to adapt to the ways robot behave. The goal of this research is to develop a personalized vision system that can estimate the intensity of facial expression in real-time and be able to support human-robot interaction.

14.1.1 Related work

A number of vision systems have been developed to estimate the intensity of facial expressions from still images using off-line systems. They can be broadly divided into two main approaches: holistic approaches and local feature approaches.

In the holistic approach, all the information from a face image is taken into account. Kimura and Yachida [117] developed a system to recognize facial expressions by extracting a variation from expressionless faces while considering the face area as a whole pattern. A variation in facial expression is represented as motion vectors of an elastically deformed net from a facial edge image. PCA is used to map the net motion into a low dimension emotion space, and the degree of expression is estimated in that space. Expression models for happiness, anger, and surprise are then constructed. The change of emotion degree is shown in form of locus on a 2D plane. However, the system cannot work in real-time.

Lisetti and Rumelhart [143] use a three-layer neural network to recognize difference in levels of expressiveness from two emotions: happy and neutral. They hypothesize that there are three areas of the face capable of independent muscular movement and subnetworks are built to deal with different portions of face images.

Fasel and Luttin [62] extract facial expressions from different images by projecting them into a subspace using PCA/ICA and then applying nearest neighbor classification. The recognition is based on the FACS (Facial Action Coding System) standard and is tested on one subject. The system can estimate two intensity levels for the left and right side of the face.

Chandrasiri, et al. [40] use multidimensional scaling to map face images as points in a low dimension that preserves their relative positions in terms of Euclidean distances, from which the trajectory of the face image sequence can be derived.

The second major popular approach is the local feature approach, which tracks the position of the features (such as eyes and mouth) on the face and hypothesizes that the relative motion of these features are related to the intensity of the expression.

Kaiser and Wehrle [114] developed a neural-network-based system to automatically code facial expression intensities according to FACS for asymmetric facial expressions. However, this system requires that markers are put on the subjects' faces.

Wang, et al. [247] developed a system to recognize the degree of continuous facial expression change. They track facial features using a labeled graph matching technique, and the expression change model is constructed by fitting B-splines to the tracked facial points. Facial expression recognition and degree estimation are achieved by making a comparison between the trajectory of the feature points and the expression change model. Estimations can be made only after the facial expression trajectory is completed.

Hong, et al. [98] achieved online expression recognition by using personalized galleries that feature face images with different expression intensities. In this case, Gabor wavelets are utilized for feature tracking and elastic graph matching is applied to match a given test face against the pictures in the gallery.

Lien, et al. [141] used three methods to estimate the facial expression intensities based on FACS. (1) A coarse-to-fine pyramid method is used to track facial feature points from which the expression intensity is estimated by nonlinearly mapping the displacement of the points. (2) Dense flow tracking and PCA are used to measure the motion of each pixel on the entire face. The expression intensity of a frame in a sequence is estimated by using the correlation property of PCA. The expression intensity of individual frames in each training sequence is determined *a priori* by experts. The intensity of a testing frame can be estimated by applying a minimum distance criterion against the frames in the training set. (3) Edge detectors are applied on different facial regions to detect high frequency furrows. The sum of square difference criterion is employed to find a close estimation of furrow expression intensity.

Both the holistic and local feature approaches have different advantages and disadvantages. In the holistic approach, all the information in the face image is preserved, which allows the classifier to discover the relevant features in the data. However, the normalization step usually involves the whole image and it generally takes a long time. Processing all image pixels is computationally expensive and a large memory space is required. This method is relatively insensitive to subtle motion and the result can be easily affected by a change in lighting.

By using the local feature method, the number of input dimensions is significantly reduced, so the computational load of the classification step can be reduced and the processing time can be sped up. In this case, the reliability of the tracker becomes very important.

A number of systems are based on FACS [141], [62], [114], in which each frame is coded independently and is not based on temporal and detailed spatial information. Classification tasks consist of determining the action units (AU) for each test image as well as the corresponding intensity of that unit. Most false classifications were obtained due to weak AU activities that were classified as being neutral. It is difficult to estimate expression intensity using FACS because only five action units allow the possibility to mark intensity on a three-level scale. The other AUs can only express two states (on/off).

Moreover, each person has his/her own way of expressing a particular emotion and the maximum intensity of an emotion.

As seen from the discussion above, preceding systems have made important contributions in analyzing facial expression. However, improvements and new techniques are required in order to develop systems that can understand human expressions more effectively. Firstly, most systems cannot analyze facial information in real-time, and this is crucial in efficiently supporting the interaction. Secondly, a number of systems extract the expression from a single face image only. We argue that the temporal change of expression is also important to the estimation of expression. Thirdly, the expression intensities of the training face images are determined by experts *a priori*. This process is tedious and time-consuming. The libraries of facial images occupy large amounts of storage space. One issue that we would like to investigate is how to automatically generate the spectrum of expression intensity based on a continuous facial change between extreme expression states. It would be desirable if the system was able to extract the intensity of the facial expression from the face data without asking humans to mark the intensity of each individual image in the training sequence. Fourthly, neural networks have been applied effectively in different areas of facial analysis [143], [199], [47], but only networks of fixed sizes and structures are used. We believe that both structure and parameters have a place in the neural network learning process. Therefore, we propose using cascade neural networks for the learning of intensity estimation model.

14.1.2 System overview

We have developed a vision and learning system that can estimate the intensity of facial expression in real-time. There are two steps in the process. The first step is to train the system so that it is capable of relating facial feature positions to expression intensities (Figure 14.1). The facial feature data are normalized and then processed using isometric feature mapping (Isomap) in order to extract the one-dimensional manifold of expression intensity. The extracted intensities and the normalized facial feature trajectories are then used in the training of cascade neural networks (CNN) and support vector machines (SVM) for expression intensity modeling. The second step is real-time intensity estimation. The new feature points on the human face are tracked and normalized in real-time and the estimates of the expression intensity is produced by the trained intensity model (Figure 14.2).

14.1.3 Extraction of facial motion data

We use a feature-based method to extract facial information, because there are many good natural features on the face for robust tracking. In terms of speed, the feature tracking approach is faster than the holistic approach in general. Moreover, facial feature point tracking is very sensitive to subtle

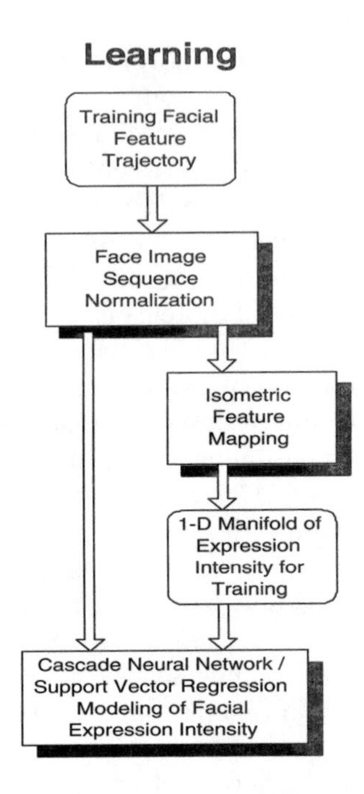

FIGURE 14.1: Learning of facial expression intensity.

facial motions of small intensity. A pyramidal implementation of the Lucas-Kanade optical flow algorithm [146] is employed to track the feature points on the face. Nineteen different feature points are placed around a subject's facial features, including the eyes, the brows, the nose, and the mouth (Figure 14.3). The search window size of the feature points ranges from 6×6 pixels to 25×25 pixels, depending on the position of each feature point. Currently, the feature points are interactively marked on the face using a computer mouse. The tracking algorithm is then capable of tracking the feature points in real-time with sub-pixel accuracy.

In-plane head motion and small out-of-plane motion are allowed, and image normalization is performed in order to separate the non-rigid facial expression from rigid head motion. Since the geometry of the facial features is different across different subjects, affine transformation is applied to each of the images in the sequences (Figure 14.4). This transformation performs rotation, translation, and scaling adjustment to the images such that all the variances in the transition sequences not caused by the facial expression changes are removed. It also provides the basis of comparison for the expressions produced by different individuals. The size of each image in the sequence is 320×240 (row $\times$ column) pixels and the images will be transformed such that in the normalized images the positions of the left inner eye corner, right inner eye corner and

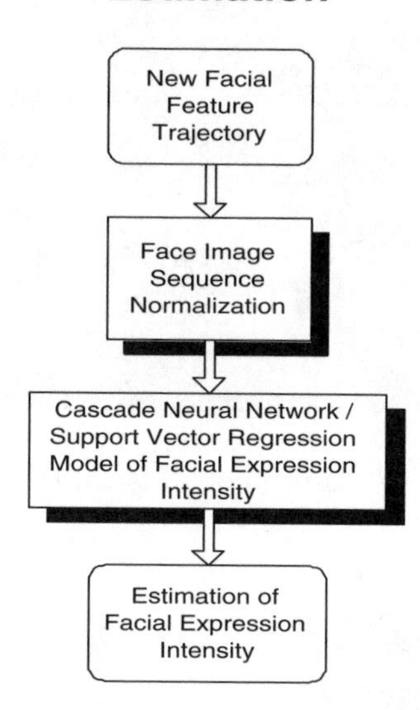

FIGURE 14.2: Real-time estimation of facial expression intensity.

Philtrum point are (122.5,115), (157.5,115), and (140,160), respectively.

14.1.4 Automatic extraction of facial expression intensity from transition

We consider the continuous change of facial expression for a particular emotion as the sequence of possible states of deformation of an object. Inspired by the fact that neural population activity is constrained in low-dimensional manifold [213], we formulate expression intensity extraction as a process of low-dimensional manifold construction. We achieve dimensionality reduction based on representing the data manifold by a low-dimensional Euclidean space where the distances between observed facial feature points are preserved. The goal of the intensity extraction process is to discover the low-dimensional representation with coordinates that capture the intensity of the expression given the high-dimensional inputs of facial feature point positions. We want to automatically extract the intensity of expression from a video training sequence that displays the transition from an expressionless face to a face with extreme emotion expression.

The aim of this distance preserving method is to find a collection of points $\{y_n\}_{n=1}^N$ in a metric space $Y \subset \Re^L$ associated with the high dimension sample of facial feature points such that the distance between two points x_n and x_m is approximately the same as the distance between their associated

FIGURE 14.3: Position of 19 facial feature points.

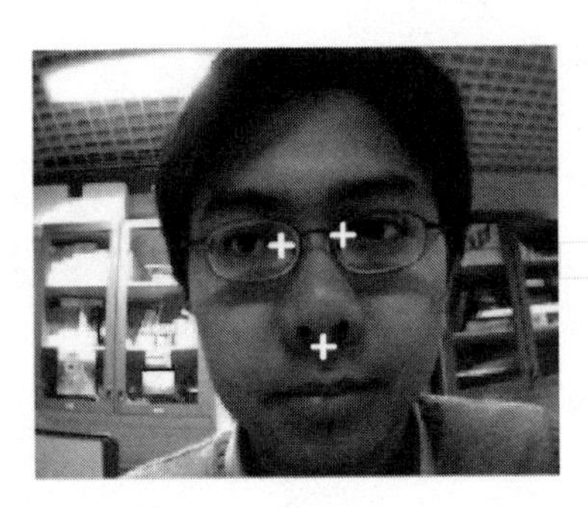

FIGURE 14.4: Normalization of feature positions by affine transformation.

low-dimensional points y_n and y_m. We applied isometric feature mapping (Isomap) [236] to achieve intensity extraction. This is a method that combines a topology-preserving network algorithm with multidimensional scaling for manifold modeling. There are three steps in Isomap:

1. *Representation of manifold*
   We construct the neighborhood graph G from input space X and determine the edge length $d_x(i, j)$ for all the data point pairs i and j. We then connect each point to all points within a fixed radius ϵ.

2. *Computation of the shortest distances in the manifold*
   In this step, the geodesic distances $d_m(i, j)$ in the low-dimensional manifold M between all point pairs are estimated by computing the shortest path distance $d_G(i, j)$ in the graph G using Floyd's algorithm.

3. *Construction of embedding*

To construct a one-dimensional embedding in Euclidean space Y, multidimensional scaling is applied to D_G. The D_G matrix contains the shortest path between all points in G.

The result of this procedure is that we can produce a one-dimensional manifold, which represents the intensity of expression, from the high-dimensional feature trajectory space. The continuous intensity levels are normalized to be within -1 and 1, in which -1 represents the expressionless face and 1 represents the extreme expression of the emotion concerned.

14.1.5 The learning of facial expression intensity

In order to capture the characteristics of facial expression changes into computational models, we propose applying cascade neural networks (CNN) and support vector regression (SVR), which are both excellent approaches for modeling continuous data.

Cascade neural network

Neural network is a powerful tool for function approximation. If the characteristics of facial expression changes are to be captured efficiently into a computational model, we need to specify the structure of the model as well as its parameters. All of the previous neural-network-based facial modeling techniques only addressed the parameter issue [143], [199], [47], and we want to point out that both are equally important in modeling facial expression.

The cascade neural network (CNN) differs from other neural network models in a number of aspects. The network structure and size are not fixed, but they can be dynamically adjusted during the training process (Figure 14.5). New hidden units receive connections from all input units as well as all previous hidden units, which is why the network is referred to as a cascade neural network.

Assuming that a cascade network has n_{in} input units (including bias unit), n_h hidden units, and n_o output units, the total number of connections in the network n_w is:

$$n_w = n_{in}n_o + n_h(n_{in} + n_o) + (n_h + 1)n_h/2 \tag{14.1}$$

For a given network node, the kind of activation function that gives the best performance will be installed. During the training process, the kind of activation function that reduces the RMS error the most is selected. The available types of activation functions include Sigmoid, Gaussian, Bessel, and sinusoids of various frequencies.

Node-decoupled extended Kalman filtering (NDEKF) is used for network training. Since the network gradually organizes its structure to optimize performance, no *a priori* model assumption about facial expression intensity is required. Network size is minimized as it grows from a simple two-layer network. In NDEKF training, only part of the network is trained at any one time; thus, it converges within a smaller number of training epochs and is able to achieve smaller RMS error compared to the quick-prop algorithm.

In general extended Kalman filtering, there is a conditional error covariance matrix that stores the interdependence of each pair of weights in the network. NDEKF simplifies the algorithm by considering only the interdependence of weights that feed into the same node. By iteratively training one hidden unit at a time and then freezing that unit's weights, the detrimental effect of node-decoupling is minimized.

The NDEKF weight-update recursion is given by:

$$\omega_{k+1}^i = \omega_k^i + \{(\psi_k^i)^T (A_k \xi_k)\}\phi_k^i \tag{14.2}$$

where ξ_k is the n_o-dimensional error vector for the current training pattern, ψ_k^i is the n_o-dimensional vector of partial-derivatives of the network's output unit signal with respect to the ith unit's net input, and ω_k^i is the input-side weight vector of length n_w^i at iteration k for unit $i \in \{0, 1, ..., n_o\}$. $i = 0$ corresponds to the current hidden unit being trained, and $i \in \{1, ..., n_o\}$ corresponds to the ith output unit, and,

$$\phi_k^i = P_k^i \xi_k^i \tag{14.3}$$

$$A_k = \left[I + \sum_{i=0}^{n_o} \{(\xi_k^i)^T \phi_k^i\} \left[\psi_k^i (\psi_k^i)^T\right] \right]^{-1} \tag{14.4}$$

$$P_{k+1}^i = P_k^i - \{(\psi_k^i)^T (A_k \psi_k^i)\} \left[\phi_k^i (\phi_k^i)^T\right] + \eta_Q I \tag{14.5}$$

$$P_0^i = (1/\eta_P) I \tag{14.6}$$

where ξ_k^i is the n_w^i-dimensional input vector for the ith unit, and P_k^i is the $n_w^i \times n_w^i$ approximate conditional error covariance matrix for the ith unit. Parameter η_Q (0.0001) is included in equation (14.5) to alleviate singularity problems for error covariance matrices [190].

In our research group, cascade neural networks [171] have been successfully applied in modeling various human functions including human control strategies in simulated vehicle driving [166], human skill in controlling a single wheel robot [254], and modeling human sensation in virtual environments [134]. This architecture has demonstrated good performance in modeling continuous human skills data, so it is suitable for our application.

Support vector machine

Support vector machine (SVM) [211] is a recently developed statistical learning technique that has found applications in various areas. We have applied SVM in applications of robot control skills learning [179] and on-line fault diagnosis of intelligent manufacturing systems [253]. SVM can be considered as a linear method in a high-dimensional feature space nonlinearly related to the

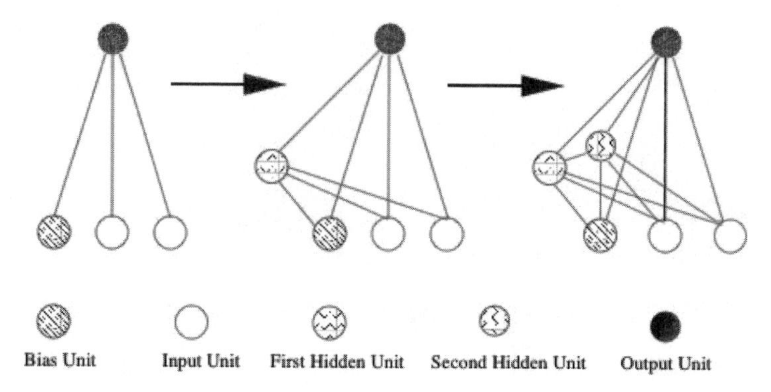

FIGURE 14.5: Training of cascade neural network.

input space. By the use of kernels, all input data are mapped nonlinearly into a high-dimensional feature space. Separating hyperplanes are then constructed with maximum margins that yields a nonlinear decision boundary in the input space. Using appropriate kernel functions, it is possible to compute the separating hyperplanes without explicitly carrying out the mapping into the feature space and the problem will turn into one of quadratic programming in terms of kernels.

Assume we have a set of training samples, S_f, which are independent and identically distributed: $S_f = \{(x_1, y_1), (x_2, y_2), ..., (x_n, y_n)\}$, where $x \in \Re^N$ denotes the input vector and $y \in \Re$ is the output value.

The nonlinear support vector regression (SVR) estimate has the form

$$y = f(x) = \sum_{i=1}^{n} (\alpha_i - \alpha_i^*) \cdot k(x_i, x) + b \tag{14.7}$$

where α_i, α_i^* are coefficients and b is a bias term that are computed as solutions of the following quadratic optimization problem:

$$\min_{\alpha_i,\alpha_i^*} - \sum_{i=1}^{n}(\alpha_i - \alpha_i^*)y_i + \tag{14.8}$$

$$\frac{1}{2}\sum_{i,j=1}^{n}(\alpha_i - \alpha_i^*)(\alpha_j - \alpha_j^*)k(x_i, x_j) \tag{14.9}$$

$$+\epsilon\sum_{i=1}^{n}(\alpha_i + \alpha_i^*)$$

$$\text{such that} \sum_{i=1}^{n}(\alpha_i - \alpha_i^*) = 0,$$

$$0 \leq \alpha_i, \alpha_i^* < C,$$

$$i = 1, 2, ..., n$$

where n is the number of vectors in the input set, ϵ is the expected level of precision, and C is a constant that provides a trade-off between the flatness of function f and the amount to which output deviation is larger than ϵ. There are only a few parameters α that take non-zero values and the corresponding x_i is known as the support vectors. Two types of kernels $k(\cdot)$ will be used in the experiment: polynomial function $k(x, y) = ((x \cdot y) + 1)^d$ and radial basis function $k(x, y) = exp(-\gamma|x - y|^2)$. Various values of polynomial order d and RBF coefficient γ will be applied for performance comparison.

14.1.6 Experiment

First, we captured sequences of facial expression change for a subject. Three types of emotional expression change were recorded and analyzed, including *happiness, anger, and sadness* (Figure 14.6). During each recording, the subject displayed the transition of one type of emotional expression. The face changed from expressionless to an expression with maximal intensity, and then changed back to an expressionless face. This transition was repeated a number of times in order to capture the personal variation in the display of the expression changes (three cycles are shown). After the video sequences were captured and each of the image frames was normalized by using affine transformation, we obtained the trajectory of the facial feature points by using the Lucas-Kanade optical flow tracking algorithm. The feature input to the learning models were the displacements of the facial feature points, each of which was calculated by subtracting the normalized positions in the first image of the sequence from normalized positions in the current image.

The intensity of the facial expression was then extracted from the trajectory of the feature points using isometric feature mapping. The Isomap-generated intensity and the trajectory of the facial feature points were utilized in the training of cascade neural networks and support vector machines. Separate models for each emotion generated by a subject were developed. After learning the facial expression intensity using CNN and SVM, the models were validated using new sequences of facial expression transition data.

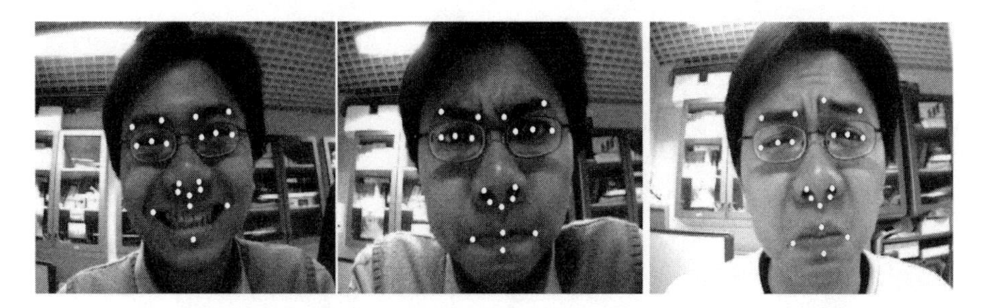

FIGURE 14.6: Expressions of happiness, anger, and sadness.

The expression intensities generated from the training facial feature trajectories using Isomap for the three expression transitions are shown in Figure 14.9. Figure 14.7 shows the residual variance of the reduced dimension using isometric feature mapping. The intensities of the expression account for 95.0%, 76.7%, and 70.6% of the total variance in the transition of expressions in happiness, anger, and sadness, respectively.

Different types of models were utilized for learning, including SVMs with polynomial kernel (Poly) of orders 2 to 5, SVMs with radial basis function kernel (RBF) of γ values 1, 5, and 10, and CNN with hidden unit numbers 5, 10, and 15. The average RMS errors between the expected intensity trajectories and the model-generated intensity trajectories for expression transition of happiness, anger, and sadness are shown in tables 14.1 and 14.2. The best model value for each emotion is shown in *italics.* SVMs with polynomial k-ernels are shown to be more efficient in estimating facial emotion intensity than the other tested models. For model validation, the best performance from CNN is shown in Figure 14.10, and the best performance from SVM is shown in Figure 14.11. The solid lines represent the intensity extracted using Isomap and the dashed lines show the real-time output of the trained facial expression intensity models.

Using a USB camera and Pentium III 1GHz laptop PC, the intensity estimation step can run at 7Hz. Based on the experimental results, the trained CNN and SVM models can successfully reveal the general trend of facial expression transition in real-time.

To demonstrate the usefulness of this approach, we have integrated the expression estimation system with a humanoid robot in our lab for a simple application example. A template of initial feature point positions is used to automatically place all the tracking points on the human's face. During interaction and conversation with humans, the humanoid can detect the de-

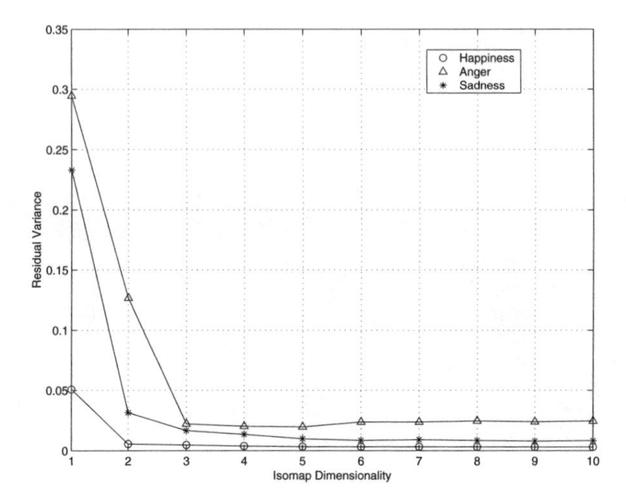

FIGURE 14.7: Residual variance of extraction of the expression intensities using Isomap.

gree of happiness and generate the corresponding body motion and speech in response (Figure 14.8).

14.1.7 Discussion

In general, the performance of the expression intensity model for sadness was not as good as that of other models. During the stage of intensity extraction using Isomap, the first degree of freedom of the sadness expression leaves some variance in the feature trajectory data not accounted for. There is more variation in how the subject expresses sadness than in expressing happiness and anger. Therefore, two more degrees of freedom may be required in order to further reduce the residual variance of Isomap in modeling sadness.

SVM provides the benefit of being able to model data patterns with small sample sizes. But it also leads to the phenomenon that when the testing expressions are very different from the training expressions, erroneous intensity values will be generated. The less efficient modeling effect of the emotion of sadness may also be caused by the possibility that the person is not able to produce a consistent expression of sadness. For the expression of happiness, there are fewer variations in the way that people smile.

For the emotion of sadness, the intensity peak positions output by the models seem to shift away from the expected peaks produced by Isomap. This can be explained by the fact that the expression of sadness displayed by humans actually involves small motions of the upper and lower parts of the face, whereas for happiness, the expression is dominated by the large motion

Table 14.1: Performance of different SVM emotion expression models (RMS error between expected intensity and model generated intensity)

	Poly ($d = 2$)	Poly ($d = 3$)	Poly ($d = 4$)	Poly ($d = 5$)	RBF ($\gamma = 1$)	RBF ($\gamma = 5$)	RBF ($\gamma = 10$)
Happiness	*0.028*	0.071	0.124	0.093	0.140	0.170	0.092
Anger	*0.085*	0.117	0.108	0.132	0.218	0.183	0.235
Sadness	0.236	*0.144*	0.214	0.270	0.340	0.390	0.370

Table 14.2: Performance of different CNN emotion expression models (RMS error between expected intensity and model generated intensity)

	CNN (HU=5)	CNN (HU=10)	CNN (HU=15)
Happiness	0.081	0.162	0.138
Anger	0.097	0.253	0.175
Sadness	0.214	0.341	0.224

of the mouth (lower part of the face). When the person displays expressions of sadness, the upper and lower part of the face may reach their motion peaks at slightly different times or at different orders. This causes the output of the models to produce a shifted peak and sometimes multiple peaks in a cycle.

The architectures of CNN and SVM are similar to each other in the sense that their outputs are both formed by the linear combination of the nodes in the layer(s) in the middle. In addition, instead of having predetermined node connections, the structure of both learning systems are learnt and constructed during the training process based on the sample patterns.

However, the difference in performance between CNN and SVM may lie in the weaknesses of CNN. The basic concept of CNN is based on that of normal neural networks; that means a large amount of training sample is usually required. A large training sample size will minimize the empirical risk but reduce the level of confidence, which may finally lead to weak a generalization capacity and cause the over-fitting problem to occur. Also, there are many local minima in the space represented by the CNN and standard optimization processes only guarantee the convergence to one of them. The final result of CNN is highly influenced by the selection of the initial set of weight matrix, which is usually generated randomly. For SVM, there is no such a need.

The performance of the system is largely dependent on the efficiency of the feature tracker. If the feature points are tracking on the wrong positions on the face, incorrect intensity values will result and the feature positions will be reinitialized to restore correct tracking.

The placement of the feature points is also a critical issue. At the current feature configuration, certain points play a more important role in determining the intensity level than others. The number of feature points and their positions are not yet optimized. This is a future research issue on which we will work.

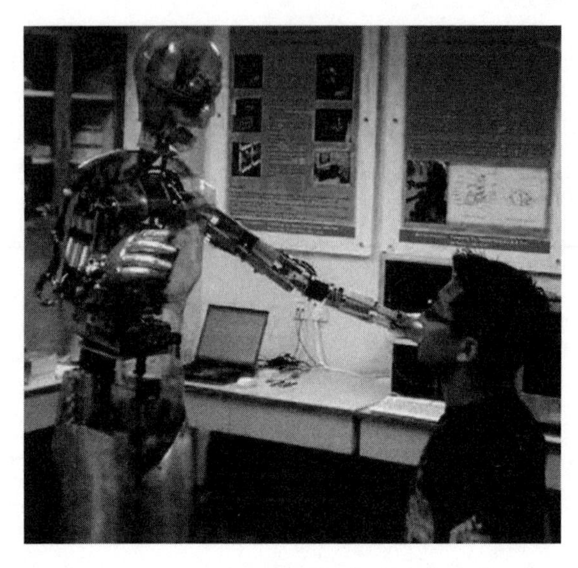

FIGURE 14.8: Robot reacts to human according to facial expression intensity.

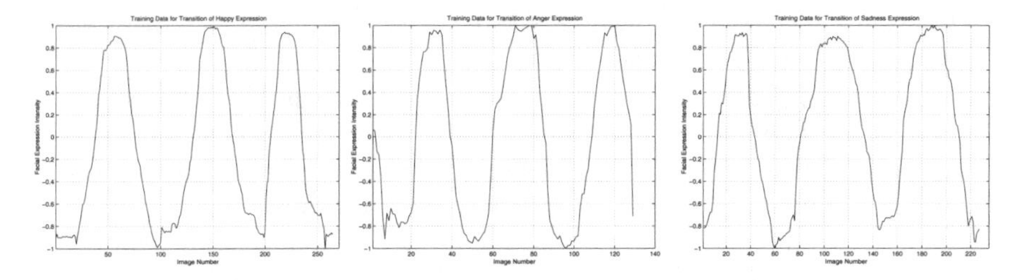

FIGURE 14.9: Training intensity data for happiness, anger, and sadness.

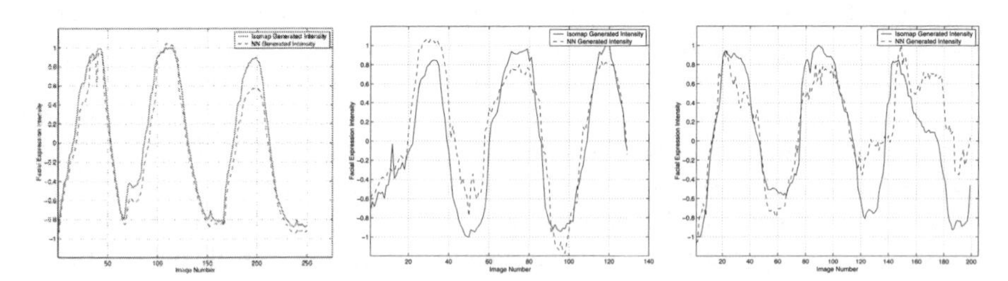

FIGURE 14.10: Validation intensity data for happiness, anger, and sadness using cascade neural network.

14.1.8 Conclusion

We have developed a personalizable vision system that can automatically estimate the intensity of human facial expression in real-time, using learning

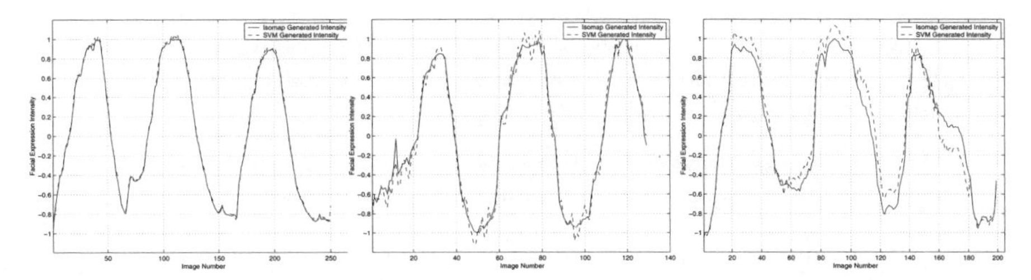

FIGURE 14.11: Validation intensity data for happiness, anger, and sadness using support vector machine.

models including cascade neural networks and support vector machines. Using isometric feature mapping, we can extract the spectrum of expression intensity from a training sequence that represents the transition between an expressionless face and the extreme expression. The system allows in-plane head rotation and is useful for real-time facial interaction with robots.

For future work, we will equip the system with the ability to model more emotion transition types, and to implement human-robot interaction tasks related to facial expressions and context. We will also investigate the issue of facial feature point selection and SVM kernel selection so that the modeling process can demonstrate greater efficiency.

14.2 Full-body action modeling

In our research, we apply artificial intelligence and statistical techniques towards observation of people, leading to modeling of their actions, and understanding of their expressions and intentions. Instead of solving these problems using heuristic rules, we propose to approach them by learning from demonstration examples. The complexities of many human actions make them very difficult to be handled analytically or heuristically. Our experiences in human function modeling [134], [135] indicate that learning human actions from observation is a more viable and efficient method in modeling the details and variations involved with human behaviors. The solutions to these technical problems will create the core modules whose flexible combination will form the basis of systems that can fit into different application areas.

In the literature, different types of methods have been applied to recognize human actions, including skin tracking and rule-based state transition analysis [18], representing human activities by a set of pose and velocity vectors [27], using hidden Markov models for action classification [201], segmenting continuous human activities into separate actions and identify them using nearest neighbor analysis [2], etc.

We have developed a tracking and learning system that is able to classify full-body human actions. The system will be tested on two application examples: sports analysis and violence detection.

1. Detect whether two table tennis players are currently engaged in a game, and the type of actions that the players are performing.
2. Detect whether person-on-person violence has occurred between two people.

These two applications have not been handled using the approach of learning by demonstration in the past.

14.2.1 System overview

Figure 14.12 shows the steps involved in our motion recognition system. In our system, the motions of the people in the scene are tracked continuously. The humans are extracted from the image using background subtraction. The positions of the body centroid and the hands are extracted from the silhouettes. We achieve detection of head and hand positions by silhouette boundary shape analysis. Preprocessing steps are carried out in order to help extracting the required body parts more efficiently. The motion trajectories of the body parts are then recorded for analysis. The original sequence of data defined by the window size is compressed and features are extracted by two computational methods: principal component analysis and independent component analysis. The extracted features are then passed to support vector machines for classification learning.

In modeling the background, we subtract each image from a model of the background scene and thresholding the resulting difference image to determine the foreground pixels. The pixel intensity of a completely stationary background in the indoor can be reasonably modeled with a normal distribution [177], [252]. The model can adapt to slow changes in the scene by recursively updating the model.

Data preprocessing

The pixels in motion will be aggregated using a connected component approach [107] so that individual target human regions can be extracted. Imperfect motion detection may exist in the following form: spurious pixels, holes inside moving features and objects, interlacing effects caused by the video digitization process, and other anomalies. The foreground regions are filtered to remove the spurious features, and then the remaining blobs are preprocessed before recognition analysis is performed. The anomalies in the target region are cleaned by applying morphological dilation and erosion, such that the small holes in the target can be removed and the interlacing anomalies can be smoothed out. The outline of the target region is then extracted by a contour following algorithm [107].

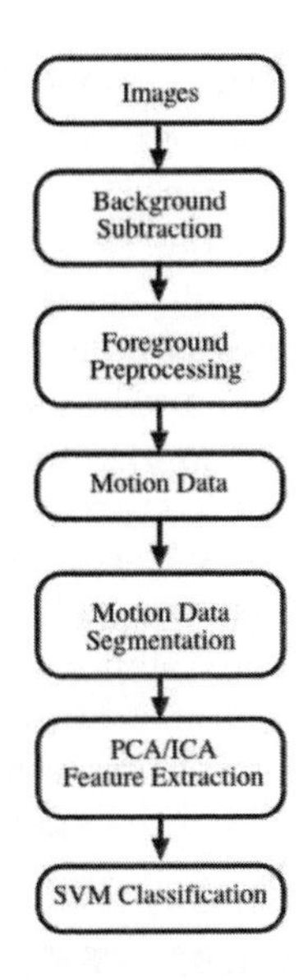

FIGURE 14.12: Steps involved in the sports analysis system.

In the next step, the positions of the human head and hands are to be located. For our application, it is not so important to locate the positions of the foot. So, instead of using skeletonization, we locate the positions of the human head and hands from the contour by using some geometric properties of the contour and convex-hull. We determine the centroid (x_c, y_c) of the human blob by using the following: $x_c = \frac{1}{N}\sum_{i=1}^{N} x_i, y_c = \frac{1}{N}\sum_{i=1}^{N} y_i$, where (x_c, y_c) is the average contour pixel position, (x_i, y_i) represent the points on the human blob contour and there are a total of N points on the contour.

The outline of the body-part tracking algorithm is as follows: (1) The contour of the detected human blob is found, (2) A recursive convex-hull algorithm is applied to locate the vertices of the contour, (3) Locate the position of the head according to the vertical axis of the blob, (4) Locate the hand positions. Our method of locating the positions of head and hands is not iterative, therefore, it is computationally simple. Also, it can be applied to human regions of different sizes and at different scales. Low pass filter is applied to the contour to average the neighboring points. Then, an ellipse is fitted within the human body contour [64] (Figure 14.13).

We assume that the head is located in the upper half of the body, close to the silhouette, and is close to the major axis of the fitted ellipse. The search for the head point is restricted to this region. Two lines are drawn upward from the centroid of the human body blob and making 25 degrees with respect to the major axis of the ellipse. The convex hull point that exists inside this region is considered to be the head. If more than one convex hull point is found within the region, the mean position of all the points found will be taken.

We locate the positions of the hands by using a similar method. During a table tennis game, the hands of the players are usually placed above the waist and near the sides. The search areas for the hands are the areas between the minor axis of the ellipse and the 25 degree lines. The average position of the convex hull points that are furthest away from the centroid will be considered as the position of the hand. If no convex hull vertex is found on the side, we then assume that the positions of the hand are too low or they are being occluded by the torso. In this case the hand points will be placed near the waist on the minor axis of the body ellipse. The search region for the head and hand points are shown in Figure 14.14. The left hand point may not be the actual position of the left hand of the person, it just represents the hand position that exists on the left side of the human body in the image.

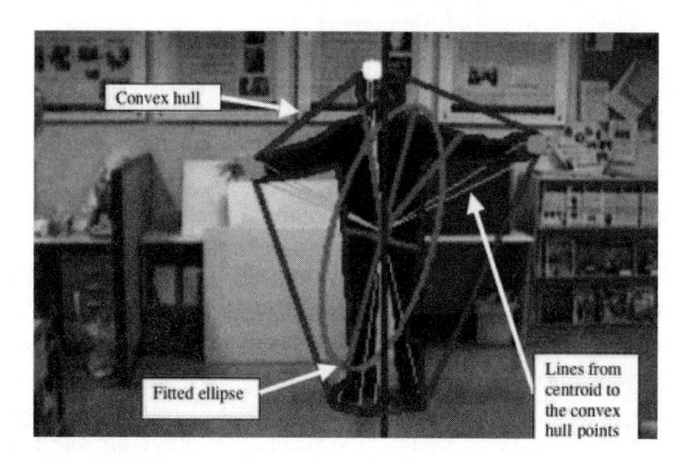

FIGURE 14.13: Convex hull and ellipse on the human image.

14.2.2 Feature extraction

We will apply principal component analysis and independent component analysis to achieve feature extraction from the body part motion trajectories.

Although the motion data presented to the learning machine can be well under control, the amount of perceptual data provided by the humans is in fact very large. In terms of human motion modeling, that means the dimension of inputs of the human motion model can be very high. Therefore it is desirable to reduce the dimension of the human motion inputs in the modeling process

FIGURE 14.14: Search regions for the head and hand points.

in order to reduce the computational complexity and memory requirements.

Principal component analysis (PCA) is a popular feature extraction technique for creating new variables which are linear combination of the original variables [111]. The new variables are uncorrelated among themselves. Geometrically, the objective of PCA is to identify a set of orthogonal axes called the principal components. The new variables (principal component scores) are formed by projecting the original variables on the principal components.

Apart from PCA, we also propose using independent component analysis (ICA) to reduce the dimension of the motion data inputs for human motion modeling. Independent component analysis is a statistical method which transforms an observed multi-dimensional vector into components that are statistically as independent as possible.

A fixed-point algorithm is employed for independent component analysis [104]. The goal of the ICA algorithm is to search for a linear combination of the prewhitened data $x'_i(t)$, where $s(t) = B^T x'(t)$, such that the negentropy (non-gaussianity) is maximized. s are the latent variables and x are the original data. w is assumed to be bounded to have norm of 1 and g' is the derivative of g. The fixed point algorithm [104] is as follows :

1. Generate an initial random vector w_{k-1}, $k = 1$
2. $w_k = E\{x' g(w_{k-1}^T x')\} - E\{g'(w_{k-1}^T x')\} w_{k-1}$
3. $w_k = w_k / \|w_k\|$
4. Stop if converged ($\|w_k - w_{k-1}\|$ is smaller than certain defined threshold). Otherwise, increment k by 1 and return to step 2.

If the process converges successfully, the vector w_k produced can be converted to one of the underlying independent components by $w_k^T x'(t), t =$

$1, 2, ..., m$. Due to the whitening process, the columns of B are orthonormal. By projecting the current estimates on the subspace orthogonal to the columns of the matrix B which are found previously, we are able to retrieve the components one after the other. Using this approach, the knowledge about the human motion model is not required.

14.2.3 Learning system based on support vector classification

We utilize support vector machines to classify the normality of the features of the trajectory points. Support vector machine [211] is a recently developed statistical learning technique which has found applications in various areas and we have successfully applied SVM in applications such as robot control skills learning [179] and on-line fault diagnosis of intelligent manufacturing systems [253]. SVM can be considered as a linear method in a high-dimensional feature space nonlinearly related to the input space. By the use of kernels, all input data are mapped nonlinearly into a high-dimensional features space. Separating hyperplanes are constructed with maximum margins which yields a nonlinear decision boundary in the input space. Using appropriate kernel functions, it is possible to compute the separating hyperplanes without explicitly carrying out the map into the feature space. The problem will turn into one of quadratic programming in terms of kernels.

The classification estimate has the form

$$f(x) = sgn(\sum_{i=1}^{l} v_i \cdot k(x_i, x) + b) \tag{14.10}$$

where $f \in \{-1, 1\}$ represents the output class, v_i are coefficients computed as solutions of the quadratic programming problem, b is the bias term, l is the number of vectors in the testing set, and x_i are the set of support vectors selected from the training data. Two types of kernels $k()$ will be used in the experiment: Polynomial $k(x, y) = ((x \cdot y) + 1)^d$ and Radial Basis Function $k(x, y) = exp(-\gamma|x - y|^2)$. Various values of polynomial order (d) and Gamma (γ) will be applied.

A single support vector machine is capable for classifying two different classes of data only. In our applications, more than one SVM is utilized and they are organized in a hierarchical manner. If there are i types of actions to be recognized, $i - 1$ SVMs will be connected together. The general form of the SVM network is shown in Figure 14.15

14.2.4 Experiment 1: Recognition of motions of table tennis players

Sports video analysis has become a popular research topic in recent years. Related work can be classified into a number of application areas, includ-

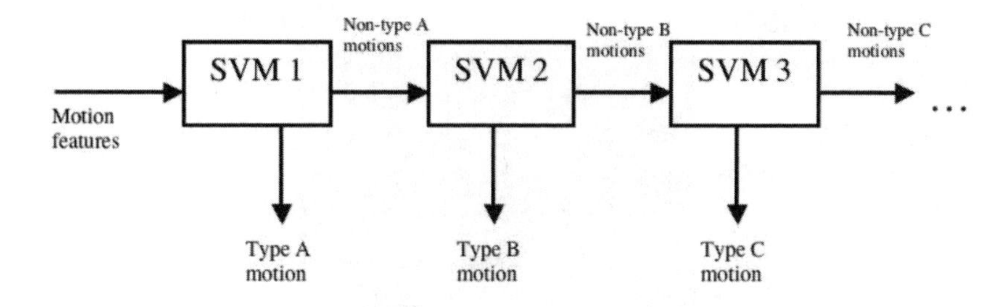

FIGURE 14.15: SVM action classifier.

ing player tracking, analysis of specific sports actions, play-break detection, scene detection, and video indexing. Specialized actions involved in certain sports have become the target of research. Tachibana, et al. [235] applied fuzzy neural network in the analysis of rowing actions, which aimed to identify the relationship between the forces/angles of on-water rowing and the performance. Neural network is used by Yoon, et al. [260] to classify various shapes of swing patterns in golf. Cheng, et al. [42] used Bayesian analysis to identify whether an athlete is running or walking.

In this experiment, we want to develop a method to detect whether table tennis players are engaged in a game. If we detect that the players are hitting the ball, we try to identify whether the player is performing a topspin or a underspin. Topspin (forward spin, Figure 14.17) is a type of spin used on most aggressive shots. When topspin is applied to the ball, the top of the ball moves away from the player. Underspin (backspin, reverse spin, Figure 14.18) is applied by drawing the face of the racket down across the ball at impact. When underspin is applied to the ball, the bottom of a ball is moving in the same direction the ball is traveling.

Trajectories of the body part motion are recorded for analysis. They include (1) x-position of body centroid, (2) x-velocity of body centroid, (3) x-position of left hand point, (4) x-velocity of left hand point, (5) x-position of right hand point, (6) x-velocity of right hand point, (7) y-position of left hand point, (8) y-velocity of left hand point, (9) y-position of right hand point, and (10) y-velocity of right hand point.

The motion of the hands are measured relative to the position of the body centroid. Figure 14.16 shows the coordinate system of the image sequence. The data is captured at a rate of 15Hz and overlapping windows are applied on the data to cut the data into segments. Each segment contains 15 frames (1 second long) and the distance between the beginning of each successive segment is 5 frames. Each segmented data frame can be considered as a matrix

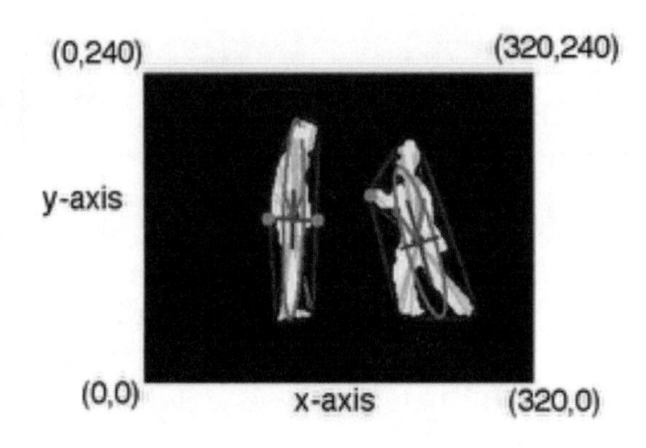

FIGURE 14.16: Coordinate system of the image.

of size 15×10. We then apply three methods to reduce the input size to the SVM for classification. The following steps are performed on each motion data segment. (1) Finding the mean for each of the input data dimension. The reduced input size will be 1×10. (2) Extracting two features from the original data matrix using PCA. Align the two PCs into one single row of size 1×20. (3) Extracting two features from the original data matrix using ICA. Align the two ICs into one single row of size 1×20.

Two SVMs are linked in the architecture for classifying the table tennis actions. The first SVM detects whether the players are actually engaged in a table tennis game. If the players are playing, the second classifier then gives us information on whether the player is doing a topspin or an underspin. Two SVM networks are employed in this application. Each human in the scene has a dedicated SVM network to handle the motions.

For SVM training, there are 25 data segments that represent topspin motion, 25 data segments that represent underspin motion, and 50 data segments that represent non-gaming behaviors. For testing, there are 10 data segments that represent topspin motion, 10 data segments that represent underspin motion, and 10 data segments that represent non-gaming behaviors. Two kinds of kernel functions are used in the SVM: polynomial functions of orders 2, 3, and 4, and radial basis functions with γ values 1, 5, and 10. After two principal components are extracted from the original data segment, more than 99% of the energy is retained. The summary of the results of the various input data reduction methods is shown in Figure 14.26. The best approach in this example is found to be the SVM with radial basis function as kernel ($\gamma = 1$) that uses ICA for feature extraction.

FIGURE 14.17: Topspin action.

FIGURE 14.18: Underspin action.

FIGURE 14.19: Person on the right attacking person on the left.

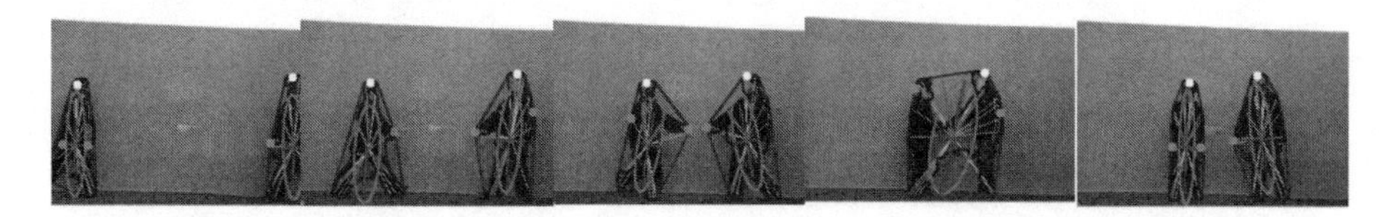

FIGURE 14.20: Two persons shake hands with each other.

14.2.5 Experiment 2: Detection of fighting actions

Human motion analysis can be a very useful tool in surveillance applications. Violence detection is one such application that has received little attention in the past. Statistical models have been proposed by Vasconcelos and Lippman [242] to detect violence scenes in video data. Transformation is performed to map image data into low-dimension feature space where categorization by degree of action can be achieved. However, the system can only recognize high-action motion sequences, such as violent actions and fast sports actions.

Denver, et al. [55] developed a system to recognize the holdup position of armed robbery. The arm positions of the people in the scene are tracked to see if one of the persons has the arms raised and the other person is pointing the arm horizontally at the first person. Datta, et al. [51] proposed to detect fighting motions between two people by analyzing the trajectory information of the limbs, including motion magnitude, motion direction, and jerk of the motion. An analytical measure is developed to calculate the phenomenon that if a person has been moving in a direction for a number of frames and suddenly changes direction and magnitude of motion, then this person is a candidate for being hit.

In this experiment, we try to recognize whether there is any fighting action occurring between the two persons in the video sequence.

The following data trajectories are recorded for analysis: (1) centroid x-position, (2) centroid x-velocity, (3) hand x-position, (4) hand x-velocity, (5) hand y-position, (6) hand y-velocity, (7) upper body contour length, and (8) upper body contour length change.

Three SVMs are linked in the architecture for classifying fighting actions. The first SVM detects whether the two persons are actually engaged in contact between their hands or just walking past each other. If the people contact each other by hands, the second classifier then gives us information on whether the players are doing hand shake or fighting. If one of the persons is being hit, the third SVM recognizes who is the attacker.

In the analysis, we mainly focus on the motions that occur between the time period after one of the persons raises a hand above the centroid position and before the two persons come into contact with each other. The data segments have variable lengths. One SVM network is used to handle the motions of both humans in the image sequence. Only the hand motions that occur between two body centroids will be considered.

The original data in each motion segment has 16 dimensions. Using PCA and ICA, two components are retrieved. The 2×16 features retrieved are aligned to form a 1×32 vector as input to the SVM network. For the statistical averaging method, the mean of the data in each dimension is taken to form a 1×16 input vector for SVM classification.

Figure 14.19 shows a sequence of motion that is classified by the system as the person on the left being hit by the person on the right. Figure 14.20 is classified by the system as a sequence of motion in which the two persons shake hands. Figures 14.21 to 14.25 show some typical motion trajectories that the person on the left is being hit by the person on the right. The punching action occurs at around frame number 20.

The results for SVM classification of fighting action are shown in Figure 14.27. The best result is produced by the SVM with radial basis function as its kernel ($\gamma = 1$) that uses ICA as the feature extraction method.

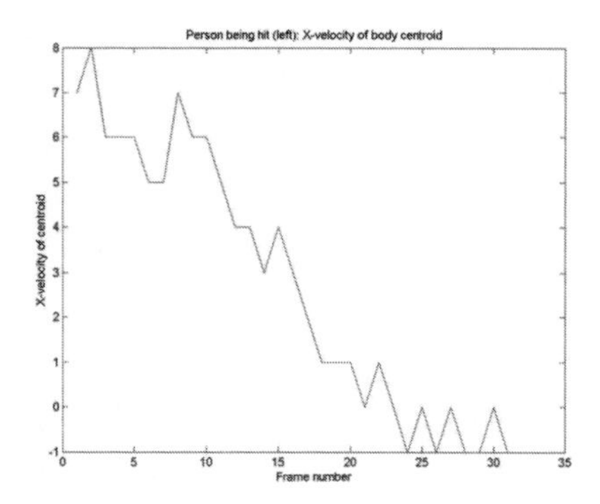

FIGURE 14.21: Person being hit (left): x-velocity of body centroid.

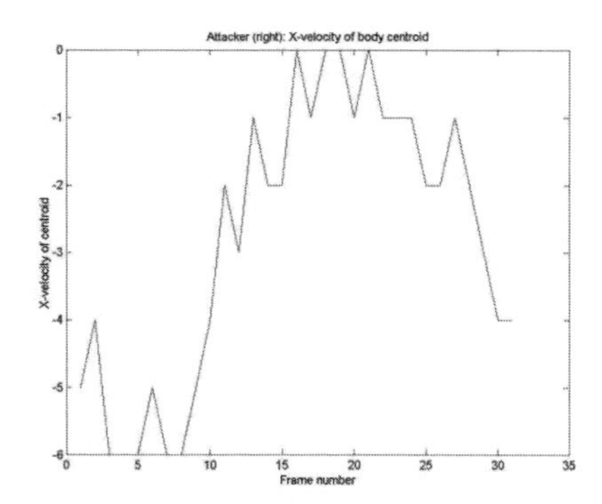

FIGURE 14.22: Attacker (right): x-velocity of body centroid.

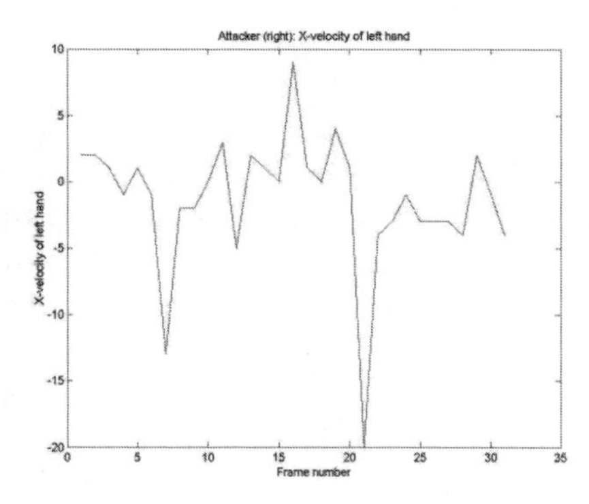

FIGURE 14.23: Attacker (right): x-velocity of attacking hand.

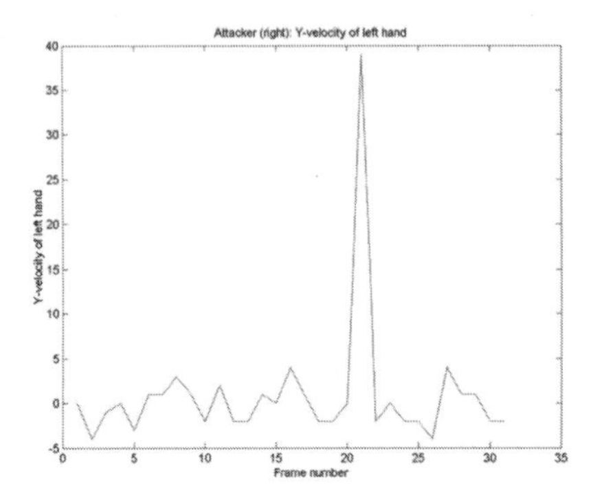

FIGURE 14.24: Attacker (right): y-velocity of attacking hand.

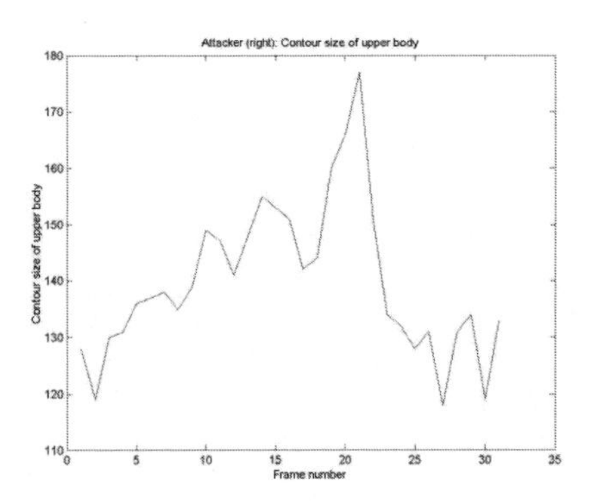

FIGURE 14.25: Attacker (right): Contour size of upper body.

14.2.6 Discussion

Based on the results from our two applications, SVMs with radial basis function perform better than those of polynomial functions. In terms of feature extraction techniques, ICA performs better than pure PCA, and statistical averaging gives the worst classification results. PCA searches for orthogonal directions of greatest variance in the data, whereas ICA component maps may not be orthogonal. In general, there is no reason to believe that the ICs forming the motion data inputs to be spatially orthogonal to one another. Although PCA is often used for input reduction, it is not always useful, because the variance of a signal is not necessarily related to the importance of the variable. The features retrieved may have no importance to the output. The performance of both computational methods are better than the case where only the mean values are used. PCA is often good to use but it works bad if there are too many isotropically distributed clusters or if there are meaningless variables (outliners) with a high noise level.

Kernel Feat. Ext.	Poly. Order/ Gamma	Number of SV		Number of errors			Success rate
		SVM1	SVM2	Not playing	Topspin	Underspin	
Polynomial/ Mean	2	14	16	5	7	6	40%
	3	18	16	6	9	8	23%
	4	12	18	6	7	8	30%
Polynomial/ PCA	2	9	7	3	2	3	73%
	3	11	9	5	5	9	40%
	4	13	12	6	8	6	33%
Polynomial/ ICA	2	4	6	3	2	2	76%
	3	6	8	2	5	4	63%
	4	6	9	3	3	4	66%
RBF/ Mean	1	13	19	7	8	7	26%
	5	8	13	9	9	8	13%
	10	13	16	8	8	9	16%
RBF/ PCA	1	14	17	3	7	4	53%
	5	19	12	7	5	7	36%
	10	15	13	7	9	6	26%
RBF/ ICA	1	7	9	2	2	2	80%
	5	6	13	5	4	4	56%
	10	9	11	4	3	7	53%

FIGURE 14.26: SVM classification results for table tennis actions.

Kernel Feat. Ext.	Poly. Order/ Gamma	Number of SV			Number of errors				Success rate
		SVM1	SVM2	SVM3	Walkpass	Shakehands	Left attack	Right attack	
Polynomial/ Mean	2	15	12	16	5	4	4	2	50%
	3	13	18	8	4	5	2	2	53%
	4	14	16	7	6	5	5	4	33%
Polynomial/ PCA	2	12	14	8	1	7	5	6	36%
	3	14	19	9	1	6	5	4	46%
	4	11	13	7	2	5	4	4	59%
Polynomial/ ICA	2	8	14	6	1	4	4	3	60%
	3	7	12	5	0	4	3	3	66%
	4	8	11	5	2	2	5	4	56%
RBF/ Mean	1	16	18	17	5	4	5	3	43%
	5	17	16	15	5	4	6	7	26%
	10	15	19	20	6	5	6	7	20%
RBF/ PCA	1	7	5	4	2	4	4	4	53%
	5	5	9	11	0	3	5	4	60%
	10	11	9	8	3	7	8	8	13%
RBF/ ICA	1	11	13	8	0	2	4	3	70%
	5	12	12	7	0	3	4	4	63%
	10	12	14	6	1	7	8	7	23%

FIGURE 14.27: Classification results for fighting actions.

The feature extraction method based on ICA is found to be able to give the best data reduction results compared to the other two processing methods presented. This method requires a relatively straightforward transformation and statistical test to be performed in order to achieve model-free input reduction. By reducing the redundancy in the input data, the training process of the human motion model becomes more efficient. After the unrequired information is removed from the inputs, not only the key characteristics of the human motion data can be retained, but also the modeling power of the system is actually improved. It is apparent that ICA provides a useful tool for reducing the human motion input size and improving the performance of

15

Conclusions

In this book we have presented a series of methods to model and transfer human action skills and reaction skills.

In the first part of this book, we present a coherent framework for learning, validating, evaluating, optimizing, and transferring discrete-time models of human control strategy. We summarize the original contributions of this work below.

- We developed a neural-network-based algorithm that combines flexible cascade neural networks with extended Kalman filtering. We show that the resulting learning architecture achieves better convergence in faster time than alternative neural-network paradigms for modeling both known continuous functions and dynamic systems, as well as for modeling human control strategies from real-time human data. We also demonstrate the fundamental problem of modeling discontinuous control strategies with a continuous function approximator.

- We developed a statistical, discontinuous framework for modeling discontinuous human control strategies. The approach models control actions as probabilistic events and choose a specific control action based on a stochastic selection criterion. We demonstrate that the resulting learning architecture is much better able to approximate discontinuous control strategies than continuous function approximators.

- As a model validation tool, we developed a stochastic similarity measured – based on hidden Markov models – that measures the level of similarity (or dissimilarity) between multi-dimensional, stochastic trajectories. We demonstrate and derive important properties of the similarity measure.

- We developed a series performance criteria to evaluate human control strategy models. Although model validation measure shows the trained HCS has the same characteristic as the human driving data, it is difficult to compare different HCS models. This in turn requires that performance criteria be developed, since few if any theoretical guarantees exist for these models. We developed a series performance criteria based on the event analysis as well as the inherence analysis including obstacle avoidance by deceleration or by lane change, tight-turning performance criteria, feeling of the passenger, and smoothness criterion of domain frequency and so on. Based on the performance criteria, we can compare the different nonexplicit physical models, as HCS models.

- We developed an iterative algorithm to optimize the HCS models. The method is based on simultaneously perturbed stochastic approximation (SPSA). This algorithm is used to improve the performance criterion of learnt human control strategy model. The algorithm keeps the overall structure of the learnt models in tact, but tunes the parameters (i.e., weights) in the model to achieve better performance. It requires no analytic formulation of performance, only two experimental measurements of a defined performance criterion per iteration. We have demonstrated the viability of the approach for the task of human driving, where we model the human control strategy through cascade neural networks. While performance improvements vary between HCS models, the optimization algorithm always settles to stable, improved performance after only a few iterations. Furthermore, the optimized models retain important characteristics of the original human control strategy.

- We developed two methods to transfer HCS model from expert to apprentice. One algorithm is also based on simultaneously perturbed stochastic approximation (SPSA), to improve the similarity measure between apprentice and expert HCS models. This transferring algorithm includes two aspects. One is structure learning algorithm which keeps the apprentice getting the same structure as expert. The other aspect is called parameter learning, which tunes the parameters (i.e., weights) in the model to achieve better similarity measure with expert. Only two experimental HMM similarity measurements between apprentice and expert HCS model are needed per iteration. We have demonstrated the apprentice performance improved while the similarity improved with the help of expert, for the new apprentice HCS model gets the high similarity with the expert HCS model, the latter shows the better performance. The other transferring learning algorithm is based on the model compensation. This algorithm need not assume expert HCS model and the apprentice HCS model have the same structure. By introducing a compensation cascade neural network, which is trained from the driving data and the "gap" of expert and apprentice HCS models; the new apprentice HCS model, which is the combination of the original apprentice HCS model and the compensation HCS model, gets the "characteristic" of expert HCS model.

- A novel navigation/localization learning methodology is presented to abstract and transfer the human sequential navigational skill to a robotic wheelchair by showing the platform how to respond in different local environments along a demonstrated, designated route using a lookup-table representation.

The central idea of the second part of this book is the formulation of action learning as a dimension reduction problem. The most important theoretical contribution, however, is a new nonparametric method for fitting trajectories to phase space data. In addition to these, the book contributes a comparison of previously existing methods which may be applied to dimension reduction for action learning.

- Action learning is formulated as the characterization of the lower dimensional manifold or constraint surface, within the much higher-dimensional state space of possible actions, upon which human action states tend to lie during performance of a given task. While it is useful to characterize the regions of configuration space visited during typical example performances, the thesis shows that it is even more useful to characterize the regions of phase space visited, as this comprises a more complete description of the performance's state space.

  Once such a description has been learnt, it is a valuable tool for recognizing and classifying particular observed performances, for performing motion analysis, for skill transfer applications, and for creating computer animations.

- We argue that the "best-fit trajectory" is the best one-dimensional model for a set of action data. To build such a model from a set of position and velocity data sampled from multiple examples of task-execution, this thesis develops two new mathematical tools:

  - a spline smoother is derived for fitting phase space data sets, and
  - the principal curves algorithm is adapted to use this new smoother for fitting trajectories in phase space without an *a priori* parameterization.

  Together, these provide a practical solution for an important general problem: *What is the best-fit path through a sampled vector field?*

  The spline smoother fits a curve which balances a measure of its smoothness against the errors in its approximation of a given set of parameterized position and velocity data. The optimal curve is found by solving a linear system. When used with this smoother in phase space, the principal curves algorithm is transformed from a method for nonlinear regression into a valuable tool for modeling the motion of a dynamic system.

- The use of parametric and nonparametric methods for mapping raw performance data to and from a lower-dimensional feature space is demonstrated on an example data set, and the nonparametric methods are shown to present greater promise.

  Linear parametric models such as PCA, while easy to interpret, are limited in the range and complexity of actions they can represent efficiently.

Nonlinear parametric models such as NLPCA are capable of efficiently representing a more general range of actions, but result in opaque and often suboptimal mappings.

Nonparametric methods combine a set of local models which are individually easy to analyze and interpret into a global model which is capable of representing complex actions. The form of the local models can be designed explicitly for the purpose of modeling action data, and as we have seen, can be designed to work simultaneously with both position and velocity data.

- We develop an intelligent visual surveillance system that can classify normal and abnormal human walking trajectories in outdoor environments by learning from demonstration. The system takes into account both the local and global characteristics of the observed trajectories and is able to identify their normality in real-time. By utilizing support vector learning and a similarity measure based on hidden Markov models, the developed system has produced satisfactory results on real-life data during testing. Moreover, we utilize the approach of longest common subsequence (LCSS) in determining the similarity between different types of walking trajectories. In order to establish the position and speed boundaries required for the similarity measure, we compare the performance of a number of approaches, including fixed boundary values, variable boundary values, learning boundary by support vector regression, and learning boundary by cascade neural networks.

- We have developed a system which can automatically estimate the intensity of facial expression in real-time. Based on isometric feature mapping, the intensity of expression is extracted from training facial transition sequences. Then, intelligent models including cascade neural networks and support vector machines are applied to model the relationship between the trajectories of facial feature points and expression intensity level. We have implemented a vision system which can estimate the expression intensities of happiness, anger, and sadness in real-time.

- We have developed a tracking and learning system that is capable of classifying full-body actions occur in sport videos and detecting the actions of person-on-person violence. A tracker is developed to locate the positions of human head and hands by using background subtraction and silhouette analysis. The motion data is then compressed by using principle component analysis and independent component analysis. The motions performed by the people in the scene can then be classified using support vector classification.

A

Appendix A: Human Control Data

In this appendix, we describe the human control data sets which we use throughout the thesis for learning and validation experiments. We use the dynamic driving simulator of Chapter 3 to collect human control data from six individuals* – (1) Larry, (2) Curly, (3) Moe, (4) Groucho, (5) Harpo, and (6) Zeppo. In order to become accustomed to the simulator's dynamics, we first allow each individual to practice "driving" in the simulator for up to 15 minutes prior to recording any actual data. We then ask each person to drive over three different, randomly generated 20km roads – roads #1, #2, and #3 in Figure 3.2 – as fast as possible without veering off the road. Between runs, we allow a short break for each operator.

Sections A.1 through A.6 plot the instantaneous velocity v (mi/h), the lateral offset from the road median d_ξ (m), the steering angle δ (rad), and the acceleration command ϕ (N) for each human control data set. Table A.1 reports corresponding aggregate statistics for each of the 18 runs.

*All human subjects are males in their mid-20s.

Table A.1: Aggregate statistics for human driving data[a]

	Run	*v (mi/h)*	ω *(rad/s)*[b]	δ *(rad)*[b]	ϕ (× 10^3 *N)*	*off road %*
Larry	$X^{(1,1)}$	71.8 (8.1)	(0.183)	(0.064)	1.70 (2.41)	1.31
	$X^{(1,2)}$	71.1 (7.2)	(0.194)	(0.072)	1.81 (2.35)	0.80
	$X^{(1,3)}$	73.7 (8.0)	(0.200)	(0.081)	2.04 (2.35)	2.05
Curly	$X^{(2,1)}$	63.1 (12.2)	(0.174)	(0.057)	1.38 (2.43)	2.94
	$X^{(2,2)}$	62.7 (9.5)	(0.174)	(0.056)	1.31 (1.85)	2.33
	$X^{(2,3)}$	64.0 (8.6)	(0.178)	(0.056)	1.29 (1.37)	2.43
Moe	$X^{(3,1)}$	70.8 (8.3)	(0.201)	(0.073)	1.90 (3.26)	1.75
	$X^{(3,2)}$	69.1 (7.7)	(0.194)	(0.073)	1.85 (3.34)	1.19
	$X^{(3,3)}$	71.5 (7.7)	(0.200)	(0.077)	1.97 (3.14)	0.59
Groucho	$X^{(4,1)}$	73.1 (9.5)	(0.244)	(0.092)	2.19 (2.77)	2.04
	$X^{(4,2)}$	71.9 (9.0)	(0.249)	(0.095)	2.24 (2.62)	1.02
	$X^{(4,3)}$	74.5 (9.4)	(0.285)	(0.114)	2.57 (2.65)	2.41
Harpo	$X^{(5,1)}$	66.8 (12.4)	(0.181)	(0.084)	1.85 (3.83)	4.02
	$X^{(5,2)}$	65.1 (13.2)	(0.208)	(0.095	1.94 (3.98)	5.27
	$X^{(5,3)}$	69.8 (12.3)	(0.226)	(0.111)	2.29 (3.76)	4.69
Zeppo	$X^{(6,1)}$	52.3 (12.2)	(0.171)	(0.053)	0.89 (1.48)	7.16
	$X^{(6,2)}$	51.7 (4.2)	(0.158)	(0.043)	0.70 (0.25)	1.36
	$X^{(6,3)}$	56.1 (5.7)	(0.204)	(0.058)	1.01 (0.34)	4.50

a. Numbers in parentheses are standard deviations.
b. Means for all runs is 0.000.

A.1 Larry

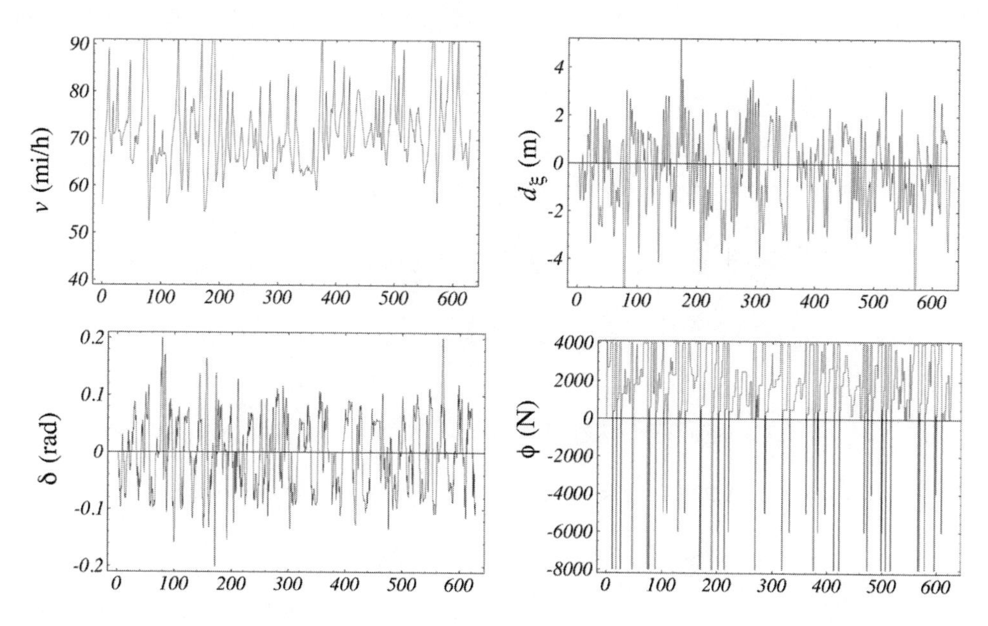

FIGURE A.1: Larry's run over road #1 as a function of time (sec).

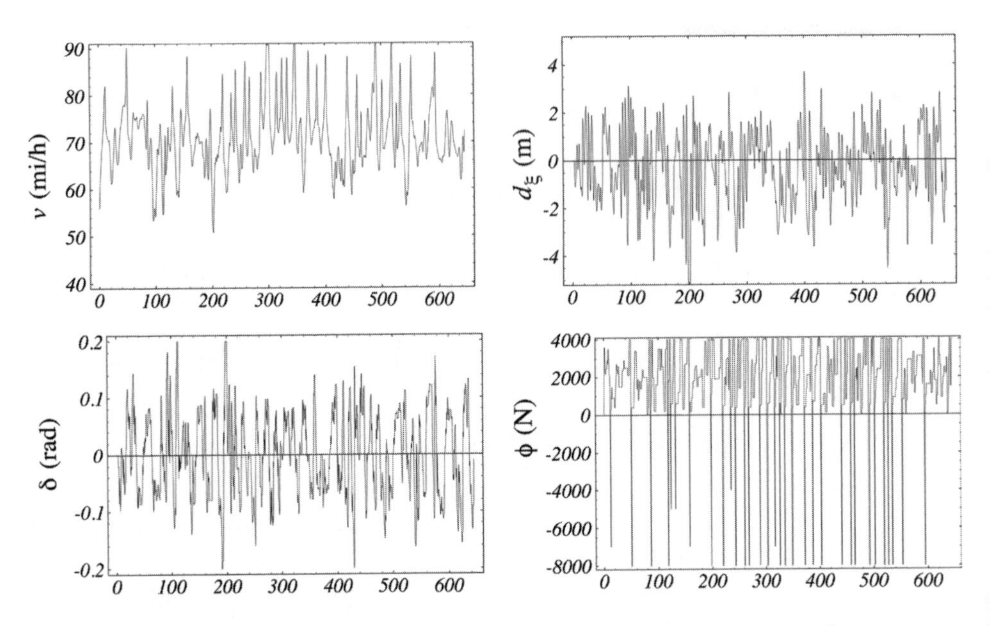

FIGURE A.2: Larry's run over road #2 as a function of time (sec).

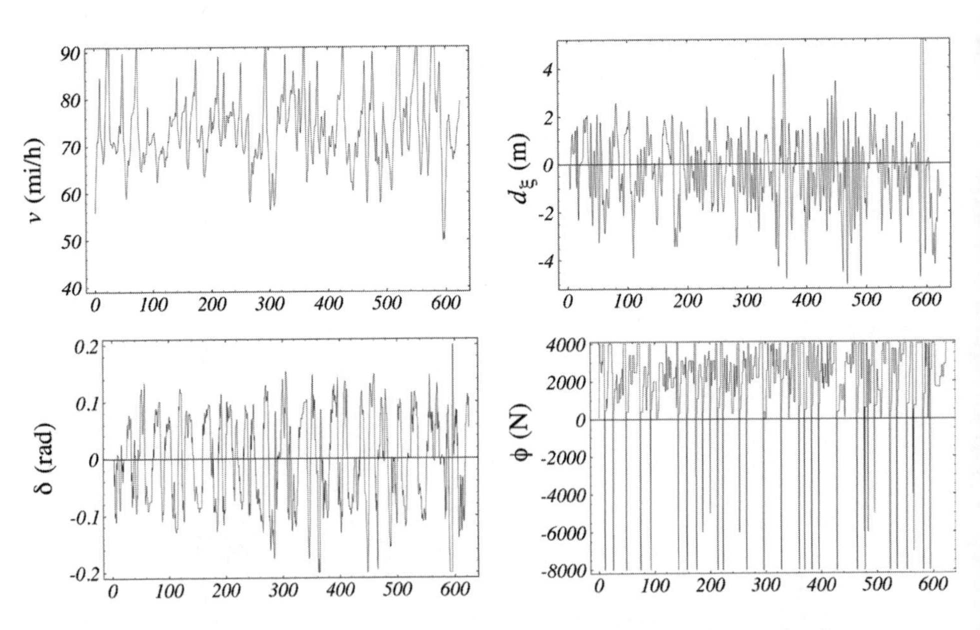

FIGURE A.3: Larry's run over road #3 as a function of time (sec).

A.2 Curly

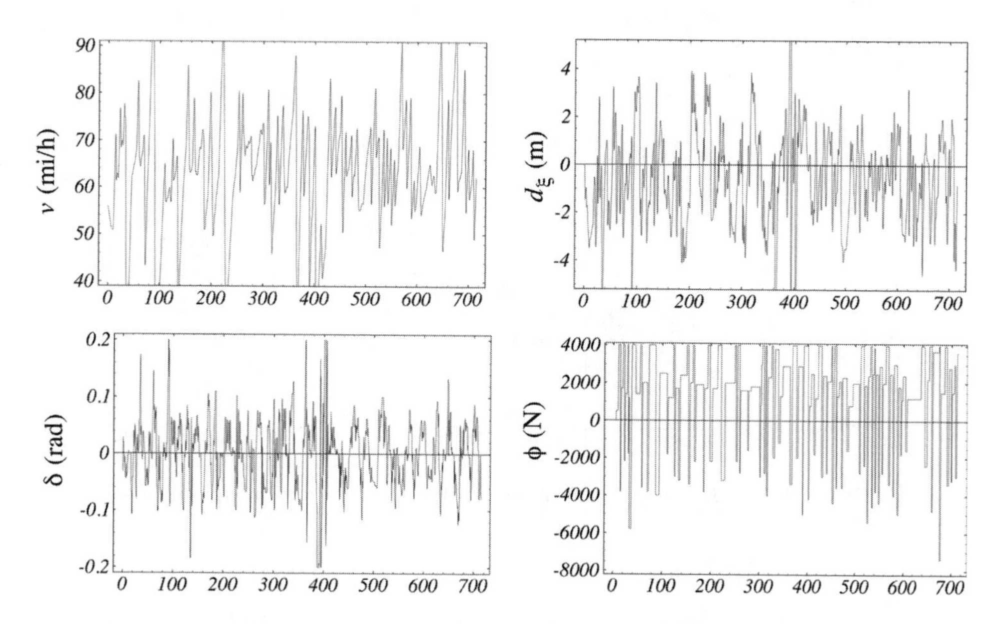

FIGURE A.4: Curly's run over road #1 as a function of time (sec).

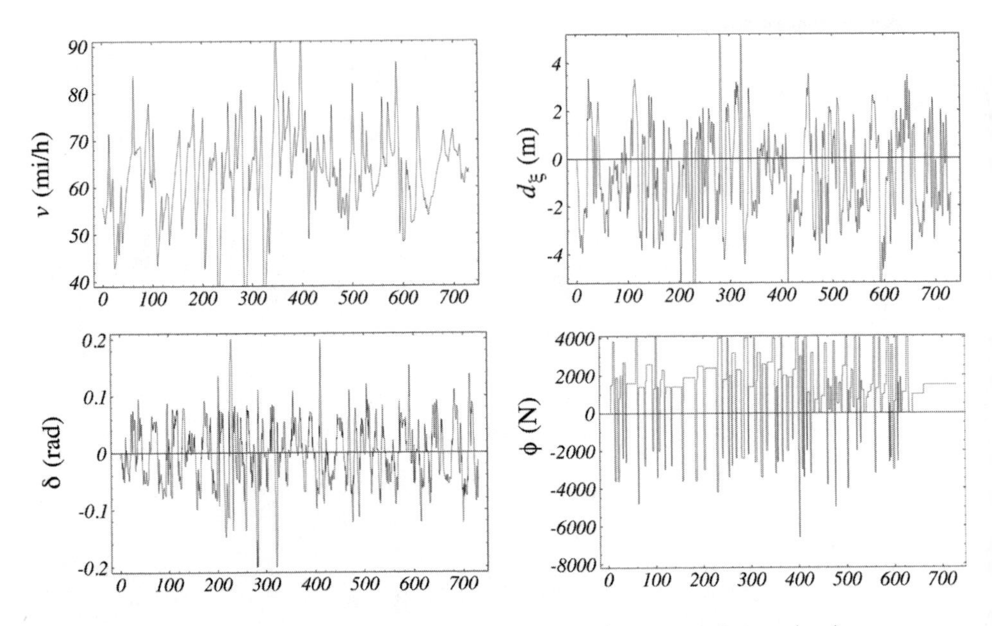

FIGURE A.5: Curly's run over road #2 as a function of time (sec).

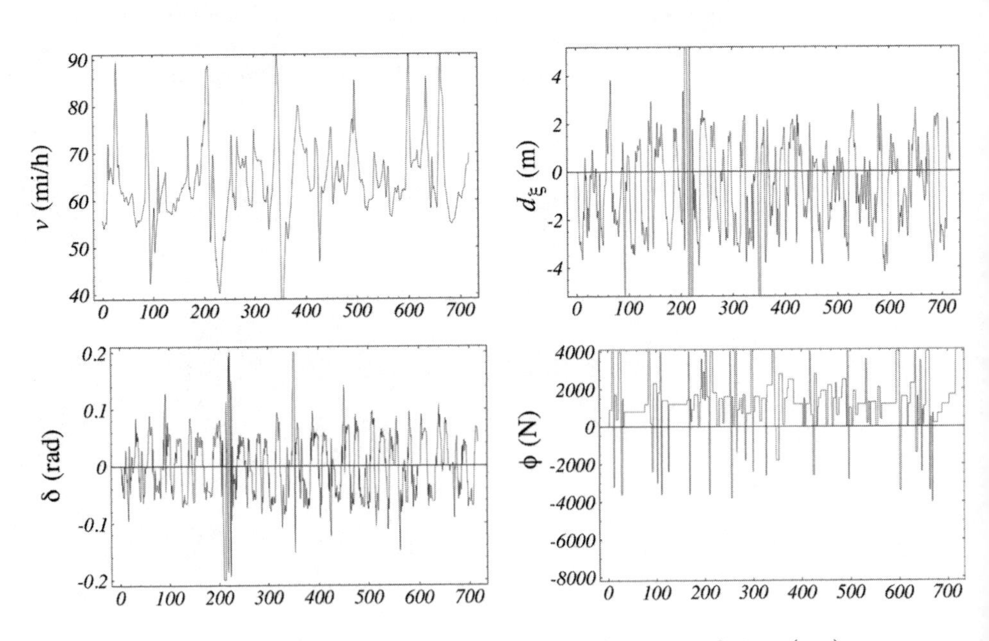

FIGURE A.6: Curly's run over road #3 as a function of time (sec).

A.3 Moe

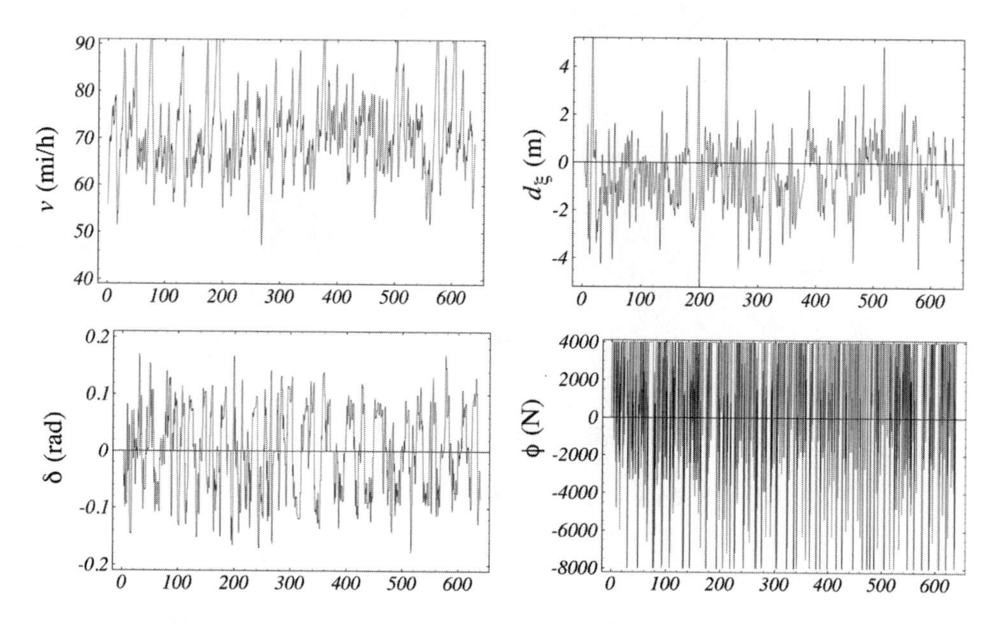

FIGURE A.7: Moe's run over road #1 as a function of time (sec).

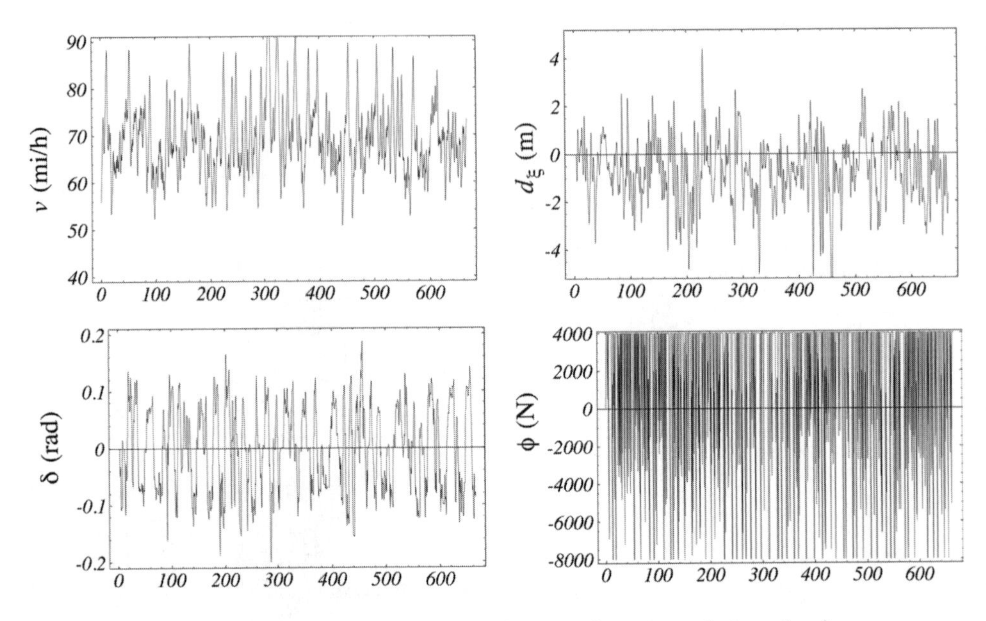

FIGURE A.8: Moe's run over road #2 as a function of time (sec).

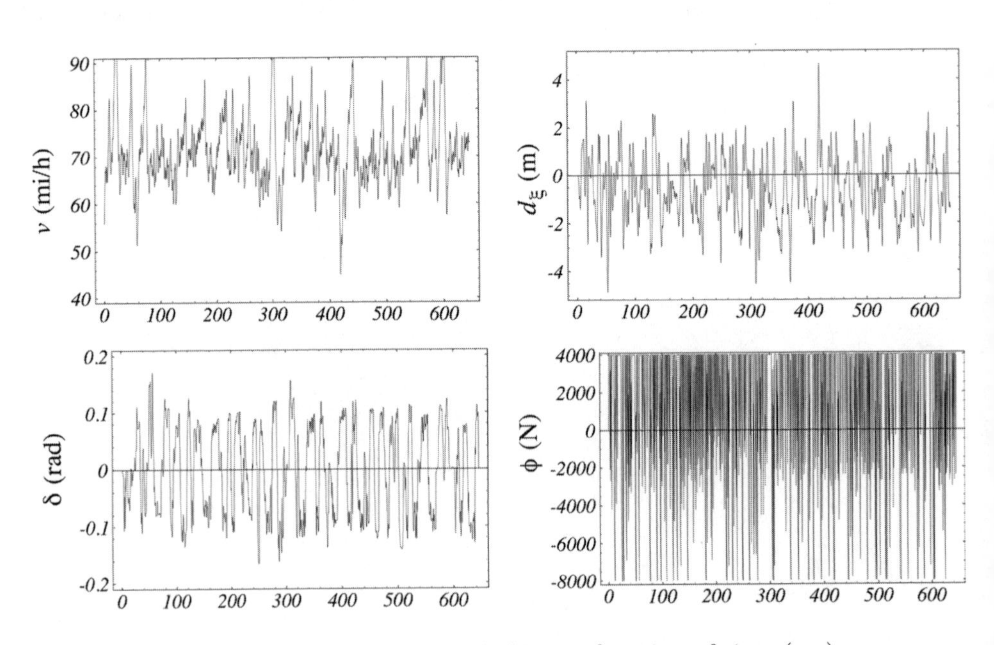

FIGURE A.9: Moe's run over road #3 as a function of time (sec).

A.4 Groucho

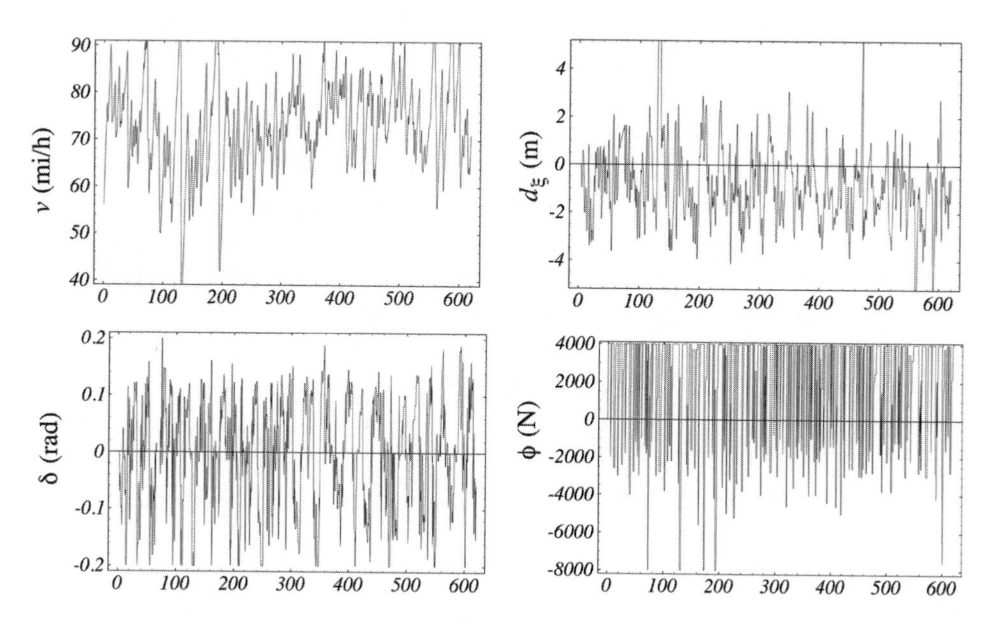

FIGURE A.10: Groucho's run over road #1 as a function of time (sec).

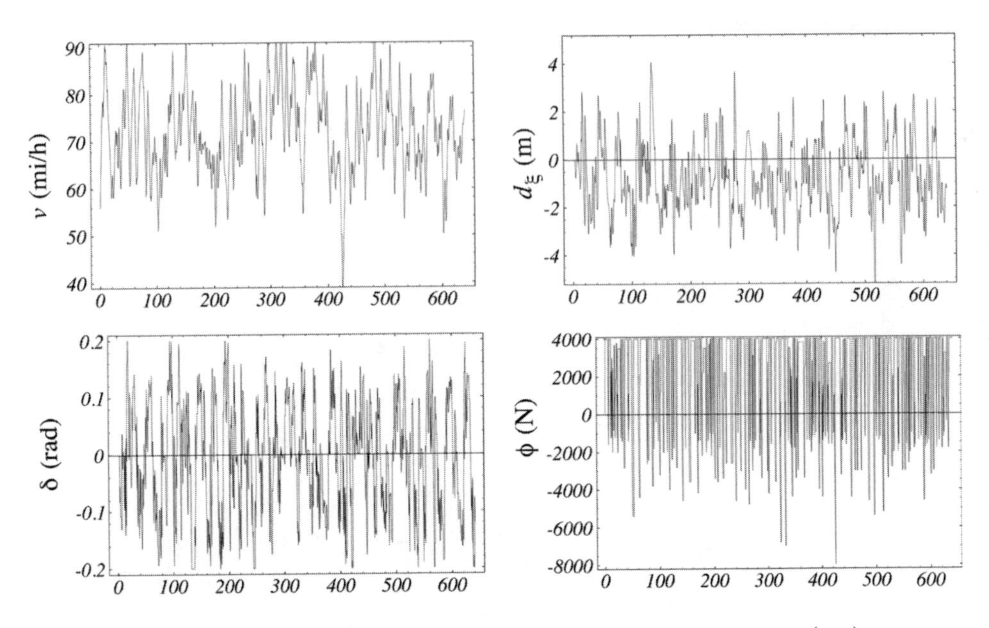

FIGURE A.11: Groucho's run over road #2 as a function of time (sec).

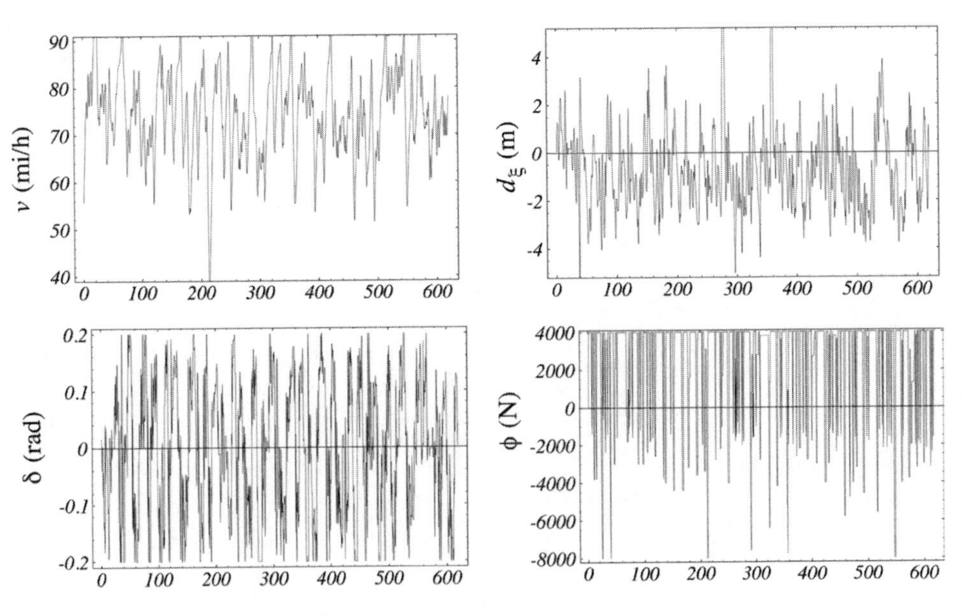

FIGURE A.12: Groucho's run over road #3 as a function of time (sec).

A.5 Harpo

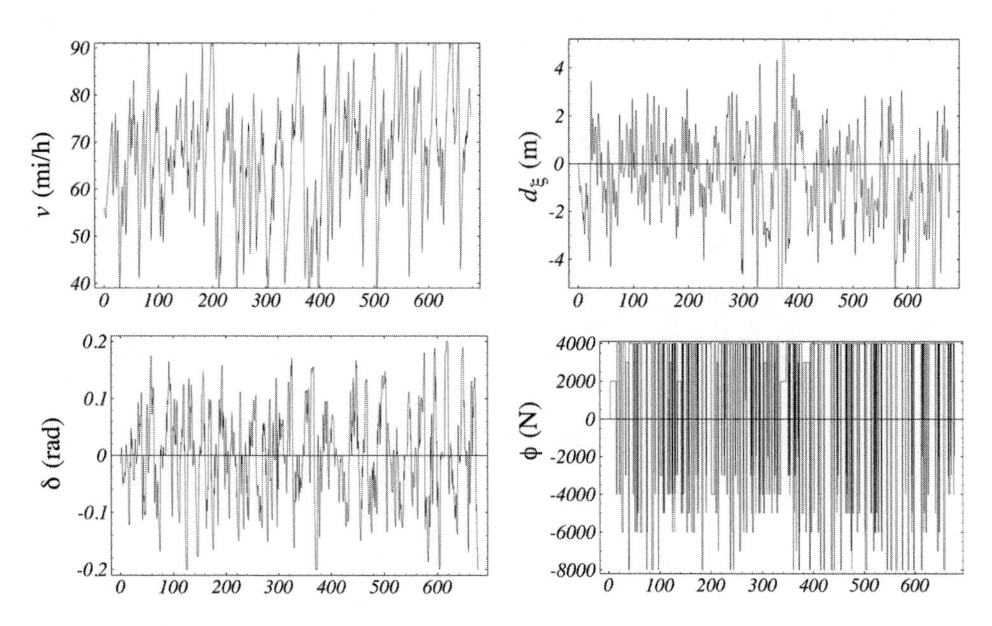

FIGURE A.13: Harpo's run over road #1 as a function of time (sec).

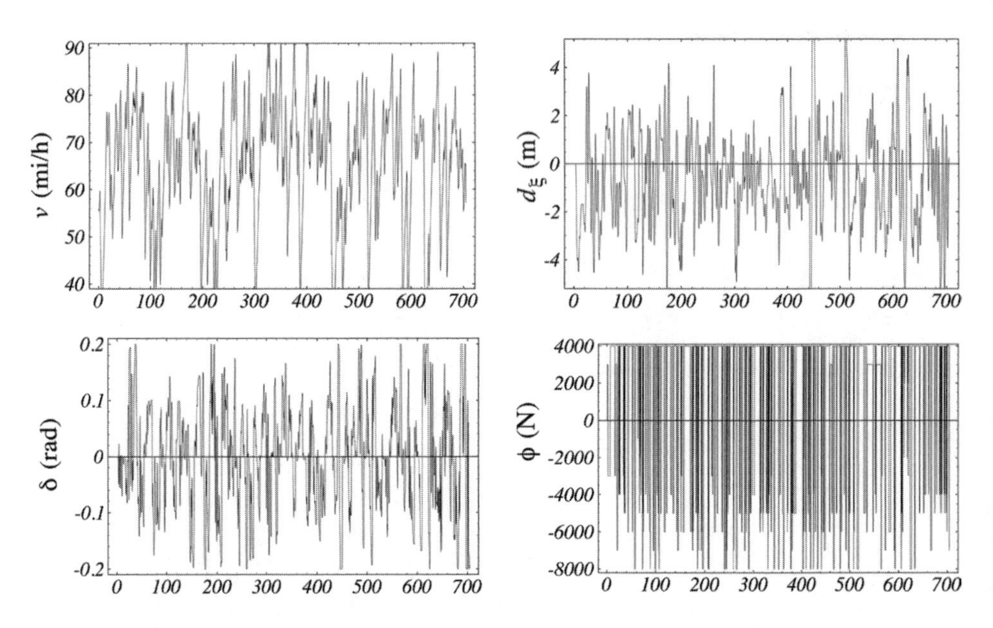

FIGURE A.14: Harpo's run over road #2 as a function of time (sec).

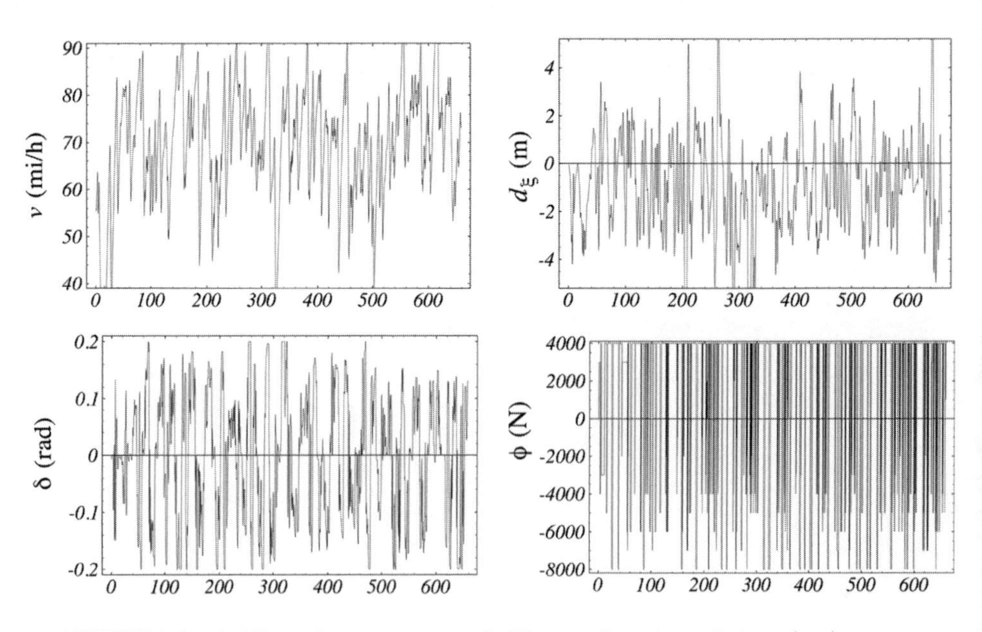

FIGURE A.15: Harpo's run over road #3 as a function of time (sec).

A.6 Zeppo

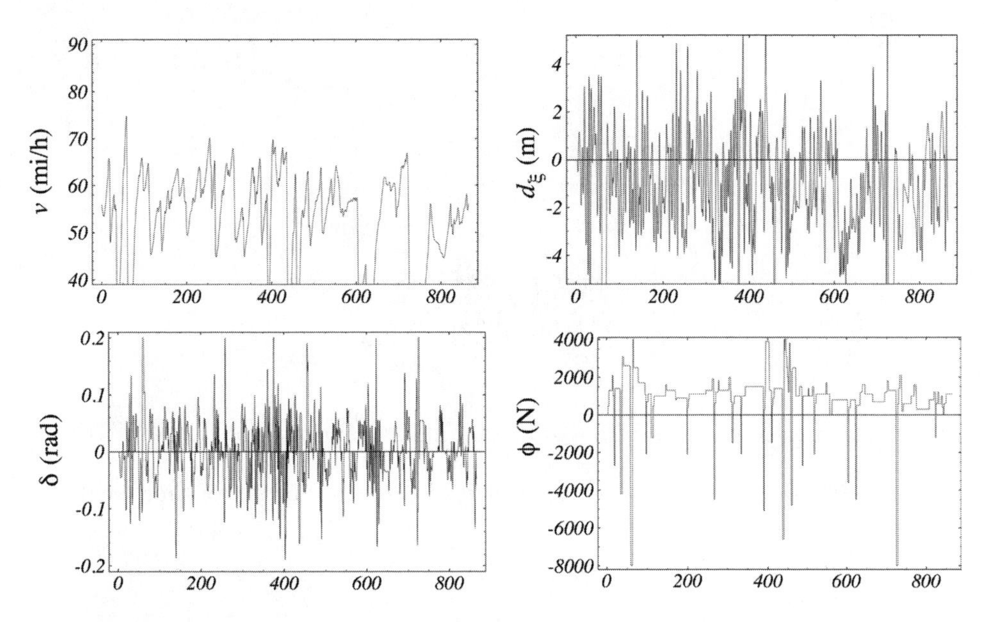

FIGURE A.16: Zeppo's run over road #1 as a function of time (sec).

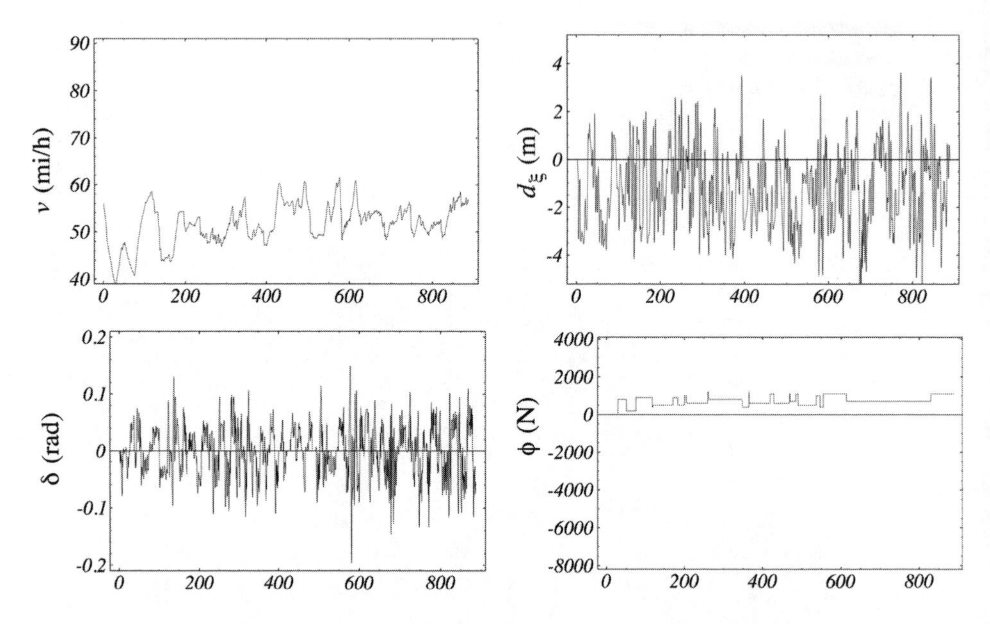

FIGURE A.17: Zeppo's run over road #2 as a function of time (sec).

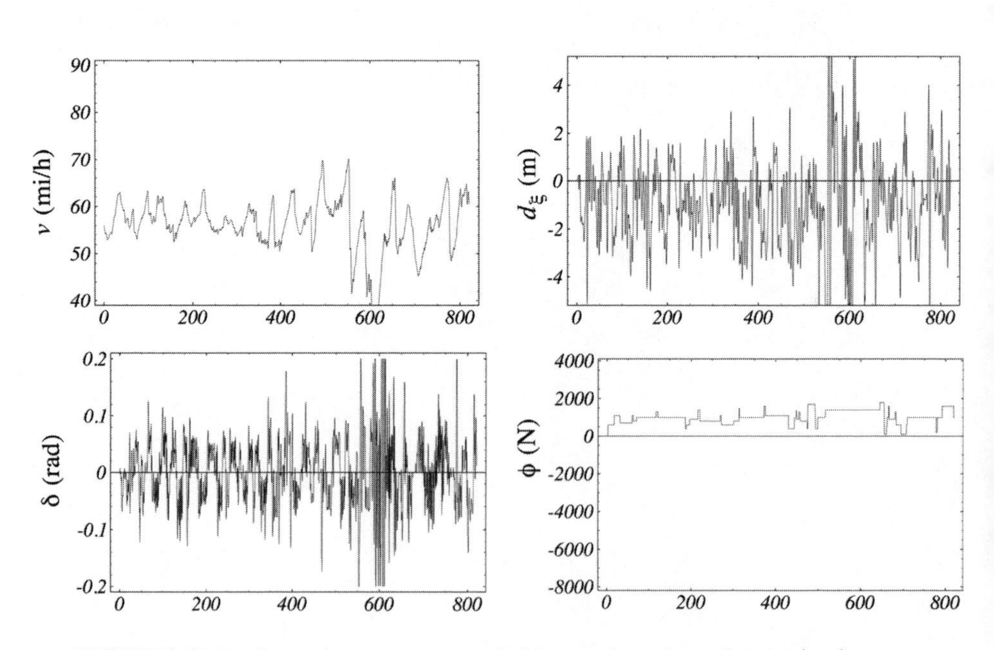

FIGURE A.18: Zeppo's run over road #3 as a function of time (sec).

References

[1] Abelson, H., et al., "Revised Report on the Algorithmic Language Scheme," *ACM Lisp Pointers IV*, vol. 4, no. 3, July-September 1991.

[2] Ali, A. and Aggarwal, J. K., "Segmentation and Recognition of Continuous Human Activity," in *Proc. IEEE Workshop on Detection and Recognition of Events in Video*, pp. 28-35, 2001.

[3] Atkeson, C. G., Moore, A. W. and Schaal, S. A., "Locally Weighted Learning," *Artificial Intelligence Review*, vol. 11, pp. 11-73, 1997.

[4] Atkeson, C. G., Moore, A. W. and Schaal, S. A., "Locally Weighted Learning for Control," *Artificial Intelligence Review*, vol. 11(15):75-113, 1997.

[5] Albus, J., "A New Approach to Manipulator Control: The Cerebellar Model Articulation Controller (Cmac)," *Trans. ASME Journal of Dynamic Systems, Measurement, and Control*, vol. 97, pp. 220-227, 1975.

[6] Anderson, B. D. O. and Moore, J. B., *Optimal Filtering*, Prentice Hall, Englewood Cliffs, New Jersey, 1979.

[7] Antsaklis, P. J., guest ed., Special Issue on Neural Networks in Control Systems, *IEEE Control System Magazine*, vol. 10, no. 3, pp. 3-87, 1990.

[8] Antsaklis, P. J., guest ed., Special Issue on Neural Networks in Control Systems, *IEEE Control System Magazine*, vol. 12, no. 2, pp. 8-57, 1992.

[9] Archibald, C. and Petriu, E., "Skills-Oriented Robot Programming," In F.C.A. Groen, S. Hirose, and C.E. Thorpe, editors, *Proc. of the International Conference on Intelligent Autonomous Systems IAS3*, pp. 104-115, Pittsburgh, PA, 1993. IOS Press.

[10] Arkin, R. C., *Behavior-Based Robotics*, The MIT Press, Cambridge, MA, 1998.

[11] Aronszajn, N., "Theory of Reproducing Kernels," *Transactions of the American Mathematical Society*, vol. 68, pp. 337-404, 1950.

[12] Asada, H. and Liu, S., "Transfer of Human Skills to Neural Net Robot Controllers," *Proc. IEEE Int. Conf. on Robotics and Automation*, vol. 3, pp. 2442-2447, 1991.

[13] Asada, H. and Yang, B., "Skill Acquisition from Human Experts Through Pattern Processing of Teaching Data," *Proc. IEEE Int. Conf. on Robotics and Automation*, vol. 3, pp. 1302-1307, 1989.

[14] Ash, T., "Dynamic Node Creation in Backpropagation Networks," *Connection Science*, vol. 1, no. 4, pp. 365-375, 1989.

[15] Astrom, K. J. and McAvoy, T. J., "Intelligent Control: An Overview and Evaluation," *Handbook of Intelligent Control: Neural, Fuzzy, and Adaptive Approaches*, D. A. White and D. A. Sofge, eds., pp. 3-34, Multiscience Press, New York, 1992.

[16] Atkeson, C. G., Moore, A. W. and Schaal, S., "Locally Weighted Learning," *Artificial Intelligence Review*, vol. 11, no. 1-5, pp. 11-73, 1997.

[17] Atkeson, C. G., Moore, A. W. and Schaal, S., "Locally Weighted Learning for Control," *Artificial Intelligence Review*, vol. 11, no. 1-5, pp. 75-113, 1997.

[18] Ayers, D. and Shah, M., "Recognizing Human Actions in a Static Room," *Proc. IEEE Workshop on Applications of Computer Vision*, pp. 42-47, 1998.

[19] Azoz, Y., Devi, L. and Sharma, R., "Reliable Tracking of Human Arm Dynamics by Multiple Cue Integration and Constraint Fusion," *Proc. of the IEEE Conference on Computer Vision and Pattern Recognition (CVPR)*, pp. 905-910, Santa Barbara, CA, June 1998.

[20] Baluja, S. and Caruana, R., "Removing Genetics from the Standard Genetic Algorithm," *Proc. of the 12th Int. Conf. on Machine Learning*, vol. 1, pp. 38-46, 1995.

[21] Bar-Joseph, Z., Gerber, G., Gifford, D. K. and Jaakkola, T. S., "A New Approach to Analyzing Gene Expression Time Series Data," *Proc. 6th Annual Int. Conf. on Research in Comp. Molecular Biology*, 2002.

[22] Basri, R. and Weinshall, D., "Distance Metric Between 3D Models and 2D Images for Recognition and Classification," *IEEE Trans. Pattern Analysis and Machine Intelligence*, vol. 43, no. 4, pp. 465-479, 1996.

[23] Bartlett, E. B., "Dynamic Node Architecture Learning: An Information Theoretic Approach," *Neural Networks*, vol. 7, no. 1, pp. 129-140, 1994.

[24] Barto, A. G., Sutton, R. S. and Watkins, C. J., "Learning and Sequential Decision Making," *Learning and Computational Neuroscience*, ed. M. Gabriel and J. W. Moore, MIT Press, Cambridge, pp. 539-602, 1990.

[25] Batavia, P. H., "Driver Adaptive Warning Systems," *Technical Report, CMU-RI-TR-98-07*, Carnegie Mellon University, 1998.

[26] Baum, L. E., Petrie, T., Soules, G. and Weiss, N., "A Maximization Technique Occurring in the Statistical Analysis of Probabilistic Functions of Markov Chains," *Ann. Mathematical Statistics*, vol. 41, no. 1, pp. 164-171, 1970.

[27] Ben-Arie, J. and Wang, Z., "Human Activity Recognition Using Multidimensional Indexing," *IEEE Trans. on Pattern Analysis and Machine Intelligence*, vol. 24, no. 8, pp. 1091-1104, August 2002.

[28] Bentley, J. L., "Multidimensional Binary Search Trees Used for Associative Searching," *Communications of the ACM*, vol. 19, no. 9, pp. 509-517, 1975.

[29] Bhat, N. V. and McAvoy, T. J., "Determining Model Structure for Neural Models by Network Stripping," *Computers and Chemical Engineering*, vol. 16, no. 4, pp. 271-281, 1992.

[30] Bishop, C. M., *Neural Networks for Pattern Recognition*, Oxford University Press, Oxford, United Kingdom, 1995.

[31] Boninsegna, M. and Rossi, M., "Similarity Measures in Computer Vision," *Pattern Recognition Letters*, vol. 15, no. 12, pp. 1255-1260, 1994.

[32] Borenstein, J., Everett, H. R. and Feng, L., "Where am I? Sensors and Methods for Mobile Robot Positioning," University of Michigan, *Technical Report*, 1996.

[33] Boussinot, F. and De Simone, R., "The Esterel Language," *Proc. of the IEEE*, vol. 79, pp. 1293-1304, 1991.

[34] Bregler, C. and Omohundro, S. M., "Surface Learning with Applications to Lipreading," In J. D. Cowan, G. Tesauro, and J. Alspector, editors, *Advances in Neural Information Processing Systems*, vol. 6, pp. 43-50. Morgan Kaufmann, San Mateo, CA, 1994.

[35] Brockwell, P. J. and Davis, R. A., *Time Series: Theory and Methods*, 2nd. ed., Springer-Verlag, New York, 1991.

[36] Burrascano, P., "A Pruning Technique Maximizing Generalization," *Proc. Int. Joint Conf. on Neural Networks*, vol. 1, pp. 347-350, 1993.

[37] Byrd, R. H., Lu, P., Nocedal Jorge and Zhu Ciyou, "A Limited Memory Algorithm for Bound Constrained Optimization," *SIAM Journal of Scientific Computing*, vol. 16, no. 5, pp. 1190-1208, 1995.

[38] Castellano, G., Fanelli, A. M. and Pelillo, M., "An Empirical Comparison of Node Pruning Methods for Layered Feed-forward Neural Networks," *Proc. Int. Joint Conf. on Neural Networks*, vol. 1, pp. 321-326, 1993.

[39] Cater, J. P., "Successfully Using Peak Learning Rates of 10 (and Greater) in Back-Propagation Networks with the Heuristic Learning

Algorithm," *IEEE First Int. Conf. on Neural Networks*, vol. 2, pp. 645-651, 1987.

[40] Chandrasiri, N. P., Park, M. C. and Naemura, T. and Harashima, H., "Personal Facial Expression Space based on Multidimensional Scaling for the Recognition Improvement", *Proc. IEEE Int. Symp. on Signal Processing and its Applications*, pp. 943-946, Brisbane, Australia, August 1999.

[41] Chen, K. and Ervin, R. D., "Worldwide IVHS Activities: A Comparative Overview," *Proc. CONVERGENCE' 92 – Int. Congress on Transportation Electronics*, pp. 339-349, 1992.

[42] Cheng, F., Christmas, J. and Kittler, J., "Recognising Human Running Behaviour in Sports Video Sequences," *Proc. 16th Int. Conf. on Pattern Recognition*, pp. 1017-1020, 2002.

[43] Chow, H. N. and Xu, Y., "Human Inspired Approach to Navigation Control," *International Journal of Robotics and Automation*, vol. 17, no. 4, pp. 171-177, 2002.

[44] Cleveland, W. S., "Robust Locally Weighted Regression and Smoothing Scatterplots," *Journal of the American Statistical Association*, vol. 74, no. 368, pp. 829-836, December 1979.

[45] Cleveland, W. S. and Loader, C., "Smoothing by Local Regression: Principles and Methods," *Technical Report 95.3*, AT&T Bell Laboratories, Statistics Department, Murray Hill, NJ, 1994.

[46] Collier, W. C. and Weiland, R. J., "Smart Cars, Smart Highways," *IEEE Spectrum*, vol. 31, No. 4, pp. 27-33, 1994.

[47] Cottrell, G. and Metcalfe, J., "Empath: Face, Gender, and Emotion Recognition Using Holons," in R. Lippman, J. Moody, and D. Touretzky (Eds), *Advances in Neural Information Processing Systems 3*, pp. 564-571. San Mateo: Morgan Kaufmann, 1991.

[48] Craven, P. and Wahba, G., "Smoothing Noisy Data with Spline Functions-Estimating the Correct Degree of Smoothing by the Method of Generalized Cross-Validation," *Numerische Mathematik*, vol. 31, no. 4, pp. 377-403, 1979.

[49] Crowley, J. L., Wallner, F. and Schiele, B., "Position Estimation Using Principal Components of Range Data," *Proc. 1998 IEEE International Conference on Robotics and Automation*, vol. 4, pp. 3121-3128, 1998.

[50] Cybenko, G., "Approximation by Superposition of a Sigmoidal Function," *Mathematics of Control, Signals, and Systems*, vol. 2, no. 4, pp. 303-314, 1989.

[51] Datta, A., Shah, M. and Lobo, N., "Person-on-Person Violence Detection in Video Data," *Proc. IEEE Int. Conf. on Pattern Recognition*, pp. 433-438, 2002.

[52] Davis, J. W., "Recognizing Movement Using Motion Histograms," *Perceptual Computing Section 487*, MIT Media Laboratory, 1999.

[53] De Boor, C., *A Practical Guide to Splines*, Springer-Verlag, New York, 1978.

[54] de Madrid, A.P., Dormido, S. and Morilla, F., "Reduction of the Dimensionality of Dynamic Programming: A Case Study," *Proc. 1999 American Control Conference*, vol. 4, pp. 2852-2856, 1999.

[55] Dever, J., Lobo, N. and Shah, M., "Automatic Visual Recognition of Armed Robbery," *Proc. IEEE Int. Conf. on Pattern Recognition*, 2002.

[56] Dong, D. and McAvoy, Thomas J., "Nonlinear Principal Component Analysis-Based on Principal Curves and Neural Networks," *Computers and Chemical Engineering*, vol. 20, no. 1, pp. 65-78, January 1996.

[57] Douglas, A. and Xu, Y., "Real-time Shared Control System for Space Telerobotics," *Journal of Intelligent and Robotic Systems: Theory and Applications*, vol. 13, no. 3, pp. 247-262, July 1995.

[58] Duda, R. O. and Hart, P. E., *Pattern Classification and Scene Analysis*, John Wiley & Sons, New York, 1973.

[59] Fahlman, S. E., "An Empirical Study of Learning Speed in Back-Propagation Networks," *Technical Report, CMU-CS-TR-88-162*, Carnegie Mellon University, 1988.

[60] Fahlman, S. E. and Lebiere, C., "The Cascade-Correlation Learning Architecture," *Technical Report, CMU-CS-TR-91-100*, Carnegie Mellon University, 1991.

[61] Fahlman, S. E., Baker, L. D. and Boyan, J. A., "The Cascade 2 Learning Architecture," *Technical Report, CMU-CS-TR-96-184*, Carnegie Mellon University, 1996.

[62] Fasel, B. and Luettin, J., "Recognition of Asymmetric Facial Action Unit Activities and Intensities", *Proc. of Int. Conf. on Pattern Recognition*, vol. I, pp. 1100-1103, Barcelona, Spain, 2000.

[63] Fels, S. S. and Hinton, G. E., "Glove-talk: A Neural Network Interface between a Data-glove and a Speech Synthesizer," *IEEE Transations on Neural Networks*, vol. 4, no. 1, pp. 2-8, January 1994.

[64] Fitzgibbon, A., Pilu, M. and Fisher, R. B., "Direct Least Squares Fitting of Ellipses," *IEEE Transactions on Pattern Analysis and Machine Intelligence*, vol. 21, no. 5, pp. 476-480, 1999.

[65] Fix, E. and Armstrong, H. G., "Modeling Human Performance with Neural Networks," *Proc. Int. Joint Conf. on Neural Networks*, vol. 1, pp. 247-252, 1990.

[66] Fix, E. and Armstrong, H. G., "Neural Network Based Human Performance Modeling," *Proc. IEEE National Aerospace and Electronics Conf.*, vol. 3, pp. 1162-1165, 1990.

[67] Foresti, G. L., Mahonen, P. and Regazzoni, C., Eds., *Multimedia Video-based Surveillance Systems: Requirements, Issues and Solutions*, Kluwer Academic Publishers, Boston, 2000.

[68] Frank, I. E. and Friedman, J. H., "A Statistical View of Some Chemometrics Regression Tools," *Technometrics*, vol. 35, no. 2, pp. 109-135, May 1993.

[69] Friedman, D. and Wise, D., "Reference Counting Can Manage the Circular Environments of Mutual Recursion," *Information Processing Letters*, vol. 8, no. 1, pp. 41-45, January 1979.

[70] Friedman, J., Bentley, J. L. and Finkel, R. A., "An Algorithm for Finding Best Matches in Logarithmic Expected Time," *ACM Transactions on Mathematical Software*, vol. 3, no. 3, pp. 209-226, September 1977.

[71] Foresti, G. L. and Roli, F., "Learning and Classification of Suspicious Events for Advanced Visual-based Surveillance," in Foresti, G. L., Mahonen, P. and Regazzoni, C., Eds., *Multimedia Video-based Surveillance Systems: Requirements, Issues and Solutions*, Kluwer Academic Publishers, Boston, 2000.

[72] Fraile, R. and Maybank, S. J., "Vehicle Trajectory Approximation and Classification," *Proc. of British Machine Vision Conference*, 1998.

[73] Friedrich, H., Kaiser, M. and Dillman, R., "What Can Robots Learn From Humans?," *Annual Reviews in Control*, vol. 20, pp. 167-172, 1996.

[74] Funahashi, K., "On the Approximate Realization of Continuous Mappings by Neural Networks," *Neural Net.*, vol. 2, no. 3, pp. 183-192, 1989.

[75] Ge, M. and Xu, Y., "A Novel Intelligent Manufacturing Process Monitor," a chapter in the book *Robotic Welding, Intelligence and Automation*, Springer-Verlag, 2003.

[76] Gersho, A., "On the Structure of Vector Quantizers," *IEEE Transactions on Information Theory*, IT-28, vol. 2, pp. 157-166, 1982.

[77] Gertz, M., Stewart, D. and Khosla, P., "A Software Architecture-based Human-Machine Interface for Reconfigurable Sensor-Based Control Systems," *Proc. of 8th IEEE International Symposium on Intelligent Control*, pp. 75-80, Chicago, IL, August 1993. IEEE.

[78] Geva, S. and Sitte, J., "A Cartpole Experiment Benchmark for Trainable Controllers," IEEE Control Systems Magazine, vol. 13, no. 5, pp. 40-51, 1993.

[79] Gingrich, C. G., Kuespert, D. R. and McAvoy, T. J., "Modeling Human Operators Using Neural Networks," *ISA Trans.*, vol. 31, no. 3, pp. 81-90, 1992.

[80] Goldberg, D. E., *Genetic Algorithms in Search, Optimization, and Machine Learning*, Addison-Wesley, New York, 1989.

[81] Golub, G. H. and Van Loan, C. F., *Matrix Computations*, Johns Hopkins Studies in the Mathematical Sciences, The Johns Hopkins University Press, Baltimore, MD, third edition, 1996.

[82] Gopher, D., Weil, M. and Bareket, T., "The Transfer of Skill from a Computer Game Trainer to Actual Flight," *Proc. Human Factors Society 36th Annual Meeting*, vol. 2, pp. 1285-1290, 1992.

[83] Guez, A. and Selinsky, J., "A Trainable Neuromorphic Controller," *Journal of Robotic Systems*, vol. 5, no. 4, pp. 363-388, 1988.

[84] Guibas, L., Motwani, R. and Raghavan, P., "The Robot Localization Problem," In Golderg, K., Halperin, D., Latombe, J.-C. and Wilson, R. (Editors), *Algorithmic Foundations of Robotics*, pp. 269-282, 1995.

[85] Gullapalli, V., Franklin, J. A. and Benbrahim, H., "Acquiring Robot Skills Via Reinforcement Learning," *IEEE Control Systems Magazine*, vol. 14, no. 1, pp. 13-24, 1994.

[86] Hagiwara, M., "Removal of Hidden Units and Weights for Back Propagation Networks," *Proc. Int. Joint Conf. on Neural Networks*, vol. 1, pp. 351-354, 1993.

[87] Hannaford, B. and Lee, P., "Hidden Markov Model Analysis of Force/Torque Information in Telemanipulation," *Int. Journal Robotics Research*, vol. 10, no. 5, pp. 528-539, 1991.

[88] Haritaoglu, I., Harwood, D. and Davis, L., "W4: Who, When, Where, What: A Real Time System for Detecting and Tracking People," *Proc. 3rd Int. Conf. on Automatic Face and Gesture*, Nara, April 1998.

[89] Hasegawa, T., Suehiro, T. and Takase, K., "A Model-based Manipulation System with Skill-based Execution," *IEEE Transactions on Robotics and Automation*, vol. 8, no. 5, pp. 535-544, October 1992.

[90] Hastie, T. and Stuetzle, W., "Principal Curves," *Journal of the American Statistical Society*, vol. 84, no. 406, pp. 502-516, June 1989.

[91] Hatwal, H. and Mikulcik, E. C., "Some Inverse Solutions to an Automobile Path-Tracking Problem with Input Control of Steering and Brakes," *Vehicle System Dynamics*, vol. 15, pp. 61-71, 1986.

[92] Hertz, J., Krogh, A. and Palmer, R. G., *Introduction to the Theory of Neural Computation*, Addison-Wesley Publishing, Redwood City, 1991.

[93] Hess, U., et al., "The Intensity of Emotional Facial Expressions and Decoding Accuracy," *Journal of Nonverbal Behavior*, vol. 21, no. 4, pp. 241-257, Winter 1997.

[94] Hess, R. A., "Human-in-the-Loop Control," *The Control Handbook*, ed. W. S. Levine, CRC Press, pp. 1497-1505, 1996.

[95] Hightower, J. and Borriello, G., "A Survey and Taxonomy of Location Sensing Systems for Ubiquitous Computing," University of Washington, *Technical Report UW CSE 01-08-03*, 2001.

[96] Hiroshe, Y., Yamashita, K. and Hijiya, S., "Backpropagation Algorithm Which Varies the Number of Hidden Units," *Neural Networks*, vol. 4, no. 1, pp. 61-66, 1991.

[97] Hirzinger, G., Brunner, B., Dietrich, J. and Heindl Johan, "Sensor-based Space Robotics-ROTEX and its Telerobotic Features," *IEEE Transactions on Robotics and Automation*, vol. 9, no. 5, pp. 649-663, October 1993.

[98] Hong, H., Neven, H. and Malsburg, C., "Online Facial Expression Recognition Based on Personalized Gallery", *Proc. Int. Conf. on Automatic Face and Gesture Recognition*, pp. 354-359, Nara, Japan, 1998.

[99] Hovland, G. E., Sikka, P. and MacCarragher, B. J., "Skill Acquisition from Human Demonstration Using a Hidden Markov Model," *Proc. IEEE Int. Conf. on Robotics and Automation*, vol. 3, pp. 2706-2711, 1997.

[100] Hu, Y. H., "From Pattern Classification to Active Learning," *IEEE Signal Processing Magazine*, vol. 14, issue 6, pp. 39-43, 1997.

[101] Huang, X. D., Ariki, Y. and Jack, M. A., *Hidden Markov Models for Speech Recognition*, Edinburgh University Press, Edinburgh, 1990.

[102] Hutchinson, M. F. and de Hoog, F. R., "Smoothing Noisy Data with Spline Functions," *Numerische Mathematik*, vol. 47, pp. 99-106, 1985.

[103] Hunt, K. J., *et. al.*, "Neural Networks for Control Systems - A Survey," *Automatica*, vol. 28, no. 6, pp. 1083-1112, 1992.

[104] Hyvarinen, A., "Fast and Robust Fixed-Point Algorithms for Independent Component Analysis," *IEEE Transactions on Neural Networks*, vol. 10, no. 3, pp. 626-634, 1999.

[105] Iba, W., "Modeling the Acquisition and Improvement of Motor Skills," *Machine Learning: Proc. Eighth Int. Workshop on Machine Learning*, vol. 1, pp. 60-64, 1991.

[106] Ikeuchi, K. and Suehiro, T., "Toward an Assembly Plan from Observation; Part I: Task Recognition with Polyhedral Objects," *IEEE Transactions on Robotics and Automation*, vol. 10, no. 3, June 1994.

[107] *Open Source Computer Vision Library Reference Manual*, Intel Corporation, 2001.

[108] Ivanov, Y., Stauffer, C., Bobick, A. and Grimson, W. E. L., "Video Surveillance of Interactions," *Proc. Workshop on Visual Surveillance*, June 1999.

[109] Jain, R., Murty, S. N. J., *et. al.*, "Similarity Measures for Image Databases," *Proc. IEEE Int. Conf. on Fuzzy Systems*, vol. 3, pp. 1247-1254, 1995.

[110] Johnson, N. and Hogg, D. C., "Learning the Distribution of Object Trajectories for Event Recognition," *Image and Vision Computing*, vol. 14, pp. 609-615, 1996.

[111] Jolliffe, I. T., *Principal Component Analysis*, Springer-Verlag, New York, 1986.

[112] Juang, B. H. and Rabiner, L. R., "A Probabilistic Distance Measure for Hidden Markov Models," *AT&T Technical Journal*, vol. 64, no. 2, pp. 391-408, 1985.

[113] Kaiser, M., "Transfer of Elementary Skills via Human-Robot Interaction," *Adaptive Behavior*, vol. 5, no. 3-4, pp. 249-280, 1997.

[114] Kaiser, S. and Wehrle, T., "Automated Coding of Facial Behavior in Human-Computer Interactions with FACS," *Journal of Nonverbal Behavior*, vol. 16, no. 2, pp. 67-83, 1992.

[115] Kang, S. B., "Automatic Robot Instruction from Human Demonstration," Ph.D. Thesis, The Robotics Institute, Carnegie Mellon University, 1994.

[116] Kidono, K., Miura, J. and Shirai, Y., "Autonomous Visual Navigation of a Mobile Robot using a Human-Guided Experience," *Proc. of the 6th International Symposium on Intelligent Autonomous Systems*, pp. 620-627, Venice, Italy, 2000.

[117] Kimura, S. and Yachida, M., "Facial Expression Recognition and its Degree Estimation", *Proc. IEEE Computer Society Conf. on Computer Vision and Pattern Recognition*, pp. 295-300, 1997.

[118] Koeppe, R. and Hirzinger, G., "Learning Compliant Motions by Task Demonstration in Virtual Environments," In *Fourth International Symposium on Experimental Robotics*, ISER'95, Lecture notes in control and information sciences, Stanford, CA, June-July 1995. SpringerVerlag.

[119] Kollias, S. and Anastassiou, D., "An Adaptive Least Squares Algorithm for the Efficient Training of Artificial Neural Networks," *IEEE Trans. on Circuits and Systems*, vol. 36, no. 8, pp. 1092-1101, 1989.

[120] Kosuge, K., Fukuda, T. and Asada, H., "Acquisition of Human Skills for Robotic Systems," *Proc. IEEE Int. Symp. on Intelligent Control*, pp. 469-474, 1991.

[121] Kovacs-Vajna, Z., "A Fingerprint Verification System Based on Triangular Matching and Dynamic Time Warping," *IEEE Trans. on Pattern Analysis and Machine Intelligence*, vol. 22, no. 1, pp. 1266-1276, 2000.

[122] Kraiss, K. F. and Kuettelwesch, H., "Teaching Neural Networks to Guide a Vehicle Through an Obstacle Course by Emulating a Human Teacher," *Proc. of the 1990 International Joint Conference on Neural Networks*, pp. 333-337, San Diego, CA, 1990.

[123] Kramer, U., "On the Application of Fuzzy Sets to the Analysis of the System-Driver-Vehicle Environment," *Automatica*, vol. 21, no. 1, pp. 101-107, 1985.

[124] Kramer, M. A., "Nonlinear Principal Component Analysis using Autoassociative Neural Networks," *AIChe Journal*, vol. 37, no. 2, pp. 233-243, February 1991.

[125] Kundu, A., Chen, G. C. and Persons, C. E., "Transient Sonar Signal Classification Using Hidden Markov Models and Neural Nets," *IEEE Jour. Oceanic Engineering*, vol. 19, no. 1, pp. 87-99, 1994.

[126] Kupeev, K. Y. and Wolfson, H. J., "On Shape Similarity," *Proc. of 12th IAPR Int. Conf. on Pattern Recognition*, vol. 1, pp. 227-231, 1994.

[127] Latombe, J. C., *Robot Motion Planning*, Kluwer Academic Publishers, 1991.

[128] LeBlanc, M. and Tibshirani, R., "Adaptive Principal Surfaces," *Journal of the American Statistical Society*, vol. 89, no. 425, pp. 53-64, March 1994.

[129] Lee, C. and Xu, Y., "(DM) 2 : A Modular Solution for Robotic Lunar Missions," *International Journal of Space Technology*, vol. 16, no. 1, pp. 49-58, 1996.

[130] Lee, C. and Xu, Y., "Online, Interactive Learning of Gestures for Human/Robot Interfaces," In *1996 IEEE International Conference on Robotics and Automation*, vol. 4, pp. 2982-2987, Minneapolis, MN, 1996.

[131] Lee, C. C., "Fuzzy Logic in Control Systems: Fuzzy Logic Controller – Part I," *IEEE Trans. on Systems, Man and Cybernetics*, vol. 20, no. 2, pp. 404-418, 1990.

[132] Lee, C. C., "Fuzzy Logic in Control Systems: Fuzzy Logic Controller – Part II," *IEEE Trans. on Systems, Man and Cybernetics*, vol. 20, no. 2, pp. 419-435, 1990.

[133] Lee, C., "Transferring Human Skills to Robots via Task Demonstrations in Virtual Environments," Ph.D. Thesis Proposal, Carnegie Mellon University, 1997.

[134] Lee, K. K. and Xu, Y., "Human Sensation Modeling in Virtual Environments", *Proc. 2000 IEEE/RSJ Int. Conf. on Intelligent Robots and Systems*, vol. 1, pp. 151-156, 2000.

[135] Lee, K. K. and Xu, Y., "Input Reduction in Human Sensation Modeling Using Independent Component Analysis," *Proc. IEEE/RSJ Int. Conf. on Intelligent Robots and Systems*, pp. 1854-1859, Vol. 4, 2001.

[136] Lee, S. L., Chun, S. J., Kim, D. H., Lee, J. H. and Chung, C. W., "Similarity Search for Multi-Dimensional Data Sequences," *Proc. Int. Conf. on Data Exchange*, pp. 599-608, 2000.

[137] Lee, S., Amato, N. M. and Fellers, J., "Localization Based on Visibility Sectors Using Range Sensors," *Proc. of 2000 IEEE International Conference on Robotics and Automation*, San Francisco, pp. 3505-3511, 2000.

[138] Lee, S. and Chen, J., "Skill Learning from Observations," *Proc. IEEE Int. Conf. on Robotics and Automation*, vol. 4, pp. 3245-3250, 1994.

[139] Lee, S. and Kim, M. H., "Cognitive Control of Dynamic Systems," *Proc. IEEE Int. Symp. on Intelligent Control*, pp. 455-460, 1987.

[140] Lee, S. and Kim, M. H., "Learning Expert Systems for Robot Fine Motion Control," *Proc. IEEE Int. Symp. on Intelligent Control*, pp. 534-544, 1988.

[141] Lien, J. J.-J., Kanade, T., Cohn, J.F. and Li, C.-C., "Subtly Different Facial Expression Recognition and Expression Intensity Estimation", *Proc. 1998 IEEE Computer Society Conf. on Computer Vision and Pattern Recognition*, pp. 853-859, 1998.

[142] Linde, Y., Buzo, A. and Gray, R. M., "An Algorithm for Vector Quantizer Design," *IEEE Trans. Communication*, vol. COM-28, no. 1, pp. 84-95, 1980.

[143] Lisetti, C. and Rumelhart, D., "Facial Expression Recognition Using a Neural Network", in *Proc. 11th Int. Conf. Facial Expression Recognition*, AAAI Press, 1998.

[144] Liu, S. and Asada, H., "Transferring Manipulative Skills to Robots: Representation and Acquisition of Tool Manipulative Skills Using a

Process Dynamics Model," *Trans. ASME Journal of Dynamic Systems, Measurement, and Control*, vol. 114, pp. 220-228, 1992.

[145] T. Lozano-Perez, "Task Planning," In Michael Brady, editor, *Robot Motion: Planning and Control*, MIT Press, Cambridge, MA, 1982.

[146] B. D. Lucas and T. Kanade, "An Iterative Image Registration Technique with an Application to Stereo Vision", *Proc. 7th Int. Joint Conf. on Artificial Intelligence*, 1981.

[147] Mackey, M. C. and Glass, L., "Oscillations and Chaos in Physiological Control Systems," *Science*, vol. 197, no. 4300, pp. 287-289, 1977.

[148] Malthouse, E. C., "Limitations of Nonlinear PCA as Performed with Generic Neural Networks," *IEEE Transactions on Neural Networks*, vol. 9, no. 1, pp. 165-173, January 1998.

[149] Mataric, M. J., Zordan, V. B., and Williamson, M. M., "Making Complex Articulated Agents Dance: An Analysis of Control Methods Drawn from Robotics, Animation and Biology," *Autonomous Agents and Multi-Agent Systems*, vol. 2, no. 1, pp. 23-44, March 1999.

[150] Matsumoto, D., et al., "American-Japanese Cultural Differences in Judgments of Expression Intensity and Subjective Experience", *Cognition and Emotion*, vol. 13, no. 2, pp. 201-218, 1999.

[151] McRuer, D. T. and Krendel, E. S., "Human Dynamics in Man-Machine Systems," *Automatica*, vol. 16, no. 3, pp. 237-253, 1980.

[152] Michelman, P. and Allen, P., "Forming Complex Dextrous Manipulations from Task Primitives," *Proc. 1994 IEEE International Conference on Robotics and Automation*, vol. 4, pp. 3383-3388, San Diego, CA, May 1994. IEEE Computer Society Press.

[153] Miller, W. T., Sutton, R. S. and Werbos, P. I., eds., "Neural Networks For Control," MIT Press, Cambridge, MA, 1990.

[154] Minka, T. P., "Automatic Choice of Dimensionality for PCA," *Technical Report 514*, MIT Media Laboratory, Perceptual Computing Section, Cambridge, MA, December 1999.

[155] Modjtahedzadeh, A. and Hess, R. A., "A Model of Driver Steering Control Behavior for Use in Assessing Vehicle Handling Qualities," *Trans. ASME Journal of Dynamic Systems, Measurement, and Control*, vol. 115, no. 3, pp. 456-464, 1993.

[156] Mjolsness, E. and DeCoste, D., "Machine Learning for Science: State of the Art and Future Prospects," *Science*, vol. 293, pp. 2051-2055, 2001.

[157] Moody, J. and Darken, C., "Fast Learning in Networks of Locally Tuned Processing Units," *Neural Computation*, vol. 1, no. 2, pp. 281-294, 1989.

[158] Moore, A. W. and Atkeson, C. G., "Prioritized Sweeping: Reinforcement Learning with Less Data and Less Real Time," *Machine Learning*, vol 13, no. 1, pp. 103-130, 1993.

[159] Morrow, J. D., *Sensorimotor Primitives for Programming Robotic Assembly Skills*, Ph.D. Thesis, Robotics Institute, Carnegie Mellon University, April 1997.

[160] Morrow, J. D. and Khosla, P. K., "Sensorimotor Primitives for Robotic Assembly Skills," *Proc. of the 1995 IEEE International Conference on Robotics and Automation*, vol. 2, pp. 1894-1899, May 1995.

[161] Morrow, J. D., Nelson, B. J. and Khosla, P. K., "Vision and Force Driven Sensorimotor Primitives for Robotic Assembly Skills," *Proc. of the 1995 IEEE/RSJ International Conference on Intelligent Robots and Systems*, vol. 3, pp. 234-240. IEEE Computer Society Press, August 1995.

[162] Mozer, M. C. and Smolensky, P., "Skeletonization: A Technique for Trimming the Fat From a Network Via Relevance Assessment," *Advances in Neural Information Processing Systems 1*, D. S. Touretzky, ed., Morgan Kaufmann Publishers, pp. 107-115, 1989.

[163] Nabhan, T. M. and Zomaya, A. Y., "Toward Generating Neural Network Structures for Function Approximation," *Neural Networks*, vol. 7, no. 1, pp. 89-99, 1994.

[164] Nagatani, K. and Yuta, S., "Designing Strategy and Implementation of Mobile Manipulator Control System for Opening Door," *Proc. 1996 IEEE International Conference on Robotics and Automation*, vol. 3, pp. 2828-2834, Minneapolis, MN, 1996. IEEE.

[165] Narendra, K. S. and Parthasarathy, K., "Identification and Control of Dynamical Systems Using Neural Networks," *IEEE Trans. on Neural Networks*, vol. 1, no. 1, pp. 4-27, 1990.

[166] Nechyba, M., *Learning and Validation of Human Control Strategy*, Ph.D. Thesis, Carnegie Mellon University, 1998.

[167] Nechyba, M. C. and Xu, Y., "Human Skill Transfer: Neural Networks as Learners and Teachers," *Proc. of the 1995 IEEE/RSJ International Conference on Intelligent Robots and Systems*, vol. 3, pp. 314-319. IEEE Computer Society Press, 1995.

[168] Nechyba, M. C. and Xu, Y., "Learning and Transfer of Human Real-Time Control Strategies," *Journal of Advanced Computational Intelligence*, vol. 1, no. 2, pp. 137-154, 1997.

[169] Nechyba, M. C. and Xu, Y., "Neural Network Approach to Control System Identification with Variable Activation Functions," *Proc. IEEE Int. Symp. on Intelligent Control*, vol. 1, pp. 358-363, 1994.

[170] Nechyba, M. and Xu, Y., "Stochastic Similarity for Validating Human Control Strategy Models," *IEEE Trans. on Robotics and Automation*, vol. 14, no. 3, June 1998, pp. 437-451.

[171] Nechyba, M. and Xu, Y., "Cascade Neural Networks with Node-Decoupled Extended Kalman Filtering," *Proc. IEEE Int. Symp. on Computational Intelligence in Robotics and Automation*, vol. 1, pp. 214-219, 1997.

[172] Neuser, S., Nijhuis, J., *et. al.*, "Neurocontrol for Lateral Vehicle Guidance," *IEEE Micro*, vol. 13, no. 1, pp. 57-66, 1993.

[173] Niedenthal, P. M., et al., "Emotional State and the Detection of Change in Facial Expression of Emotion," *European Journal of Social Psychology*, vol. 30, pp. 211-222, 2000.

[174] Nilsen, K., "Reliable Real-Time Garbage Collection in C++," *Computing Systems*, vol. 7, no. 4, pp. 467-504, 1994.

[175] Ogata, H. and Takahashi, T., "Robotic Assembly Operation Teaching in a Virtual Environment," *IEEE Transactions on Robotics and Automation*, vol. 10, no. 3, pp. 391-399, June 1994.

[176] O' Hare, D. and Roscoe, S., *Flight Deck Performance: The Human Factor*, Iowa State University Press, Ames, 1990.

[177] Olson, T. and Brill, F., "Moving Object Detection and Event Recognition Algorithms for Smart Cameras," in *Proc. DARPA Image Understanding Workshop*, pp. 159-175, 1997.

[178] Omohundro, S., "Bumptrees for Efficient Function, Constraint, and Classification Learning," *Advances in Neural Information Processing Systems 3*, R. P. Lippmann, J. E. Moody and D. S. Touretzky, eds., Morgan Kaufmann Publishers, pp. 693-699, 1991.

[179] Ou, Y. and Xu, Y., "Learning Human Control Strategy for Dynamically Stable Robots: Support Vector Machine Approach," *Proc. IEEE Int. Conf. on Robotics and Automation*, pp. 3455-3460, Taipei, Taiwan, September, 2003.

[180] Owens, J. and Hunter, A., "Application of the Self-Organizing Map to Trajectory Classification," *Proc. 3rd IEEE Int. Workshop on Visual Surveillance*, pp. 77-83, 2000.

[181] Park, E. H., et al., "Adaptive Learning of Human Motion by a Telerobot Using a Neural Network Model as a Teacher," *Computer and Industrial Engineering*, vol. 27, pp. 453-456, 1994.

[182] Pavlovic, V. I., Sharma, R. and Huang, T. S., "Visual Interpretation of Hand Gestures for Human-Computer Interaction: A Review," *IEEE*

Transaction on Pattern Analysis and Machine Intelligence, vol. 19, no. 7, 1997.

[183] Pentland, A. and Liu, A., "Toward Augmented Control Systems," *Proc. Intelligent Vehicles*, vol. 1, pp. 350-355, 1995.

[184] Pèrez, J. A., et al., "Continuous Mobile Robot Localization: Vision vs. Laser," *Proc. 1999 IEEE International Conference on Robotics and Automation*, Detroit, Michigan, pp. 2917-2923, 1999.

[185] Plaut, D., Nowlan, S. and Hinton, G., "Experiment on Learning by Backpropagation," *Technical Report, CMU-CS-86-126*, Carnegie Mellon University, 1986.

[186] Pomerleau, D. A. and Jochem, T., "Rapidly Adapting Machine Vision for Automated Vehicle Steering," *IEEE Expert*, vol. 11, no. 2, pp. 19-27, 1996.

[187] Pomerleau, D. A., *Neural Network Perception for Mobile Robot Guidance*, Ph.D. Thesis, School of Computer Science, Carnegie Mellon University, 1992.

[188] Pomerleau, D. A., "Reliability Estimation for Neural Network Based Autonomous Driving," *Robotics and Autonomous Systems*, vol. 12, no. 3-4, pp. 113-119, 1994.

[189] Press, W. H., et al., *Numerical Recipes in C: The Art of Scientific Computing, 2nd ed.*, Cambrige University Press, Cambridge, MA, 1992.

[190] Puskorius, G. V. and Feldkamp, L. A., "Decoupled Extended Kalman Filter Training of Feedforward Layered Networks," *Proc. Int. Joint Conf. on Neural Networks*, vol. 1, pp. 771-777, 1991.

[191] Qin, S., Su, H. and McAvoy, T. J., "Comparison of Four Neural Net Learning Methods for Dynamic System Identification," *IEEE Trans. on Neural Networks*, vol. 3, no. 1, pp. 122-130, 1992.

[192] Rabiner, L. R., "A Tutorial on Hidden Markov Models and Selected Applications in Speech Recognition," *Proc. IEEE*, vol. 77, no. 2, pp. 257-286, 1989.

[193] Rabiner, L. R., Juang, B. H., Levinson, S. E. and Sondhi, M. M., "Some Properties of Continuous Hidden Markov Model Representations," *AT&T Technical Journal*, vol. 64, no. 6, pp. 1211-1222, 1986.

[194] Radons, G., Becker, J. D., Dulfer, B. and Kruger, J., "Analysis, Classification and Coding of Multielectrode Spike Trains with Hidden Markov Models," *Biological Cybernetics*, vol. 71, no. 4, pp. 359-373, 1994.

[195] Rao, K. R. and Elliott, D. F., *Fast Transforms: Algorithms, Analyses and Applications*, Academic Press, New York, 1982.

[196] Rath, T. and Manmatha, M., "Word Image Matching Using Dynamic Time Warping," *Technical Report MM-38*, Center Intelligent Information Retrieval, University of Massachusetts Amherst, 2002.

[197] Rees, J. and Donald, B., "Program Mobile Robots in Scheme,"*Proc. of 1992 IEEE International Conference on Robotics and Automation*, pp. 2681-2688, Nice, France, May 1992. IEEE.

[198] Reinsch, C. H., "Smoothing by Spline Functions," *Numerische Mathematik*, vol. 10, pp. 177-183, 1967.

[199] Rosenblum, M., Yacoob, Y. and Davis, L. S. "Human Expression Recognition from Motion Using a Radial Basis Function Network Architecture," *IEEE Trans. on Neural Networks*, vol. 7, issue 5, pp. 1121-1138, Sept. 1996.

[200] Remagnino, P., Jones, G., Paragios, N. and Regazzoni, C., Eds., *Video-Based Surveillance Systems: Computer Vision and Distributed Processing*, Kluwer Academic Publishers, Boston, 2002.

[201] Ren, H. and Xu, G., "Human Action Recognition in Smart Classroom," *IEEE Proc. Int. Conf. on Automatic Face and Gesture Recognition*, pp. 399-404, 2002.

[202] Rumelhart, D. E., McClelland, J. L. and the PDP Research Group, *Parallel Distributed Processing: Explorations in the Microstructure of Cognition, Volume 1: Foundations*, MIT Press, Cambridge, MA, 1986.

[203] Salichs, M. A. and Moreno, L., "Navigation of Mobile Robots: Open Questions, *Robotica*, vol. 18, pp. 227-234, 2000.

[204] Samad, T., "Neurocontrol: Concepts and Applications," *Proc. IEEE Int. Conf. on Systems, Man, and Cybernetics*, vol. 1, pp. 369-274, 1992.

[205] Schaal, S., "Is Imitation Learning the Route to Humanoid Robots?," *Trends in Cognitive Sciences*, vol. 3, no. 6, pp. 233-242, 1999.

[206] Schaal, S., "Nonparametric Regression for Learning Nonlinear Transformations," In H. Ritter, O. Holland, and B. M. Ohl, eds., *Prerational Intelligence in Strategies, High-Level Processes and Collective Behavior*, vol. 2, pp. 595-621. Kluwer Academic Press, 1999.

[207] Schaal, S. and Atkeson, C. G., "Memory-Based Robot Learning," *Proc. IEEE Int. Conf. on Robotics and Automation*, vol. 4, pp. 2928-2933, 1994.

[208] Schaal, S. and Atkeson, C. G., "Robot Juggling: Implementation of Memory-Based Learning," *IEEE Control Systems Magazine*, vol. 14, no. 1, pp. 57-71, 1994.

[209] Schaal, S., Vijayakumar, S. and Atkeson, C. G., "Local Dimensionality Reduction," In M. I. Jordan, M. J. Kearns, and S. A. Solla, eds., *Advances in Neural Information Processing Systems*, vol. 10, MIT Press, Cambridge, MA, 1998.

[210] Schneider, J. G., *Robot Skill Learning Through Intelligent Experimentation*, Ph.D. Thesis, School of Computer Science, University of Rochester, 1995.

[211] Scholkopf, B. and Smola, A. J., *Learning with Kernels: Support Vector Machines, Regularization, Optimization, and Beyond*, MIT Press, Cambridge, MA, 2002.

[212] Scott, D. W., *Multivariate Density Estimation : Theory, Practice, and Visualization*, John Wiley & Sons, New York, 1992.

[213] Seung, H. S. and Lee, D. D.. "The Manifold Ways of Perception," *Science*, vol. 290, pp. 2268-2269, 2000.

[214] Shebilske, W. L. and Regian, J. W., "Video Games, Training, and Investigating Complex Skills," *Proc. Human Factors Society 36th Annual Meeting*, vol. 2, pp. 1296-1300, 1992.

[215] Sheridan, T. B., "Space Teleoperation Through Time Delay: Review and Prognosis," *IEEE Trans. on Robotics and Automation*, vol. 9, no. 5, pp. 592-606, 1993.

[216] Sheridan, T. B., *Telerobotics, Automation, and Human Supervisory Control*, Cambridge Press, Cambridge, 1992.

[217] Shibata, T., et al., "Development and Integration of Generic Components for a Teachable Vision-Based Mobile Robot," *IEEE/ASME Transactions on Mechatronics*, vol. 1, no. 3, pp. 230-236, 1996.

[218] Shimokura, K. and Liu, S., "Programming Deburring Robots Based on Human Demonstration with Direct Burr Size Measurement," *Proc. IEEE Int. Conf. on Robotics and Automation*, vol. 1, pp. 572-577, 1994.

[219] Shum, H. Y., Hebert, M. and Ikeuchi, K., "On 3D Shape Similarity," *Technical Report, CMU-CS-95-212*, Carnegie Mellon University, 1995.

[220] Silverman, B. W., "Spline Smoothing: The Equivalent Variable Kernel Method," *The Annals of Statistics*, vol. 12, no. 3, pp. 896-916, 1984.

[221] Silverman, B. W., "Some Aspects of the Spline Smoothing Approach to Non-Parametric Regression Curve Fitting," *Journal of the Royal Statistical Society*, Series B, vol. 47, no. 1, pp. 1-52, 1985.

[222] Singhal, S. and Wu, L., "Training Multilayer Perceptrons with the Extended Kalman Algorithm," *Advances in Neural Information Processing Systems 1*, ed. Touretzky, D. S., Morgan Kaufmann Publishers, pp. 133-140, 1989.

[223] Skubic, M. and Volz, R. A., "Learning Force Sensory Patterns and Skills From Human Demonstration," *Proc. IEEE Int. Conf. on Robotics and Automation*, vol. 1, pp. 284-290, 1997.

[224] Sniedovich, M., *Dynamic Programming*, Marcel Dekker, New York, 1992.

[225] Song, J., Xu, Y., Nechyba, M. C. and Yam, Y., "Two Performance Measures for Evaluating Human Control Strategy," *Proc. IEEE Int. Conference on Robotics and Automation*, May, 1998.

[226] Song, J., Xu, Y., Yam, Y. and Nechyba, M. C., "Optimization of Human Control Strategies with Simultaneously Perturbed Stochastic Approximation," *Proc. IEEE Int. Conference on Intelligent Robots and Systems*, October, 1998.

[227] Sotelino, L. G., Saerens, M. and Bersini, H., "Classification of Temporal Trajectories by Continuous-Time Recurrent Nets," *Neural Networks*, vol. 7, no. 5, pp. 767-776, 1994.

[228] Spall, J. C., "Multivariate Stochastic Approximation Using a Simultaneous Perturbation Gradient Approximation," *IEEE Trans. on Automation Control*, vol. 37, no. 3, pp. 332-341, 1992.

[229] Stone, M., "Cross Validatory Choice and Assessment of Statistical Predictions," *Journal of the Royal Statistical Society*, Series B, vol. 36, pp. 111-147, 1974.

[230] Sun, M., Burk, G. and Sclabassi, R. J., "Measurement of Signal Similarity Using the Maxima of the Wavelet Transform," *Proc. IEEE Int. Conf. on Acoustics, Speech, and Signal Processing*, vol. 3, pp. 583-586, 1993.

[231] Sussman, G., "A Computer Model of Skill Acquisition," *Number 1 in Artificial Intelligence Series*, American Elsevier Publishing Company, New York, 1975.

[232] Sutton, R. S., "Learning to Predict by the Methods of Temporal Differences," *Machine Learning*, vol. 3, no. 1, pp. 9-44, 1988.

[233] Sutton, R. and Towill, D. R., "Modeling the Helmsan in a Ship Steering System Using Fuzzy Sets," *Analysis, Design and Evaluation of Man-Machine Systems: Selected Papers from the Third IFAC Conference*, vol. 1, pp. 157-162, 1988.

[234] Tan, S. and Mavrovouniotis, M. L., "Reducing Data Dimensionality through Optimizing Neural Network Inputs," *AIChe Journal*, vol. 41, no. 6, pp. 1471-1480, June 1995.

[235] Tachibana, K., Furuhashi, T., Shimoda, M., Kawakami, Y. and Fukunaga, T., "An Application of Fuzzy Modeling to Rowing Motion Analy-

sis," *Proc. IEEE Conf. on Systems, Man, and Cybernetics*, pp. 190-195, 1999.

[236] Tenenbaum, J. B., Silva, V., and Langford, J. C.. "A Global Geometric Framework for Nonlinear Dimensionality Reduction," *Science*, vol. 290, pp. 2319-2323, 2000.

[237] Thodberg, H. H., "Improving Generalization of Neural Networks Through Pruning," *Int. Journal of Neural Systems*, vol. 1, no. 4, p-p. 317-326, 1991.

[238] Tibshirani, R., "Principal Curves Revisited," *Statistics and Computing*, vol. 2, no. 4, pp. 183-190, December 1992.

[239] Tipping, M. E. and Bishop, C. M., "Mixtures of Probabilistic Principal Component Analysis," *Technical Report NCRG/97/003*, Neural Computing Research Group, Department of Computer Science and Applied Mathematics, Aston University, June 1997.

[240] Tong, S., *Active Learning: Theory and Applications*, Ph.D. Thesis, S-tanford University, 2001.

[241] Ude, A., "Trajectory Generation from Noisy Positions of Object Features for Teaching Robot Paths," *Robotics and Autonomous Systems*, vol. 11, no. 2, pp. 113-127, 1993.

[242] Vasconcelos, N. and Lippman, A., "Towards Semantically Meaningful Feature Spaces for the Characterization of Video Content," in *Proc. of IEEE Int. Conf. Image Processing*, pp. 25-28, 1997.

[243] Vijayakumar, S. and Schaal, S., "Robust Local Learning in High Dimensional Spaces," *Neural Processing Letters*, vol. 7, pp. 139-149, 1998.

[244] Vlachos, M., Kollios, G., Gunopulos, D., "Discovering Similar Multidimensional Trajectories," *Proc. 18th Int. Conf. on Data Engineering*, San Jose, CA, pp. 673-684, 2002.

[245] Voyles, R. M., Morrow, J. D. and Khosla, P. K., "Towards Gesture-Based Programming: Shape from Motion Primordial Learning of Sensorimotor Primitives," *Robotics and Autonomous Systems*, vol. 22, no. 3-4, pp. 361-375, 1997.

[246] Wahba, G., *"Spline Models for Observational Data," Proc. CBSM-NSF Regional Conference Series in Applied Mathematics,* no 59, Society for Industrial & Applied Mathematics, Philadelphia, PA, 1998.

[247] Wang, M., Iwai, Y. and Yachida, M., "Recognizing Degree of Continuous Facial Expression Change", *Proc. 14th Int. Conf. On Pattern Recognition*, vol. 2, pp. 1188-1190, 1998.

[248] Watkins, C. J., *Learning from Delayed Rewards*, Ph.D. Thesis, King's College, University of Cambridge, 1989.

[249] Werman, M. and Weinshall, D., "Similarity and Affine Invariant Distances Between 2D Point Sets," *IEEE Trans. Pattern Analysis and Machine Intelligence*, vol. 17, no. 8, pp. 810-814, 1995.

[250] White, D. A. and Sofge, D. A., eds., *Handbook of Intelligent Control: Neural, Fuzzy, and Adaptive Approaches*, Multiscience Press, New York, 1992.

[251] Wold, H., "Soft Modeling by Latent Variables: The Non-Linear Iterative Partial Least Squares (NIPALS) Approach," In J. Gani, eds., *Perspectives in Probability and Statistics*, pp. 117-142. Applied Probability Trust, 1975.

[252] Wren, C., Azarbayejani, A., Darell, T. and Pentland, A., "Pfinder: Real-Time Tracking for the Human Body," *Proc. 3rd IEEE Int. Conf. Automatic Face and Gesture Recognition*, 1998.

[253] Xu, Y., Ge, M. and Tarn, T.-J., "A Novel Intelligent Monitor for Manufacturing Processes," in T.-J. Tarn, S.-B. Chen and C. Zhou, eds., *Robotic Welding, Intelligence and Automation*, Springer-Verlag, 2003.

[254] Xu, Y., Yu, W. and Au, K., "Modeling Human Control Strategy in a Dynamically Stabilized Robot," *Proc. 1999 IEEE/RSJ Int. Conf. on Intelligent Robots and Systems*, vol. 2, pp. 507-512, 1999.

[255] Yamato, J., Kurakake, S., Tomono, A. and Ishii, K., "Human Action Recognition Using HMM with Category Separated Vector Quantization," *Trans. Institute of Electronics, Information and Communication Engineers D-II*, vol. J77D-II, no. 7, pp. 1311-1318, 1994.

[256] Yamato, J., Ohya, J. and Ishii, K., "Recognizing Human Action in Time-Sequential Images Using Hidden Markov Models," *Trans. Institute of Electronics, Information, and Communication Engineers D-II*, vol. J76D-II, no. 12, pp. 2556-2563, 1993.

[257] Yang, B. and Asada, H., "Hybrid Linguistic/Numeric Control of Deburring Robots Based on Human Skills," *Proc. IEEE Int. Conf. on Robotics and Automation*, vol. 2, pp. 1467-1474, 1992.

[258] Yang, J., Xu, Y. and Chen, C. S., "Hidden Markov Model Approach to Skill Learning and its Application to Telerobotics," *IEEE Trans. on Robotics and Automation*, vol. 10, no. 5, pp. 621-631, 1994.

[259] Yang, J., Xu, Y. and Chen, C. S., "Human Action Learning via Hidden Markov Models," *IEEE Trans. on Systems, Man, and Cybernetics – Part A: Systems and Humans*, vol. 27, no. 1, pp. 34-44, 1997.

[260] Yoon, H. S., Bae, C. S. and Min, B. W., "Neural Networks for Golf Swing Analysis from Weight-Shift," *Proc. IEEE World Congress on Computational Intelligence* pp. 3083-3087, 1994.

[261] Zhang, J. and Knoll, A., "Learning Manipulation Skills by Combining PCA Technique and Adaptive Interpolation," In *Sensor Fusion and Decentralized Control in Robotic Systems, Proc. of the SPIE - The International Society for Optical Engineering*, vol. 3523, pp. 211-222, Boston, 1998.

Index